This is a clear, well written, sophisticated and comprehensive assessment of the EU's role in international climate change politics. Bringing together the leading experts, it explores all the various facets of this role – including how it is perceived by others – and manages to do so in a theoretically informed way that adds to our understanding of leadership in in international politics. I recommend it wholeheartedly.

Anand Menon, *Professor of European Politics and Foreign Affairs, King's College London, UK*

This book addresses a topic of high political and academic relevance, based on a highly innovative analytical framework – a standard read for anyone interested in international climate change politics.

Christoph Knill, *Professor of Empirical Theories of Politics, Ludwig-Maximilians-University, Munich, Germany*

This volume makes substantial contributions to our understanding of the EU roles in global politics, Europe's ability to provide global leadership, and the impediments to such leadership. Rather than simply assume or assert EU leadership, authors explore different leadership types and styles, enabling analyses that find clear leadership success in some cases and circumstances, while identifying failure and limitations elsewhere. The book should be read by scholars, students and European policymakers alike.

Stacy VanDeveer, *Professor of Political Science, University of New Hampshire*

The European Union in International Climate Change Politics

Climate change has emerged as an issue of central political importance while the EU has become a major player in international climate change politics. How can a 'leaderless Europe' offer leadership in international climate change politics – even in the wake of the UK's Brexit decision?

This book, which has been written by leading experts, offers a critical analysis of the EU leadership role in international climate change politics. It focuses on the main EU institutions, core EU member states and central societal actors (businesses and environmental NGOs). It also contains an external perspective of the EU's climate change leadership role with chapters on China, India and the USA as well as Norway. Four core themes addressed in the book are: leadership, multi-level and polycentric governance, policy instruments, and the green and low-carbon economy. Fundamentally, it asks why certain EU institutional actors, member states and societal actors have tried to take on a leadership role in climate change politics and how, if at all, have they managed to achieve this?

This text will be of key interest to scholars, students and practitioners in EU studies and politics, international relations, comparative politics and environmental politics.

Rüdiger K.W. Wurzel is Professor of Comparative European Politics and Jean Monnet Chair in European Union Studies in the School of Law and Politics, University of Hull.

James Connelly is Professor of Political Theory, School of Law and Politics, University of Hull.

Duncan Liefferink is an Assistant Professor in the Department of Political Sciences of the Environment, Institute for Management Research, Radboud University Nijmegen, the Netherlands.

Routledge Studies in European Foreign Policy
Series Editors: Richard Whitman
University of Kent, UK
and
Richard Youngs
University of Warwick, UK

This series addresses the standard range of conceptual and theoretical questions related to European foreign policy. At the same time, in response to the intensity of new policy developments, it endeavours to ensure that it also has a topical flavour, addressing the most important and evolving challenges to European foreign policy, in a way that will be relevant to the policy-making and think-tank communities.

1 **The European Union in International Climate Change Politics**
Still taking a lead?
Edited by Rüdiger K.W. Wurzel, James Connelly and Duncan Liefferink

The European Union in International Climate Change Politics

Still taking a lead?

Edited by Rüdiger K.W. Wurzel, James Connelly and Duncan Liefferink

Routledge
Taylor & Francis Group
LONDON AND NEW YORK

First published 2017
by Routledge
2 Park Square, Milton Park, Abingdon, Oxon OX14 4RN

and by Routledge
605 Third Avenue, New York, NY 10017

Routledge is an imprint of the Taylor & Francis Group, an informa business

British Library Cataloguing in Publication Data
A catalogue record for this book is available from the British Library

Library of Congress Cataloging in Publication Data
A catalog record for this book has been requested

ISBN: 978-1-138-64718-3 (hbk)
ISBN: 978-1-315-62719-9 (ebk)

Typeset in Times New Roman
by Wearset Ltd, Boldon, Tyne and Wear

To Alfred, Kelsey, Ruth, Tom, Annika and Rein

Contents

Figures

Tables

Contributors

Mikael Skou Andersen is Professor of Environmental Policy Analysis at Aarhus University's Department of Environmental Science. His research has been interdisciplinary, connecting policy, law and economics with insights and models from the natural sciences, aiming to inform policy making. Mikael Skou Andersen is a member of the Scientific Committee of the European Environment Agency (EEA).

Guri Bang is Research Director at the Center for International Climate and Environmental Policy, Oslo (CICERO). Her research focuses on comparative energy and climate policy, low-carbon energy transitions, international climate politics, and US climate and energy politics. She is leading research projects on domestic climate policy development in the US, EU, China, India, Brazil, Japan, and Russia for the Norwegian Center for Strategic Challenges in International Climate Policy (CICEP) and on low-carbon energy policy in India for the Norwegian Research Council.

Elin Lerum Boasson is Associate Professor in Political Science, University of Oslo, and Senior Researcher at CICERO Center for Climate and Environmental Research – Oslo. She has published extensively on EU climate policy, Norwegian climate policy and how EU rules and regulations influence Norwegian climate politics and policy.

Pierre Bocquillon is Lecturer at the University of East Anglia. He was Postdoctoral Researcher in the Centre for Environment, Energy and Natural Resource Governance (C-EENRG), University of Cambridge. His PhD, written in the Department of Politics and International Studies of the University of Cambridge, analyses EU energy policy-making. He researches and writes on French and European energy and climate change policies, the politics of renewable energy promotion, and the democratic governance of energy and climate change.

Daan Boezeman is Assistant Professor in the Department of Political Sciences of the Environment, Institute for Management Research, Radboud University Nijmegen, the Netherlands. He focuses on climate governance, in particular the construction of authoritative knowledge for adaptation, mitigation and geo-engineering policy.

Charlotte Burns is Senior Lecturer in Environmental Politics and Policy, based in the Environment Department at the University of York. She is an expert on the European Parliament's environmental behaviour.

Heleen de Coninck is Associate Professor in innovation studies at the Environmental Science Department at Radboud University's Faculty of Science. She also worked for over 10 years at the unit Policy Studies of the Energy Research Centre of the Netherlands. Her main field of work is international policy for climate mitigation and technology transfer. Heleen holds Masters degrees in Chemistry and Environmental Sciences, and conducted her PhD on the role of technology in the international climate negotiations.

James Connelly is Professor of Political Theory at the University of Hull where he teaches environmental politics, political theory and electoral systems. He is co-author of *Politics and the Environment: From Theory to Practice* (3rd edition, Routledge 2012) and has also written on environmental citizenship and virtues.

Claire Dupont is a post-doctoral researcher at the Institute for European Studies and the Political Science Department of the Vrije Universiteit Brussel. She works on EU climate and energy policy and governance under the framework of the research programme 'Evaluating Democratic Governance in Europe (EDGE)'.

Aurélien Evrard is Lecturer in political science at *Université Sorbonne Nouvelle/Paris 3* and Research Associate at *Sciences Po/Centre d'études européennes*. His main research areas concern comparative environmental and energy politics and the influence of the European Union over domestic policies.

Wyn Grant is Emeritus Professor of Politics at the University of Warwick and is a Fellow of the Academy of Social Sciences. He has written extensively on environmental issues, in particular the interface between agriculture and the environment.

Martin Jänicke is Emeritus Professor at the Free University (FU) Berlin where he was Founding Director of the Environmental Policy Research Centre (FFU). He is Senior Fellow at the Institute for Advanced Sustainability Studies (IASS) in Potsdam. He was vice-chairman of the German Council of Environmental Advisors (SRU) from 2000 to 2004 and has been a member of numerous scientific boards and councils while also acting as Review Editor of the 5th Assessment Report of the IPCC. His books on state failure, ecological modernisation, best practices in environmental policy and global environmental governance have been translated into several languages.

Karolina Jankowska is a scientific advisor in the Berlin House of Representatives, having previously worked as a researcher, consultant and lecturer. She has published widely on Polish, German and EU energy and climate change policy issues. Karolina was awarded a PhD from the Free University Berlin

(Germany) for her thesis *The forces of change: The Polish transition from coal-based manufacturing to renewable oriented society*.

Andrew Jordan is Professor of Environmental Policy at the Tyndall Centre, University of East Anglia and Chair of COST Action IS1309 – Innovations in Climate Governance (INOGOV). He has published widely on EU and UK environment and climate policy and is currently completing a book on durability and flexibility in climate policy.

Kirsten Jörgensen is Senior Lecturer at the Department of Political and Social Sciences and the Environmental Policy Research Centre (FFU) at the Free University Berlin. Her primary research interest is comparative environmental and climate policy including subnational climate governance. She initiated the German-Indian Sustainability and Climate Change student exchange and is coordinator of the Indian-European Multi-level Climate Governance Research Network. She is a member of the Indo-German Expert Group on Green and Inclusive Economy.

Bård Lahn is a research fellow at CICERO Centre for Climate and Environmental Research – Oslo. His research interests include international climate politics, REDD+, and Norway's role in the UNFCCC negotiations. He holds an MA from the TIK Centre for Technology, Innovation and Culture at the University of Oslo.

Xinlei Li is an assistant professor in the School of Political Science and Public Administration at Shandong University (China). She received her PhD from the Free University Berlin (Germany). In 2013 she was awarded the St. Gallen Wings of Excellence Award. Her research on energy transformation, climate leadership and transnational climate municipal networks has been published in *Foreign Affairs Review*, *Chinese Journal of European Studies* and *International Review*. She directs a project on 'Research on China Clean Energy Diplomacy Strategy' supported by the National Social Science Fund of China (NSSFC) (2015 to 2018).

Duncan Liefferink is Assistant Professor in the Department of Political Sciences of the Environment, Institute for Management Research (IMR), Radboud University Nijmegen, the Netherlands. His main research fields are European and comparative environmental politics, with a particular interest in the dynamic interrelationship between national and EU environmental policy making.

Elizabeth Monaghan is Lecturer in Politics at the University of Hull. She works on issues of democracy and legitimacy in the EU, with a particular focus on civil society organisations and theories of representation, and the European Citizens' Initiative.

Helle Ørsted Nielsen is a senior researcher in the Department of Environmental Science at Aarhus University. Her fields of research include implementation of climate and environmental policy in multi-level-governance settings as well as policy design and effects from a behavioural perspective, focussing on target group responses to policy.

Sebastian Oberthür is Professor for Environment and Sustainable Development at the Institute for European Studies (IES), Vrije Universiteit Brussel (VUB). Trained as a political scientist with a strong background in international law, he is an internationally leading expert on issues of international and European environmental and climate governance.

Tim Rayner is a research fellow with the Tyndall Centre for Climate Change Research, based at the University of East Anglia, UK. His interests centre on European Union climate policy and governance (both mitigation and adaptation aspects), as well as policy appraisal and evaluation and their roles in the policy process.

Miranda A. Schreurs is Professor of Environment and Climate Policy at the Bavarian School of Public Policy, Technical University of Munich. Her research focuses on climate change, low-carbon energy transitions, and environmental movements in Europe, the US, and Asia. She was Professor of Comparative Politics and Director of the Environmental Policy Research Centre (FFU) at the Freie Universität Berlin. She serves as Vice-Chair of the European Environment and Sustainable Development Advisory Councils, a network of advisory councils across Europe.

Jon Birger Skjærseth is Research Professor at the Fridtjof Nansens Institute. His research interests focus on various environmental challenges, including climate and energy policies and strategies at the corporate, national, EU and international levels. He has published extensively in these fields, including a number of books.

Israel Solorio is Associate Professor of Public Administration at the National Autonomous University of Mexico. He acts as co-convenor of the UACES Collaborative Research Network on the European Energy Policy.

John Vogler is a professorial research fellow at Keele University, UK and a member of the ESRC Centre for Climate Change Economics and Policy. He is the author of a number of works on EU external policy and notably, with Charlotte Bretherton, *The EU as a Global Actor* (Routledge 2006). His latest book is *Climate Change in World Politics* (Palgrave 2016).

Rüdiger K.W. Wurzel is Professor of Comparative European Politics and Jean Monnet Chair in EU Studies at the University of Hull where he is Director of the Centre for European Union Studies (CEUS). He has published widely on environmental policy and politics, new modes of governance, EU and European governance issues. He is co-author (with Anthony Zito and Andrew Jordan) of *Environmental Governance in Europe* (Edward Elgar, 2013).

Foreword and acknowledgements

Climate change is one of today's biggest challenges facing humankind. It is widely accepted that climate change poses a serious threat. However, the move towards low-carbon economies and societies also offers opportunities.

Shortly after the Second World War, the European Union (EU) was set up as a 'leaderless Europe' in which decision-making powers were spread among a wide range of EU institutional, member state and societal actors. It nevertheless developed into a leader in international climate change politics in the 1990s. Despite a burgeoning literature on EU climate change leadership, it is still not well understood *why*, *when* and *how* the EU, its member states, and societal actors can offer *what type* of leadership in EU and international climate change politics. Nor do we understand well how key external international climate change actors (such as the US, China and India) relate to and/or influence the EU's leadership efforts in international climate change politics. In order to get a better understanding of these issues our book differentiates between four different types of leadership (structural, entrepreneurial, cognitive and exemplary) and two different styles of leadership (humdrum/transactional and heroic/transformational). Moreover, as the global political system has changed rapidly in the twenty-first century, doubts have been raised about the EU's continued ability to offer sustained leadership in international climate change politics. This book assesses whether the EU is still a leader in international climate change politics. The victory of the Leave Campaign in the Brexit referendum in June 2016, which will lead to the UK exiting from the EU, has plunged the EU into the most serious crisis in its history, which it will be able to overcome only with skilful leadership. The outcome of the Brexit referendum necessitated changes to several chapters close to the completion deadline of the manuscript. Chapters 1, 2, 12 and 19 reflect on the likely consequences of the Brexit decision. Although leadership is the overarching theme, all chapters also address the additional main themes of multi-level and polycentric governance, policy instruments and the 'green' or low-carbon economy.

Early versions of the chapters in this book were discussed at a Centre for European Union Studies (CEUS) workshop in Hull (UK) in September 2015. The editors then commented on revised versions of the chapters in early 2016. Xinlei Li managed at phenomenal speed and with great efficiency to write the chapter on China despite being given very short notice. We are grateful to the Commission, the Faculty of Arts and Social Sciences at the University of Hull as well as the Department of Political Sciences of the Environment and the hotspot Europeanization of Policy and Law (EUROPAL) at Radboud University Nijmegen for co-funding the workshop in Hull. We would like to thank Francesca Manchini from the Commission's London office for her support. Our thanks also extend to Jeremy Moulton who helped us with the organisation of the workshop. We are grateful to anonymous referees and the series editors for very useful comments. We would like to thank Maurizio Di Lullo for his very helpful comments on Chapter 1. Sophie Iddamalgoda and Andrew Taylor have been excellent editorial assistant and editor respectively throughout the book project. Rudi Wurzel and Duncan Liefferink would like to thank the COST Action funded 'Innovations in Climate Governance' (INOGOV) programme for giving them the opportunity to continue their work on leadership in climate change politics by funding another workshop in Hull on 'Pioneers and Leaders in Polycentric Climate Governance' (PiLePoC).

Rudi owes a huge debt to Ita and our two boys for their patience, tolerance and good humour during the time-consuming completion process of this book. James would like to thank Rana for her help and support. Duncan is grateful to Riet for her enduring support. The book is dedicated to our children.

RKWW, JC and DL

Abbreviations

AAP	Aam Admi Party (India)
ACEA	European Automobile Manufacturers' Association
ADEME	*Agence de l'Environnement et de la Maitrise de l'Energie* (France)
AIIB	Asian Infrastructure Investment Bank
AITC	All India Trinamool Congress (India)
APEC	Asian-Pacific Economic Cooperation
APSEC	APEC Sustainable Energy Centre
ATTAC	*Association pour la Taxation des Transactions financière et l'Aide aux Citoyens*
BASIC	Brazil, South Africa, India and China
BEE	Bureau of Energy Efficiency (India)
BJD	Biju Janata Dal (India)
BJP	Bharatiya Janata Party (India)
BMU	*Bundesministerium für Umwelt, Naturschutz und Reaktorsicherheit* (Federal Ministry for the Environment, Nature Protection and Nuclear Safety, Germany)
BMUB	*Bundesministerium für Umwelt, Naturschutz, Bau und Reaktorsicherheit* (Federal Ministry for the Environment, Nature Protection, Construction and Nuclear Safety, Germany)
BMWi	*Bundesministerium für Wirtschaft und Energie* (Federal Ministry for Economics and Energy, Germany)
BREXIT	British exit from the EU
C	Celsius
CAN-Europe	Climate Action Network Europe
CAP	Common Agricultural Policy
CBDR	Common But Differentiated Responsibilities
CCAs	Climate Change Agreements (United Kingdom)
CCL	Climate Change Levy (United Kingdom)
CCS	carbon capture and storage
CDM	Clean Development Mechanism
CDU	*Christlich Soziale Union* (Christian Democratic Union, Germany)
CEC	Commission of the European Communities

CEES	Central and Eastern European States
CEO	chief executive officer
CEPS	Centre for European Policy Studies
CER	certified emission reduction
CEU	Commission of the European Union
CFSP	Common Foreign and Security Policy
CMA	China Meteorological Administration
CO_2	carbon dioxide
CO_2-eq.	carbon dioxide equivalent
COMECON	Council for Mutual Economic Assistance
COP	Conference of the Parties (to the Kyoto Protocol)
CPP	Clean Power Plan (United States)
CRESP	China Renewable Energy Scale-up Program
CSU	*Christlich Soziale Union* (Christian Social Union, Germany)
DECC	Department of Energy and Climate Change (United Kingdom)
DEFRA	Department of Environment, Food and Rural Affairs (United Kingdom)
DG	Directorate General
DGEC	Directorate General for Energy and Climate (France)
DGEMP	Directorate General for Energy and Raw Materials (France)
DMK	Dravida Munnetra Kazhagam (India)
EA	Energy Agreement for Sustainable Growth (the Netherlands)
EC2	Europe–China Clean Energy Center
ECF	European Climate Foundation
EEA	European Environment Agency
EEAA	European Economic Area Agreement
EEAS	European External Action Service
EEB	European Environmental Bureau
ECCP	European Climate Change Programme
ECF	European Climate Foundation
EnBW	*Energie Baden-Württemberg AG*
ENGOs	environmental non-governmental organisations
EP	European Parliament
EPA	Environmental Protection Agency (US)
EPHA	European Public Health Alliance
ERT	European Round Table for Industrialists
ESD	Effort Sharing Decision (EU)
ETS	emissions trading scheme
EU	European Union
FoE-Europe	Friends of the Earth Europe
FYP	Five-Year Plan (China)
GEI	Global Environmental Institute
GBI	generation-based incentives
GHG	greenhouse gas
GHGE	greenhouse gas emissions

GIES	*Groupe Interministériel sur l'Effet de Serre* (France)
GRI	Global Reporting Initiative
HEAL	Health and Environment Alliance
HFCs	hydrofluorocarbons
ICARE	China–EU Institute for Clean and Renewable Energy
ICAO	International Civil Aviation Organization
IDDR	Institute for Sustainable Development and International Relations
IEA	International Energy Agency
IMA	Inter-ministerial Working Group (*Interministerielle Arbeitsgruppe*, Germany)
INC	Indian National Congress Party
INDCs	intended nationally determined contributions
IPCC	Intergovernmental Panel on Climate Change
IRENA	International Renewable Energy Agency
JI	joint implementation
JNNSM	Jawaharlal Nehru National Solar Mission
LCTPi	Low Carbon Technology Partnership Initiative
LRTAP	Long Range Transboundary Air Pollution
LTAs	long-term agreements
LULUCF	Land Use, Land-Use Changes and Forestry
MEEDDAT	Ministry of Ecology, Energy, Sustainable Development and Planning (France)
MIES	inter-ministerial mission (France)
MLG	multi-level governance
MSR	Market Stability Reserve
NAO	National Audit Office (United Kingdom)
NAP	National Allocation Plan
NAPCC	National Action Plan on Climate Change
NCSC	National Center for Climate Change Strategy and International Cooperation (China)
NDCs	nationally determined contributions
NDRC	National Development and Reform Commission (Denmark)
NEPIs	New Environmental Policy Instruments
NEPP	National Environmental Policy Plan (the Netherlands)
NFI	Naturefriends International
NIMBY	not in my backyard
OCC	Office of Climate Change (United Kingdom)
OECC	*Oficina Española de Cambio Climático* (Spanish Office on Climate Change)
OECD	Organization of the Petroleum Exporting Countries
OPEC	Oil Producing and Exporting Countries
PAT	Perform Achieve and Trade
PGE	*Polska Grupa Energetyczna* (Polish Energy Group)
PiS	*Prawo i Sprawiedliwość* (Law and Justice Party, Poland)

PNLCC	*Plan National de Lutte Contre le Changement Climatique* (France)
PO	*Platforma Obywatelska* (Civic platform, Poland)
POPE	*Programme d'Orientation de la Politique Energétique Française* (France)
PP	*Partido Popular* (Popular Party, Spain)
PSL	*Polskie Stronnictwo Ludowe* (Polish People's Party)
PSOE	Partido Socialista Obrero Español (Spanish Socialist Workers Party)
PV	photovoltaic
QMV	qualified majority voting
RCEP	Royal Commission on Environmental Pollution (United Kingdom)
RDD&D	Research, Development, Demonstration and Deployment
RED	Renewable Energy Directive
REDD	reducing emissions from deforestation and forest degradation
REDP	China Renewable Energy Development Project
REIO	Regional Economic Integration Organisations
REN21	Renewable Energy Policy Network for the 21st Century
RGGI	Regional Greenhouse Gas Initiative
ROCs	Renewable Obligation Certificates (United Kingdom)
RPO	Renewable Energy Purchase Obligations (India)
RWE	Rheinisch-Westfälisches Elektrizitätswerk AG
SDPC	State Development and Planning Commission (China)
SDSN	Sustainable Development Solutions Network
SEA	Single European Act
SEM	Single European Market
SEPA	State Environmental Protection Administration
SER	Social-Economic Council (the Netherlands)
SET Plan	Strategic European Energy Technology Plan
SPD	*Sozialdemokratische Partei Deutschlands* (Social Democratic Party of Germany)
SRIs	socially responsible investors
TEPOS	*Territories à Energies Positives* (Positive Energy Territories, France)
TFEU	Treaty on the Functioning of the European Union
T&E	Transport and Environment
TWh	terawatt hours
UN	United Nations
UNEP	United Nations Environment Programme
UNFCCC	United Nations Framework Convention on Climate Change
US	United States of America
USCAN	United States Climate Action Network Europe
Visegrad	Poland, Hungary, Slovakia and the Czech Republic
WBCSD	World Business Council for Sustainable Development
WTO	World Trade Organisation
WWF	Word Wide Fund for Nature

Part I

Introduction

1 Introduction

European Union climate leadership

Rüdiger K.W. Wurzel,[1] *Duncan Liefferink and James Connelly*

Introduction

There is no shortage of would-be leaders in EU climate change politics. The EU institutions (e.g. European Council, Council of the EU, Commission and the European Parliament (EP)), member states and societal actors have all, though to varying degrees and at different time periods, tried to offer leadership in EU and international climate change politics. Importantly, public support for EU environmental policy in general, and climate change policy in particular, has been consistently high (e.g. Eurobarometer 2015). The economic recession which followed the 2008 financial crises triggered only a moderate drop in public support for EU action on climate change, although considerable variation exists between member states.

As will be discussed in more detail below, at first sight the EU seems ill equipped to offer political leadership because decision-making powers are dispersed among a wide range of political actors including EU institutional and member state actors thus making the EU a 'leaderless Europe' (Hayward 2008). In the 1950s, when the EU was founded, 'efforts were made in "taming" the "beast" ... of leadership' (Blondel 1987: 3). The dispersal of the EU's decision-making powers has, however, led to the emergence of a wide range of veto actors which have repeatedly led the EU into political stalemate and 'joint decision traps' (Scharpf 1988) from which it is able to escape, typically, only after lengthy periods of arduous negotiations and by adopting complex compromises and sub-optimal policy solutions. The victory of the Leave Campaign in the Brexit referendum in June 2016, which will lead to the UK exiting from the EU (see Chapter 12), has plunged the EU into the most serious crisis in its history, which it will be able to overcome only with skilful leadership.

This chapter will first give a short historical overview of EU climate change politics in a global context, followed by an introduction to the key analytical concepts and themes of this book.

A short history of EU climate change politics

As a brief introduction to the chapters which follow, this section provides a thumbnail guide to core events and actions in EU internal and external climate change politics.

The following five phases of EU climate change policy can broadly be identified: (1) late 1980s to 1992: formation and formulation phase; (2) 1992–2001: Kyoto Protocol negotiation phase; (3) 2001–2005: Kyoto Protocol rescue phase; (4) 2005–2015: Kyoto Protocol implementation and negotiation of a follow up agreement phase; and, (5) since 2015: 2015 Paris Agreement ratification and implementation phase.

Various EU institutional actors (Chapters 2–5), member states (Chapters 6–12) and societal actors (Chapters 13–14) as well as non-EU countries (Chapters 13 and 16–18) reacted somewhat differently to the challenges of climate change and the changing opportunity structures of EU and/or global climate change politics. As will become clear in the following chapters, from the perspective of different EU institutional, member state, societal and non-EU actors the phases can appear differently.

Late 1980s–1992: formation and formulation phase

In 1986 the EP became the first EU institution to request a common climate change policy (see Table 1.1 and Chapter 4). Two years later the Commission issued a communication on climate change. In 1990, the European Council adopted a resolution which demanded the early adoption of GHGE reduction targets on the UN level. In the same year a joint Environment and Energy Council adopted a political agreement on the stabilisation of the EU's CO_2 emissions by 2000 (compared to 1990) which, however, was conditional on other highly developed countries taking similar steps. According to Haigh (1996: 162) it enabled the EU 'to take a strong and leading role, particularly in relation to the United States'. While the US had acted as a leader on the Montreal Protocol on ozone layer depleting substances, in climate change politics it was the EU which started to take over the leadership role from the US (see Chapters 2 and 17).

The EU signed the UN framework convention on climate change (UNFCCC) at the 1992 UN Rio 'Earth summit'. Because it had not yet adopted adequate common policy measures to implement its commitments under the UNFCCC, the EU created a 'capability-expectation gap' (Hill 1993) which it was able to close only with the adoption of legally binding climate policy measures within the framework of the 2000 and 2005 European Climate Change Programmes (ECCPs).

In early 1992, at a time of high public environmental awareness and relatively strong support for deeper European integration, the Commission proposed an EU-wide carbon dioxide (CO_2)/energy tax which was, however, vetoed by the UK on sovereignty grounds (see Chapter 12). The Council adopted the Commission's proposals for a Framework Directive on energy efficiency measures by

Table 1.1 Main phases of EU climate change politics

Late 1980s–1992: formation and formulation phase

- 1986: EP climate change policy resolution
- 1988: Commission communication
- 1990: European Council for early adoption of targets. Joint Environmental and Energy Council agreed CO_2 stabilisation of by 2000 (at 1990 levels)
- 1991: Commission's proposal for EU climate change policy

1992–2001: Kyoto Protocol negotiation phase

- 1992: UN Rio conference adopted UNFCCC: EU accepted CO_2 stabilisation by 2000 (compared to 1990)
- 1997: Kyoto Protocol negotiations: EU proposed 15% reduction of three GHGs by 2010 (compared to 1990) but settled for 8% reduction of six GHGs by 2008–12 (compared to 1990/1995)
- 1998: Burden sharing agreement
- 2000: First ECCP

2001–2005: Kyoto Protocol rescue phase

- 2001: US dropped out of the Kyoto Protocol. EU Environmental Council and European Council for Kyoto Protocol ratification
- 2002: EP voted (540 to 4 votes) in favour of Kyoto Protocol ratification. Kyoto Protocol ratified by EU
- 2003: Commission's EU ETS proposal

2005–2015: Kyoto Protocol implementation and negotiations of a follow-up agreement

- 2005: Kyoto Protocol entered into force. EU ETS became operational.
- 2007: European Council agreed '20–20 by 2020' climate and energy package:
  - Unilateral 20% GHGE reductions by 2020 (compared to 1990)
  - Binding 20% renewable energy by 2020
  - Non-binding 20% energy efficiency improvement by 2020
  - Conditional 30% GHGE reductions by 2020 if 'comparable efforts' by other developed and 'adequate efforts' by leading developing countries
- 2008: EU adopted legally binding CO_2 limits for cars, revision of EU ETS and effort sharing decision
- 2009: EU agreed €7.2 billion fast track money for climate mitigation and adaptation in developing countries. Copenhagen climate conference (COP15) resulting in the Copenhagen Accord
- 2014: European Council adopted '2030 climate and energy package':
  - at least 40% reduction of GHGE by 2030 (compared to 1990)
  - at least 27% increase of renewables by 2030
  - indicative energy saving target of 27% by 2030

Since 2015: Paris Agreement ratification and implementation phase

- 2015: Paris Agreement:
  - Limit global temperature rise to 2.0/1.5°C
  - Peak of global emissions as soon as possible
  - Voluntary national reduction pledges (NDCs)
- 2016 onwards: Ratification process of the 2015 Paris Agreement

member states (SAVE), a Decision on renewable energy (ALTENER) and a Decision to monitor CO_2 emissions, but these were insufficient measures for reaching the EU's proposed CO_2 emissions stabilisation target.

1992–2001: Kyoto protocol negotiation phase

During the 1992 UNFCCC negotiations the EU acted largely as a symbolic leader because it did not yet have in place significant reduction measures to back up its ambitious rhetoric. In the negotiations leading to the Kyoto Protocol the EU initially offered a 15 per cent reduction in GHGE by 2010 (compared to 1990 levels) on condition that its main economic competitors (at the time the US and Japan) would accept similar reductions. Because the US accepted only a 7 per cent reduction target, the EU settled for an 8 per cent reduction target in GHGE by 2008–12. Against initial opposition from the EU the US insisted on the inclusion of the following flexible mechanisms in the Kyoto Protocol: (1) emissions trading, (2) joint implementation (JI), which allowed developed countries jointly to implement GHGE reduction projects with economies in transition, and (3) the clean development mechanism (CDM) which permitted developed countries to sponsor GHGE reduction projects in developing countries for which the former could earn saleable certified emission reduction (CER) credits.

In 1998, the EU adopted the 'burden sharing' agreement which set member states differentiated reduction targets for achieving the EU's collective 8 per cent CO_2 emissions target. Germany and the UK, which were at the time the EU's largest GHGE emitters, accepted CO_2 reductions rates of 21 and 12.5 per cent respectively. Germany benefited from 'wall fall profits' (due to the deindustrialisation of the former East Germany following the fall of the Berlin Wall in 1989), and the UK was helped by its 1980s 'dash for gas'[2] which has a lower carbon content than coal (see Chapters 8 and 12).

In October 2000, the Environmental Council accepted most of the Commission's communication *Towards a European Climate Change Programme* (CEC 2000) thus paving the way for the adoption of the first ECCP in 2000.

2001–2005: Kyoto Protocol rescue phase

On 13 March 2001, President George W. Bush announced that the US would not ratify the Kyoto Protocol which had been signed by his predecessor Bill Clinton (see Chapter 16). Because, at the time, the US was the largest emitter of GHGE, the Kyoto Protocol seemed doomed. However, a few weeks later, the Environmental Council agreed that the EU should pursue the Kyoto Protocol ratification process. After approval by the European Council and overwhelming support from the EP the EU ratified the Kyoto Protocol in May 2002 (see Chapters 4 and 5). After much lobbying from the EU, both Japan and Russia eventually ratified the Kyoto Protocol (thus reaching the required 55 per cent of the total 1990 CO_2 emissions from industrialised countries) which entered into force in 2005.

Frustrated by the veto to its CO_2/energy tax proposal and encouraged by the early experience with emissions trading in the US, the Commission's Directorate General for Environment (DG Environment) commissioned studies on emissions trading in the late 1990s (see Chapter 3). Following meetings with member governments and stakeholders, the Commission published its proposal for an EU ETS Directive in 2001 (CEC 2001). The EP and the Environmental Council speedily adopted a modified version of the Commission's EU ETS proposal in 2003 (Skjaerseth and Wettestad 2008). The EU had thus somewhat belatedly been transmogrified from an emissions trading laggard to a leader which set up the world's first supranational ETS (Skjærseth and Wettestad 2008; Wurzel 2008; see also Chapter 3).

2005–2015: Kyoto Protocol implementation and follow-up agreement negotiation phase

The EU ETS, which became operational in 2005, has become the EU's main climate change policy instrument (Skjærseth and Wettestad 2008) although, within the framework of the second EPCC, the EU also adopted Directives on energy efficiency and the promotion of renewable energy and concluded a voluntary agreement with the European, Japanese and Korean automobile manufacturers on the reduction of CO_2 which, however, was later overtaken by legislation as the voluntary agreement had failed.

In March 2007 the European Council meeting affirmed the EU's climate leadership when it adopted the '20–20–20' climate and energy package which included a binding unilateral 20 per cent CO_2 reduction target by 2020 (compared to 1990 levels), a legally binding 20 per cent renewable energy target by 2020 and a non-binding 20 per cent energy efficiency improvement by 2020. The EU also adopted a 30 per cent CO_2 emissions reduction target by 2020 which was, however, made conditional on 'comparable efforts' by other developed and 'adequate efforts' by leading developing countries. After arduous negotiations the EU adopted the effort sharing decision (which replaced the burden sharing agreement) and agreed on a review of the flagging EU ETS at the European Council in December 2008. In order to strengthen its international climate leader position the EU offered €7.22 billion in 'fast track' funding for climate adaptation measures in developing countries and tried to form alliances on the international level (see Chapter 2).

However, although the EU had adopted relatively ambitious internal climate change policy measures and pledged significant climate funds for developing countries, it was not able to significantly influence the Copenhagen Accord which 'was taken note of' rather than adopted (because Venezuela and other countries were opposed to its adoption) (written communication, EU official, 2 June 2016) – at the 2009 Copenhagen climate conference (COP15). The Copenhagen Accord, widely seen as a weak and vague agreement, was largely negotiated by the US and Brazil, South Africa, India and China (BASIC countries) without direct EU involvement at the crucial negotiating phase (see Chapters 2 and 5). There are complex reasons why the 2009 Copenhagen climate conference arguably constituted a low point (if not the lowest point) for EU international climate leadership (see Chapters 2, 5, 6 and 17). However, to a large degree the EU recovered its status as a leader in international climate change politics at the 2010 Cancún (COP16), 2011 Durban (COP17) and 2014 Lima (COP20) climate conferences which eventually paved the way for a Kyoto Protocol follow-up agreement in the form of the 2015 Paris Agreement.

In October 2014 the European Council adopted the 2030 climate and energy package which stipulated a legally binding GHGE reduction target of at least 40 per cent by 2030 (compared to 1990), a legally binding target to increase the share of renewables to 27 per cent by 2030 (although without binding national targets) and an indicative energy saving target of 27 per cent by 2030 (see Chapter 5). However, unlike in the run-up to the 2009 Copenhagen climate conference (COP15), the EU did not adopt an effort sharing decision prior to the 2015 Paris Agreement.

Since 2015: Paris Agreement ratification and implementation phase

In the climate change agreement, adopted in Paris in December 2015, 195 countries committed themselves to achieving as soon as possible a peak of global GHGE while trying to limit the global temperature rise to well below 2°C and to undertake efforts to limit the temperature rise to 1.5°C above pre-industrial levels. The inclusion of the relatively ambitious 1.5°C goal in the Paris Agreement came as a surprise to EU policy makers and even ENGOs (Interviews 2015–2016). The Paris Agreement will enter into force once it has been ratified by at least 55 countries accounting for at least 55 per cent of global GHGE.

In the run up to the 2015 Paris climate conference (COP21) countries had to submit national voluntary pledges (INDCs). The Paris Agreement turned the INDCs into Nationally Determined Contributions (NDCs) which, however, are likely to limit the global temperature rise merely to about 3° C, unless they can be made more ambitions during the Paris Agreement implementation phase (Interview, EU official, 2016). Whether the NDCs can be considered as legally binding is contested. A five-yearly review process, permitting only the ratcheting upwards of NDCs is meant to ensure that the 1.5°C goal will be achieved. Moreover, a $100 billion global climate fund was set up to assist developing countries in adapting to climate change.

The 2015 Paris Agreement constitutes a departure from the 1997 Kyoto Protocol because it adopted a bottom-up approach which allowed countries to put forward voluntary national reduction pledges. The EU had initially favoured the continuation of the Kyoto Protocol's top-down approach of internationally agreed legally binding reduction targets and deadlines. It finally accepted the new approach when it became apparent that several key countries (including the US, China and India) were only prepared to accept a bottom-up approach. However, the EU succeeded with its demand that the national voluntary pledges should be enshrined in a legally binding UN climate change treaty which stipulates a five-yearly review and transparency measures (Interview, EU official, 2016; Oberthür and Bodle 2016).

The EU played a significant role in the negotiations by brokering a 'high-ambition coalition' which eventually was made up of more than 100 developed and developing countries (including the US but not China and India). As mentioned above, in its 2030 climate and energy package, the EU had already 'pledged' a legally binding unilateral 40 per cent GHGE reduction target for 2030 (compared to 1990 levels) (see Chapters 2, 3–5 and 7). The EU's approach has been referred to as 'leadiating' (Bäckstrand and Elgström 2013; see also Chapter 5) because it consisted of acting both as leader and mediator. Such an approach helped the EU to regain much of its international climate politics leader status which it had lost at the 2009 Copenhagen climate conference (COP15). However, ENGOs have argued that the 2015 Paris Agreement's 1.5°C goal would necessitate a significant ratcheting up of the EU's 40 per cent GHGE emission reduction target for 2030 (see Chapter 15).

Core themes and overall analytical framework

This book adopts an actor-centred approach to the analysis of EU climate change politics. Its chapters all address, from the perspective of the main actor(s) on which they focus, the following four key themes: (1) leadership, (2) multi-level and polycentric governance, (3) policy instruments, and (4) green and low-carbon economy. Of these, leadership is the over-arching theme.

Leadership

The EU has widely been portrayed as a leader in international climate change politics (e.g. Bäckstrand and Elgström 2013; Grubb and Gupta 2000; Jordan *et al.* 2010; Oberthür and Roche Kelly 2008; Wurzel and Connelly 2011a). However, despite a burgeoning literature on EU climate change leadership it is still not well understood *why*, *how* and *when* the EU, its member states, and societal actors can offer *what type* of leadership in EU and international climate change politics.

Despite the surge in scholarly interest in the EU's leadership role in international climate change politics, leadership remains a contested concept which is used in differing ways in different studies (cf. Nye 2008; Young 1991). Our Introduction explains the leadership concept as applied by the authors of the chapters in this book.

Many studies of environmental politics have identified different *types* of leadership, although their respective classifications vary between different studies. Drawing primarily on Young (1991), Underdal (1994), Grubb and Gupta (2000), Wurzel and Connelly (2011a) and Liefferink and Wurzel (2016), we distinguish four main *types* of leadership – structural, entrepreneurial, cognitive and exemplary – and two *styles* of leadership, namely humdrum/transactional and heroic/transformational.[3]

Structural leadership relates to an actor's hard power, which depends on material resources (e.g. military and economic resources). As the relevance of military power tends to be low for solving environmental problems – not even the world's most powerful country could prevent climate change by military action – structural leadership rests primarily on economic power. It is the Single European Market (SEM) – the world's largest internal market – which gives the EU the economic power resources required for structural leadership. The EU may allow, for example, the import of certain products into its SEM only if they comply with minimum environmental standards. Similarly, member states (especially large ones) and societal actors (e.g. businesses and environmental NGOs – ENGOs) can provide structural leadership (Wurzel and Connelly 2011b). As the UK was the EU's second largest GHG emitter and the world's fifth largest economy at the time of the 2016 Brexit decision, the EU's structural leadership is likely to be diminished once the UK has left the EU.

Second, *entrepreneurial* leadership involves diplomatic, negotiating and bargaining efforts which are necessary for finding compromise solutions in climate change negotiations. Entrepreneurial leadership enables the adoption of package deals which offer benefits to all parties involved. Successful French diplomatic efforts in the run-up to and during the 2015 Paris climate conference (COP21) provide good examples (see Chapter 7). In international climate change politics it has traditionally been the Commission (see Chapter 3) and the rotating six-monthly EU Council Presidency (see Chapter 5) which have offered entrepreneurial leadership. However, especially large member states – France, Germany and, until it voted to leave the EU, the UK – took on the role of 'lead negotiators' which stay in place well beyond the rotating EU Presidency whose influence has declined in international climate change negotiations (Interviews 2013–2015; see Chapters 2 and 5). The European External Action Service (EEAS), which was created in 2011, has potentially provided the EU with additional entrepreneurial leadership capacity in the form of a diplomatic service which can be activated on international climate change issues although in practice it is underutilised. Considering the UK's relatively progressive stance on climate change and its important role in the EU's climate diplomacy (see Chapter 12), the UK's vote in June 2016 to leave the EU may appreciably weaken the former and the latter's entrepreneurial leadership capacity.

Third, *cognitive* leadership involves the definition and/or redefinition of interests, problem perceptions and conceivable solutions through concepts such as the 'green' or low-carbon economy, or ecological modernisation, which propounds the view that ambitious environmental measures are beneficial for both the environment and economy (Jänicke 1993). Cognitive environmental leadership often relies on scientific expertise and practical implementation knowledge (Liefferink and Wurzel 2016). The EU has been portrayed as a 'normative power' (Manners 2002), which suggests that it relies more heavily on cognitive than on structural leadership. The Commission and other actors have tried to reconceptualise climate change from being a purely environmental issue, to one entailing both a security dimension (e.g. energy security and 'climate refugees') and an economic dimension (e.g. 'green' jobs in the low-carbon economy) (cf. Wurzel and Connelly 2011b: 277–8).

Lacking significant military powers and only infrequently mobilising its structural 'market power' (Damro 2012) when exerting climate leadership, the EU has specialised in providing entrepreneurial and particularly cognitive leadership. Small member states which are keen on ambitious EU and/or international climate change policy measures but have few power resources often use similar strategies. For example, Denmark and the Netherlands have gained disproportionate influence (by comparison with their population sizes) on EU environmental policy (e.g. Liefferink and Andersen 1998).

Fourth, our book will also focus on *exemplary* leadership or leadership by example (Liefferink and Wurzel 2016). Exemplary leadership is similar but not identical to directional leadership as defined by Grubb and Gupta (2000). While directional leadership assumes an *intention* to set a good example for others – actors which use directional leadership want to attract followers – our definition of exemplary leadership includes both (1) *intentional* example setting with the aim of attracting followers (i.e. directional leadership) and (2) *unintentional* example setting. In the latter case, actors adopt ambitious climate change measures primarily for internal reasons without wanting to influence others. However, *unintentional* example setting may nevertheless be emulated by others. Denmark is often held up as a member state which is primarily interested in adopting ambitious domestic environmental policy measures rather than in attracting followers (Andersen and Liefferink 1997; see also Chapter 6). The policy diffusion and transfer literature offers many instances of intentional and unintentional transnational example setting (e.g. Tews *et al.* 2003).

In this book we also draw on different leadership *styles* which has two major analytical advantages (Wurzel and Connelly 2011b; Liefferink and Wurzel 2016). First, it allows for the analysis of *how* leaders and pioneers act. Second, it enables us to introduce a *time dimension* (e.g. short or long term) for assessing the actions of leaders and pioneers. For our analytical leadership styles dimension we draw on Hayward (1975, 2008), who differentiated *humdrum* from *heroic* leadership, and Burns (1978, 2003) who distinguished between *transactional* and *transformational* leadership (see Liefferink and Wurzel 2016). Following Lindblom's (1959) concept of 'muddling through', Hayward (1975, 2008: 6) defined a humdrum leadership style as one where there is no 'explicit, overriding, long-term objective and action is incremental, departing only slightly from existing policies as circumstances require'. In contrast, a heroic style, which normally can be used only occasionally, 'sets explicit long-term objectives to be pursued by maximum coordination of public policies and by an ambitious assertion of political will' (Hayward 2008: 7).

For Burns (2003: 375), *transactional* leadership amounts to reactive leadership which adjusts to external circumstances and aims at achieving 'short-term expedient goals rather than long-term political strategy' (Burns 2003: 5). In contrast, *transformational* leadership aims to bring about profound or even 'revolutionary' change, which usually requires the pursuit of long-term objectives.

Importantly, although transactional leadership usually fosters only incremental piecemeal changes, '[c]ontinual transaction over a long period of time can produce transformation' (Burns 2003: 25). Transactional and transformational styles should therefore be perceived as part of a continuum; the same applies to humdrum and heroic leadership (Wurzel and Connelly 2011b; Liefferink and Wurzel 2016).

Although subtle differences exist between humdrum and transactional leadership styles, and between heroic and transformational leadership styles (Liefferink and Wurzel 2016), for the purposes of the argument of this book they are sufficiently close to permit the merging of humdrum with transactional, and heroic with transformational leadership styles.

Table 1.2 presents summary definitions of the leadership types and styles as used in the chapters which follow.

Actors' *internal* ambitions (to adopt progressive domestic climate change measures) have to be distinguished from their *external* ambitions (to lead others on climate change policy). Importantly, ambitions may change over time (e.g. the UK developed from an environmental laggard to a climate change leader) and vary between policy areas (e.g. Germany has high ambitions for renewable energy and low ambitions for phasing out coal (see Chapters 8 and 12 respectively).

Actors with *high internal* and *low external ambitions* are *pioneers* which unilaterally adopt more progressive internal climate change policy measures without regard for other actors. Pioneers set a good example without intending to attract followers. However, others may nevertheless follow their example. Leaders with *high internal* and *high external ambitions* are *pushers* which actively seek to attract followers. *Symbolic leaders* are characterised by low internal and high

Table 1.2 Types and styles of leadership

Types of leadership	*Styles of leadership*
1 *Structural leadership:* • Relies on material resources (e.g. economic or military strength) derived from hard power. 2 *Entrepreneurial leadership:* • Diplomatic, negotiating and bargaining skills. 3 *Cognitive leadership:* • Definition/redefinition of interests through ideas (e.g. low carbon economy). 4 *Exemplary leadership:* • *Intentional* and *unintentional* example setting.	a *Humdrum/transactional leadership:* • *Humdrum* leadership is short-term and incremental, leading to marginal adjustments of existing policies. *Transactional leadership* is reactive and aims at short-term expedient goals without the provision of long-term strategies. b *Heroic/transformational leadership:* • *Heroic* leadership relies on long-term objectives, strong policy coordination and the assertion of political will. *Transformational* leadership aims to bring about radical or 'revolutionary' change.

Sources: adapted from Young (1991), Burns (1978, 2003), Hayward (1975, 2008), Nye (2008), Wurzel and Connelly (2011a/b) and Liefferink and Wurzel (2016).

external ambitions. *Laggards* have neither internal nor external ambitions (Liefferink and Wurzel 2016).

The early literature on leaders and pioneers in environmental/climate change politics focused almost exclusively on states (e.g. Andersen and Liefferink 1997; Underdal 1994; Young 1991). However, this has changed with non-state actors, such as the EU, ENGOs, businesses and cities, having also been identified as potential environmental/climate policy leaders and pioneers. The reasons why non-state actors have increasingly been seen as offering climate change leadership are complex and contested. The list of possible reasons includes the 2008 global financial crises and resultant global recession which pushed environmental/climate policy actors onto the defensive, the political salience of climate change becoming somewhat less important relative to other political issues (e.g. economic or migration issues), and state actor-dominated top-down *government* allegedly giving way to non-hierarchical forms of *governance* in which non-state actors play a central role. Although the reasons are contested, there is widespread agreement in the academic literature and among practitioners (Interviews 2013–2015) that some of the EU's traditional environmental leaders (e.g. the Netherlands, see Chapter 9) have become more cost-conscious and less willing to provide climate change leadership.

Multi-level and polycentric climate governance

Arguably, to a greater extent than most policy areas, climate change policy permeates all levels of governance and requires the involvement of a wide variety of actors ranging from international organisations to individual citizens. This is one important reason why this book pays attention not only to the EU institutions, and (member) states but also to businesses and ENGOs. Multi-level governance (MLG) and polycentric governance therefore constitute important research themes in this book.

According to Stephenson (2013: 817) MLG 'has been used to try to provide a simplified notion of what is pluralistic and highly dispersed policy-making activity, where multiple actors (individuals and institutions) participate, at various political levels, from the supranational to the sub-national or local level', while Algica and Tarko (2012: 237) have defined polycentricity as 'a social system of many decision centres having limited and autonomous prerogative and operating under an overarching set of rules'. MLG and polycentricity share certain core presuppositions, although conceptually they are *not* identical. Importantly, by comparison with polycentricity, MLG studies usually assume a stronger involvement of governmental actors in both the setting and implementation of the rules of the game. However, compared with state-centred approaches, MLG places more emphasis on sub-national governmental and non-state actors. Polycentricity, on the other hand, attributes a stronger degree of autonomy to subnational local actors and non-governmental actors, including individuals. Most MLG-inspired EU studies emphasise the mutual dependency of EU and governmental actors (including national and subnational governmental actors) as well as (although to a somewhat lesser degree) non-governmental actors. Polycentric studies are primarily interested in the discovery of cases of societal self-coordination at various levels of governance as well as the mechanisms and rules of the game which underpin successful self-coordination in the face of collective action problems. It is possible to argue that polycentricity is closer to the operation of markets (see Chapter 14) than networks in which governmental actors (including EU institutional actors) play an important, if not dominant, role as most MLG studies promulgate.

The concept of MLG has long been widely used in the assessment of the EU, especially its structural policy (e.g. Hooghe 1996; Marks 1993) and (though to a lesser degree) its environmental policy (e.g. Fairbrass and Jordan 2004). The concept of polycentricity has rarely been used to study EU climate change policy, although there are exceptions (Jänicke 2015; Jordan and Huitema 2014; Rayner and Jordan 2013). This is despite the plea by Elinor Ostrom (2012: 82), one of the pioneers of the concept of polycentricity: 'Indeed, I argue very strongly for the need for polycentric institutions to cope with climate change'.

Broadly speaking, the EU has consistently demanded an MLG-type global climate governance system that relies on the UN as the negotiating forum for the adoption of legally binding GHGE reduction targets and deadlines. However, other important actors in international climate change politics (e.g. US, China and India, see Chapters 16, 17 and 18) instead favoured voluntary pledges or so-called Intended Nationally Determined Contributions (INDCs) (see below).

Partly because the EU constitutes an unconventional MLG system rather than a conventional state, it has sometimes struggled both to *act* as a coherent collective political entity in the international climate change negotiations and, in cases where it did so, to be *recognised* by third states as such an actor. The EU's capacity to act in the international arena has occasionally been contested by both EU internal actors (e.g. member governments) and actors outside the EU (e.g. developing countries). In other words the EU's 'actorness' has been questioned: which is one important reason why most international climate change agreements are so-called mixed agreements that are signed by the EU *and* its member states (see Chapter 2).

Some of the EU's core characteristics as an MLG system help to explain its occasional inability to adopt progressive internal and/or external climate change goals and thus to provide successful climate change leadership. At first sight the EU seems ill equipped to offer climate change leadership because decision-making powers are dispersed among a wide range of actors. The EU was founded shortly after the Second World War with the aim of '"taming" the "beast" … of leadership' (Blondel 1987: 3). It is therefore not surprising that the EU has been characterised as a 'leaderless Europe' (Hayward 2008). The existence of many veto actors has repeatedly created political deadlock and 'joint decision traps' (Scharpf 1988) from which the EU had difficulties to escape.

Against this background, it is remarkable how often the EU's climate change policy has in fact managed to overcome decisional stalemate. This shows that the EU's MLG structures do not necessarily preclude it from adopting progressive internal climate policy measures and from acting as a leader in international climate change politics (Schreurs amd Tiberghien 2007). Because MLG extends beyond the boundaries of the EU, this book also offers an external perspective provided by chapters on the US (Chapter 16), China (Chapter 17) and India (Chapter 18).

Policy instruments

Environmental policy instruments can be grouped into the following three main categories: (1) regulations (or 'command-and-control' instruments); (2) market-based instruments (e.g. eco-taxes and emissions trading schemes); and, (3) voluntary agreements and informational devices (see Wurzel, Zito and Jordan 2013). Regulations set legally binding targets and deadlines which governmental actors enforce. Market-based instruments provide fiscal incentives for actors which can choose the most efficient compliance option. Voluntary agreements rely on companies or states' willingness to honour pledges while informational instruments appeal to the 'green' conscience of consumers.

Majone (1996) has characterised the EU as a 'regulatory state' because it relies heavily on traditional regulation. Until the late 1990s, this has been true also for the EU's climate change policy (e.g. Wurzel, Zito and Jordan 2013). There are three main reasons for this. First, environmental policy was initially perceived as an inherently regulatory policy. Second, the EU can make use only of a limited number of redistributive policy instruments (e.g. structural funds). Third, the EU Treaties allow member states to veto eco-taxes.

Thus, attempts to adopt market-based instruments failed because the Commission's 1992 proposal for a common CO_2/energy tax was vetoed by the UK on sovereignty grounds (see Chapter 12). Unable to overcome this stalemate, the Commission published its proposal for a Directive establishing an EU ETS in 2001 (CEC 2001). The EP and the Environmental Council adopted the Commission's EU ETS proposal speedily and with relatively few changes (see Chapter 3). When the EU ETS became operational in 2005 it was the world's first supranational ETS. This is remarkable as it was the US which had first innovated with regional emissions trading schemes (for sulphur dioxide and nitrous oxides) in the 1980s, while the EU initially opposed American efforts to include emissions trading as a possible policy instrument in the 1997 Kyoto Protocol (Skjærseth and Wettestad 2008; Wurzel 2008). Voluntary agreements and informational policy tools have often been promoted by businesses (see Chapter 14) while ENGOs have remained highly sceptical about self-regulatory tools (see Chapter 15). Such instruments play only a secondary role in EU climate change policy.

Green economy and low-carbon economy

One of the clearest statements in favour of a 'green' or low-carbon economy by an EU institutional actor can be found in the Commission's *Roadmap for Moving to a Competitive Low Carbon Economy in 2050* (CEC 2011). The document argues that ambitious EU climate change policy measures will trigger 'smart, sustainable and inclusive growth' (CEC 2011: 3) and states that

> [i]nvesting early in the low carbon economy would stimulate a gradual structural change in the economy and can create in net terms new jobs both in the short- and the medium-term.... In the longer-term, the creation and preservation of jobs will depend on the EU's ability to lead in terms of the development of new low carbon technologies through increased education, training, programmes to foster acceptability of new technologies, R&D and entrepreneurship, as well as favourable economic framework conditions for investments.
>
> (CEC 2011:12)

Although the 2050 Roadmap does not explicitly mention ecological modernisation, it is clearly compatible with the concept which has received widespread support particularly in Northern Europe (CEC 2008; Jänicke 1993; Wurzel and Connelly 2011b). The claim is that ecological modernisation can create a 'double dividend' or 'win-win' situation in which economic growth and the protection of the environment take place simultaneously (Jänicke 1993). From a low-carbon economy perspective, climate change therefore not only poses a threat but also creates opportunities for 'green' jobs. Moreover, a speedy uptake of renewable energy would decrease EU member states' dependence on energy imports (e.g. from Russia) and increase the EU's energy security. However, the 2008 financial crises and the subsequent economic recession have diminished somewhat the support for a 'green' or low-carbon economy while proponents of ecological modernisation have been put on the defensive. The immediate turmoil in the global stock markets and enduring uncertainty which followed the 2016 Brexit decision has made it even more challenging for proponents of the green economy to convince others with their arguments.

Structure of the book

The next chapter (with this Introduction constituting Part I) focuses on the EU's ability to act as a global player in international climate change politics. Part II assesses why, when, and how supranational institutional actors pushed the EU to adopt a leadership position in international climate change politics. The next two parts analyse why, when, and how member states (Part III) and important societal actors (Part IV) have offered domestic leadership and/or supported EU and international climate change leadership. Finally Part V offers an external perspective by focusing on the US, China and India.

Notes

1 Rudi Wurzel would like to thank the British Academy (SG131240) and Faculty of Arts and Social Sciences (FASS) at Hull University for funding parts of the research on which this chapter is based. He would like to thank all interviewees for giving up pressured time to be interviewed. The authors of this chapter take full responsibility for normative statements and any possibly errors.

2 In the 1980s, the Thatcher government closed many coal mines and built gas fired power stations for cost reasons and to break the political influence of the left-wing miners unions.

3 Underdal (1994) distinguishes between coercive, instrumental and unilateral leadership, Grubb and Gupta (2000) between structural, instrumental and directional leadership and Wurzel and Connelly (2011a) between structural, entrepreneurial and cognitive leadership.

References

Algica, P. and Tarko, V. (2012) 'Polycentricity: From Polanyi to Ostrom and Beyond', *Governance*, 25(2): 237–62.

Andersen, M.S. and Liefferink, D. (eds) (1997) *European Environmental Policy. The Pioneers*, Manchester: Manchester University Press.

Bäckstrand, K. and Elgström, O. (2013) 'The EU's Role in Climate Change Negotiations: From Leader to "Leadiator"', *Journal of European Public Policy*, 20(10): 1369–86.

Blondel, J. (1987) *Political Leadership. Towards a General Analysis*, London: Sage.

Burns, J.M. (1978) *Leadership*, New York: Harper & Row.

Burns, J.M. (2003) *Transforming Leadership*, New York: Grove Press.

CEC (2000) *Towards a European Climate Change Programme*, Brussels: Commission of the European Communities.

CEC (2001) *Proposal for a Directive Establishing a Scheme for Greenhouse Gas Emission Allowance Trading. COM(2001)581*, Brussels: Commission of the European Communities.

CEC (2008) *Boosting Growth and Jobs by Meeting Our Climate Change Commitments, Press Release IP/08/80, 23.1.2008*, Brussels: Commission of the European Communities.

CEC (2011) *A Roadmap for Moving to a Competitive Low Carbon Economy in 2050. COM(2011)112 Final*, Brussels: Commission of the European Union.

Damro, C. (2012) 'Market Power Europe', *Journal of European Public Policy*, 19(5): 682–99.

Eurobarometer (2015) *Climate Change. Special Eurobarometer 435*, Brussels: European Commission.

Fairbrass, J. and Jordan, A. (2004) 'Multi-level Governance and Environmental Policy', in A. Bache and M. Flinders (eds), *Multi-level Governance*, Oxford: Oxford University Press, 147–64.

Grubb, M. and Gupta, A. (2000) 'Climate Change, Leadership and the EU', in J. Gupta and M. Grubb (eds), *Climate Change and European Leadership*, Dordrecht: Kluwer, 3–14.

Haigh, N. (1996) 'Climate Change Policies and Politics in the European Community', in T. O'Riordan and J. Jäger (eds), *Politics of Climate Change*, London: Routledge, 155–86.

Hayward, J. (1975) 'Introduction', in J. Hayward (ed.), *Planning, Politics and Public Policy*, Cambridge: Cambridge University Press, 1–21.

Hayward, J. (2008) 'Introduction: Inhibited Consensual Leadership within an Interdependent Confederal Europe', in J. Hayward (ed.), *Leaderless Europe*, Oxford: Oxford University Press, 1–14.

Hill, C. (1993) 'The Capability-Expectations Gap, or Conceptualising Europe's International Role', *Journal of Common Market Studies*, 31(3): 305–28.

Hooghe, L. (ed.) (1996) *Multi-level Governance and European Integration*, Oxford: Clarendon Press.

Jänicke, M. (1993) 'Über ökologische und politische Modernisierungen', *Zeitschrift für Umweltpolitik und Umweltrecht*, 16: 159–75.

Jänicke, M. (2015) 'Horizontal and Vertical Reinforcement in Global Climate Governance', *Energies*, 8: 5782–99.

Jordan, A., Huitema, D., van Asselt, H., Rayner, T. and Berkhout, F. (2010) *Climate Change Policy in the European Union*, Cambridge: Cambridge University Press.

Jordan, A. and Huitema, D. (2014) 'Innovations in Climate Policy: the Politics of Invention, Diffusion, and Evaluation', *Environmental Politics*, 23(5): 715–34.

Liefferink, D. and Andersen, M.S. (1998) 'Strategies of the "Green" Member States in EU Environmental Policy-making', *Journal of European Public Policy*, 5(2): 254–70.

Liefferink, D. and Wurzel, R.K.W. (2016), 'Environmental Leaders and Pioneers: Agents of Change?', *Journal of European Public Policy*, DOI:10.1080/13501763.2016.1161657.

Majone, G. (1994) 'The Rise of the Regulatory State', *West European Politics*, 17(3): 77–101.

Marks, G. (1993) 'Structural Policy and Multi-level Governance in the EC', in A. Cafruny and G. Rosenthal (eds), *The State of the European Community*, Boulder: Lynne Rienner, 391–411.

Nye, J. (2008) *The Powers to Lead*, Oxford: Oxford University Press.

Oberthür, S. and Roche Kelly, C. (2008) 'EU Leadership in International Climate Policy: Achievements and Challenges', *The International Spectator*, 43(2): 35–50.

Oberthür, S. and Bodle, R. (2016) 'Perspectives on EU Implementation of the Paris Outcome', *Carbon and Climate Law Review* (forthcoming).

Ostrom, E. (2012) 'The Future of the Commons: Beyond Market Failure and Government Regulation', in E. Ostrom, C. Chang, M. Pennington and V. Tarko (eds), *The Future of the Commons*, London: The Institute of Economic Affairs, 68–83.

Rayner, T. and Jordan, A. (2013) 'The European Union: the Polycentric Climate Leader?', *WIRES (Wiley Interdisciplinary Reviews) Climate Change*, 4 (2):75–90.

Scharpf, F. (1988) 'The Joint-Decision Trap: Lessons from German Federalism and European Integration', *Public Administration*, 66: 239–78.

Schreurs, M. and Tiberghien, Y. (2007) 'Multi-level Reinforcement: Explaining European Union Leadership in Climate Change Mitigation', *Global Environmental Politics*, 7(4): 19–46.

Skjærseth, J.B and Wettestad, J. (2008) *EU Emissions Trading: Initiation, Decision-making and Implementation*, Aldershot: Ashgate.

Stephenson, P. (2013) 'Twenty Years of Multi-level Governance: "Where Does It Come From? What Is It? Where Is It Going?"', *Journal of European Public Policy*, 20(6): 817–37.

Tews, K., Busch, P. and Jörgens, H. (2003) 'The Diffusion of New Environmental Policy Instruments', *European Journal of Political Research*, 42(4): 569–600.

Underdal, A. (1994) 'Leadership Theory. Rediscovering the Arts of Management', in I.W. Zartman (ed.) *International Multilateral Negotiation*, San Francisco: Jossey-Bass, 178–97.

Wurzel, R.K.W. (2008) *The Politics of Emissions Trading in Britain and Germany*, London: Anglo-German Foundation.

Wurzel, R.K.W. and Connelly, J. (2011a) 'Introduction' in R. Wurzel and J. Connelly (eds) *The European Union as a Leader in International Climate Change Politics*, London: Routledge, 3–20.

Wurzel, R.K.W. and Connelly, J. (2011b) 'Conclusion' in R. Wurzel and J. Connelly (eds) *The European Union as a Leader in International Climate Change Politics*, London: Routledge, 271–89.

Wurzel, R.K.W, Zito, A. and Jordan, A. (2013) *Environmental Governance in Europe*, Cheltenham: Edward Elgar.

Young, Oran (1991) 'Political Leadership and Regime Formation: On the Development of Institutions in International Society', *International Organization*, 45(3): 281–308.

2 Global climate politics

Can the EU be an actor?

John Vogler

Introduction

It has become commonplace in discussions of environmental and other global issues to treat the EU as if it were a single purposive entity. The Union is urged to act, even to lead. From the early days of the UNFCCC it has, perhaps ostentatiously, ascribed such a leading role to itself. Leadership is the theme of this volume but it is logically inseparable from the capacity to act which may, inelegantly, be described as 'actorness'. Normally this point would hardly be worth making, individuals and state governments obviously perform acts of leadership. However, with the EU there is a problem that requires consideration of the definition of a political actor, something that is normally taken for granted. The problem is quite simply expressed. The Union is neither an emergent federal state nor an over-developed international organisation. In an international system where the capacity to act has conventionally been confined to sovereign states (although with some modification for certain types of international organisation) the EU is unique, an entity which is according to the international lawyers, *sui generis*.

This chapter sets up some criteria to establish the extent to which the EU may be regarded as an international actor in general and more specifically in the field of climate change policy. The analysis is based upon previous work (Vogler 1999, Bretherton and Vogler 1999, 2006, 2013) and suggests four characteristics that one might expect an international actor to exhibit. They are: autonomy, volition, negotiating capability and the ability to deploy policy instruments. These outward and visible signs of actorness rely upon various capabilities possessed by the Union but they are also dependent upon its broader 'presence' in the global political economy and upon the opportunity structure that confronted Union decision-makers at particular points in the history of the international climate regime. While this structure may have facilitated the development of EU actorness and its leadership ambitions during the 1990s and in the first decade of the twenty-first century, the failure to deliver a new climate settlement at Copenhagen in 2009 (see Chapter 6) and the relative exclusion of the Union from the negotiation of the 'Accord' were seen as an adjustment to new and changed international realities (Grubb 2010).

The status of an actor and the exercise of leadership are not automatic consequences of capability, presence and opportunity. Just as in the social life of an individual, actorness is conditional upon the expectations and constructions of third parties that serve to establish identity, reputation and credibility. The EU as actor may capitalise on such constructions, but they may equally prove to be damaging, as appears to have been the case with the Common Foreign and Security Policy and Christopher Hill's (1993) 'capability-expectations gap'. External environmental policy has been a great deal more impressive, in terms of fulfilling and even exceeding such expectations. This is particularly so in respect to climate change where the prospects for the development of actorness were not perhaps as favourable as in other areas of external environmental policy, falling more directly within Union rather than member states' competence.

Autonomy

This broaches the difficult question of the extent to which the Union is a policy actor distinguishable from its component member states. It involves formal questions of competence to act in external policy and legal recognition alongside the more subtle constructions of those who interact with the EU. The paradigm example of the EU as a single actor with evident, although circumscribed, independence from the member states is to be found in the area of trade – the Common Commercial Policy initially under Article 113 of the Treaty of Rome provided a clear Treaty basis for exclusive Community competence in trade negotiations. This was necessitated by the practicalities of setting up a single external tariff for the Customs Union. Indicative of an autonomy rarely, if ever, found in other international organisations is the fact that the Commission has the right of initiative and there is normally (with certain exceptions) qualified majority voting in the Council. Only the Commission represents the Union when matters of exclusive competence are under negotiation although the member states may agree to allow the Commission to act for them in other areas as well. In a formal recognition of autonomy and the right to sign treaties, the Union has legal personality bestowed upon it by the Treaty of Lisbon (this was the case for the European Community under the original Treaty of Rome) and is a full member of the World Trade Organization (WTO) in its own right alongside the member states. This is a comparatively rare status for the EU. It usually only enjoys observer status in international organisations and most notably at the UN (the UN Food and Agriculture Organization and Convention on Trade in Endangered Species and a number of fisheries organisations provide exceptions to this rule as well as the Regional Economic Integration Organisations (REIO) provisions for Conventions outlined below). To outside observers the Commission representing the Union at the WTO and in bilateral trade discussions does appear as a single actor embodied in the person of the Trade Commissioner. However, the Commission still functions as an agent for its member state principals in the Trade Policy Committee that convenes at various levels. Nonetheless the Commission enjoys a significant degree of autonomy and has attempted, especially since the entry into force of the Lisbon Treaty, to reproduce this arrangement in other international negotiations.

Climate change was initially defined by the European Community as a matter of environmental policy. As such there was no original treaty basis until the 1987 Single European Act under which the European Community became a party to the UNFCCC and its Kyoto Protocol. Internal rule-making provided another basis for asserting competence, exclusive of the member states, once the Community/Union had legislated on an issue. The 1971 ERTA judgement of the European Court of Justice (case 22/70) ensured that once competence had been achieved internally it was automatically extended to the conduct of external policy. This allowed the Commission to insist that it should participate alongside the member states in international environmental negotiations, sharing competence and concluding 'mixed agreements' where both the Union and member states are signatories. As has often been remarked, climate change is a quintessentially multi-sectoral problem that does not align easily with Community/Union competences and where key issues of taxation and energy policy remain within the exclusive competence of member states.

In terms of formal representation, shared competence for climate policy meant presidency leadership where the member state currently 'President in Office' represented the Union. The Commission appeared alongside the 28 member states and it had been said that the EU negotiated 'at 29'. After the entry into force of the Lisbon Treaty in 2010 the Commission sought to extend its representational remit so as to become sole negotiator in areas within shared competence. There was an argument to be made on grounds of efficiency and also on the basis of an article in the new Treaty (171(1) TFEU) stating that the Commission 'shall ensure the Union's external representation'. Member states resisted this and although they might on occasion be willing to allow the Commission such a role on an ad hoc basis, the text of the Lisbon Treaty, which mentioned climate change as a strategic concern (Art. 4(2)(e)), did so 'in a way that does not change the existing distribution of competences' (House of Lords 2008). United Nations Environment Programme (UNEP) sponsored negotiations for what became the Minimata Convention on Mercury, provided a test case during 2010 in which a Commission-Council dispute over competence led the Union into an embarrassing failure to offer any initial position. The issue was resolved through the continuation of dual representation where both Commission and Presidency sit behind a single EU nameplate. In climate negotiations this is apparent during plenary sessions where the Climate Action and Energy Commissioner sits alongside the minister of the state holding the Presidency of the Council and will even share the making of formal statements. As will be further discussed below, this does not accord with the actual distribution of negotiating responsibilities (see Chapter 5).

Even if the Union has competence for an issue, this does not necessarily mean that it will be accepted as an actor by other parties to a negotiation. The European Community confronted this problem of external recognition in the late 1970s when, on the basis of its newly acquired competences for air pollution it attempted to participate, alongside the member states, in the Long Range Transboundary Air Pollution (LRTAP) negotiations that emerged from the East-West détente process. The 1979 LRTAP established a formula that continues to be used to allow Union participation in the UNFCCC and various other global environmental regimes. Under it REIOs can participate and sign agreements alongside state parties, when they can demonstrate relevant competence and on condition that they do not acquire separate voting rights. In fact the EU is the only extant REIO and voting provisions are not significant, the UNFCCC having failed to agree its rules of procedure on precisely this issue. However, it provides an actual and symbolic advance on the observer status of the EU found elsewhere in the UN system and at UNEP.

The distribution of competences between the Union and member states suggests a limited role for the Union and perhaps, by analogy with trade or fisheries policy, diminished autonomy for the EU as an actor. In fact this does not appear to have been the case and the international actorness of the EU should not be assessed solely in terms of the extent to which the Union enjoys exclusive competence and is recognised as an REIO. Majority voting in the Council and the oversight of the European Parliament are also significant in so far as they indicate the existence of a policy actor which is not subject to individual national vetoes, as remains the case in intergovernmental areas such as the Common Foreign and Security Policy (CFSP). A degree of Union legislative autonomy is, furthermore, of importance for other Parties in negotiations who need to be reassured that the EU can deliver its collective commitments.

Volition

The ability to formulate distinct policy is the second attribute of an actor. It is a logical precursor to attempts to exercise leadership. Such leadership has always had an exemplary, and indeed transformational, character in that the EU has challenged other Parties to match its targets and timetables. There are some areas of EU external relations, such as the CFSP, where woeful inadequacies in terms of collective political will reflect major differences in political interests and orientation between the member states even over largely symbolic issues. External climate change policy is different because its credibility increasingly came to depend upon the effective delivery of internal policies. As various chapters in this book illustrate, this is reflected in the move from mainly symbolic to cognitive, exemplary and entrepreneurial leadership at the Union level. Despite its early classification as an atmospheric environmental issue climate policy must be concerned with energy and transport-related emissions where there are marked differences between the member states in terms of their respective energy mix, ranging from extensive reliance upon nuclear power generation (Chapter 7) to continued dependence upon coal (Chapter 10) or imported gas. Furthermore climate policy raises major sectoral competitiveness issues. Of course, many of the Parties to the UNFCCC face similar problems in reconciling a variety of interests in their climate policy, yet forging a common purpose among the member states of the EU sets an even more demanding challenge. Nonetheless, it is one that has been largely met. From the first phase of climate negotiations (see Chapter 1) the EU has been able to define a collective purpose usually of greater ambition than most of its interlocutors, providing the basis for entrepreneurial and cognitive leadership. The 'rescue' of the Kyoto Protocol was particularly significant being sustained by an internal burden-sharing agreement that facilitated a commitment to reduce emissions by 8 per cent. This EU 'bubble' relied upon a particularly fortuitous set of circumstances relating to the use of the 1990 baseline. Both Germany, through re-unification, and the UK through the closure of most of the British deep-mined coal industry were able to promise large, but essentially painless, cuts in their GHGE. Southern 'cohesion' member states were also allowed emissions growth targets (Ringius 1997, Vogler 2009). Matters became much more difficult after enlargement and under new co-decision procedures when choices entailing real economic costs for member states had to be made in the 2008 Climate and Energy Package that supported the Union's 20-20-20 position (see Chapters 1, 3 and 5). Chapters 3–5 of this volume document in detail the processes whereby successive internal and external climate policy positions were achieved. After the setback at Copenhagen internal policy development in, for example, the roadmap towards a decarbonised economy (Commission 2011) and willingness to engage, almost alone, in a second commitment period of the Kyoto Protocol, provided an internal basis for the Union's significant diplomatic action in sponsoring the negotiation of the Durban Platform. In the same period the failure of the Union's progressive attempt to include international aviation emissions within ETS resulted, not from a lack of volition but, rather, external resistance (Keating 2014). Before the 2015 Paris climate conference (COP15) the Union called for a new agreement that would be 'ambitious, legally binding, multilateral rules-based with global participation and informed by science' based on a renewed 2030 Climate and Energy Policy Framework with an 'intended contribution, of a 40% GHGE reduction' (European Council 2014). This indicates, once again, that the Union has displayed collective volition, under increasingly difficult circumstances, without being condemned to move at the speed of the 'slowest ship' or to resort to the 'lowest common denominator'. It also demonstrates the importance given to developing appropriate internal policies as the necessary underpinning of external exemplary leadership and negotiating credibility. This was summed up by a Malaysian official who remarked 'We know that the EU not only talks the talk, but walks the walk' (Interview, Bonn 2013).

Negotiating capability

An observer of the complexities of shared competence, in the context of a Union that enlarged from 12 to 28 member states (and 27 member states once the UK has left the EU) within the lifetime of the climate regime, could be forgiven for assuming that the EU, despite its ability to set objectives, would be incapable of effective negotiation or the exercise of entrepreneurial leadership. There are certainly a number of actual and potential impediments that include: reliance on a rotating presidency (van Schaik and Egenhofer 2003), the need for co-ordination meetings within a negotiation, the habit of the Nordic countries to hold their own meetings in advance of EU co-ordination, the need to accommodate new member states and the tendency of some Members to defy Presidency leadership and to attempt to cut deals with other Parties (The Hague COP of 2000 provides an example of such a UK attempt to circumvent the EU Presidency and negotiate a deal on sinks with the US). There is, too, ample evidence in the negotiating record of the climate regime of EU inertia and what has been described as 'Herculean problems of co-ordination' (Grubb and Yamin 2001). One consequence of the EU's policy system for international negotiations, with coordination and reference back to the Council, is that it may lack the necessary flexibility. This has been evident at critical points in the development of the climate regime, for example in its inability to respond to the US introduction of the Kyoto mechanisms and subsequently in its rigidity towards the US at the 2000 Hague COP (ibid.).

On the other hand, the Union, despite its difficulties, has been able to make strong long-term commitments to which other parties have had to respond. At Kyoto it managed to obtain two key objectives, crucial to its subsequent success. First, the retention of the 1990 baseline and second, the acceptance of the controversial EU 'bubble'. Its inflexibility and cumbersomeness can, on occasion, be virtues not normally recognised in discussions of entrepreneurial leadership and Union negotiators have used it as a negotiating ploy, stressing the difficulty of unpicking a painfully negotiated internal position (Bretherton and Vogler 2006: 34). Commitment to the Kyoto Protocol and targets and timetables, which for a period in Brussels seems almost to have acquired the status of religious belief, was in many ways admirably transformational, but hampered the development of a feasible post-2012 regime, when the US was equally strongly opposed (see Chapter 16).

The 2009 Copenhagen UNFCCC climate conference (COP15) was privately described by the President of the European Council Van Rompuy as an 'incredible disaster' in which the EU had been 'totally excluded and mistreated' (*Guardian* 2010: 6) with the, then, head of the Commission Barroso being heard to remark that there were on this occasion 'too many leaders'. Subsequent to this setback came a marked revival in the Union's climate diplomacy, culminating in its sponsorship of the negotiation of the Durban Platform. The European Council for Foreign Relations awards annual grades for the performance of the various sectors of external policy. These, especially for the CFSP, are often none too flattering – low 'Bs' and 'Cs'. In 2011 climate policy achieved a rare A-grade in recognition of the fact that here the EU had 'made diplomatic progress where none seemed likely. Although imperfect, the [Durban] deal was a significant victory for European diplomacy' (European Council for Foreign Relations 2012: 122) and provided the essential basis for the 2015 Paris Agreement.

A number of reasons may be advanced for this diplomatic revival in entrepreneurial leadership. One is certainly the talent and activism of the Union's first Commissioner for Climate Action, Connie Hedegaard. There was also a systematic attempt to mediate between the various negotiating groups in advance of Durban under what was known as the Cartagena Dialogue (van Schaik 2012). It was led by the Union in concert with Australia. Over the longer term there has been a practical institutional evolution in the way in which the Union approaches climate negotiations (see Chapters 3 and 5). This has served to overcome the difficulties of dual formal representation and the inherent instability of leadership by a Council Presidency that rotates every six months. The establishment of 'lead countries' and 'issue coordinators' allocates negotiation tasks within the EU climate delegation on a more permanent basis than the rotating Council Presidency and appears to encourage continuity of approach, full involvement of delegates and the most effective deployment of expertise in what is a sprawling and technically demanding set of negotiations (Delreux and Van den Brande 2013; see also Chapter 5).

Policy instruments

Actors in the international political system, if they are not to play a merely declaratory role, have to possess some capability to affect the decisions of others. This relates closely to concepts of leadership. Influence may be exerted through the power of example or through the dissemination of knowledge – described as cognitive or intellectual leadership (see Chapter 1). The EU has claims to both, but frequently there will be a need to back up policy positions through pressure or inducement in the exercise of structural leadership. The ultimate policy instruments for structural leadership available to state actors are military but they are hardly relevant to the problem of arranging international cooperation for the mitigation of climate change. Armed force will very likely figure in the management of the consequences of climate change and there has been spasmodic consideration in Brussels of the significance of climate change for the Common Security and Defence Policy (Council 2008).

The instruments potentially available to the EU in pursuit of its international climate cooperation strategies are, nonetheless, significant. They include the diplomatic resources of 28 states (and 27 states after Brexit), the very substantial structural inducements and penalties arising from the control of access to the Single European Market (SEM) and the Community's large aid budget. The latter was deployed by the ECOFIN Council before both the Copenhagen and Paris COPs to put climate funding contributions 'on the table' as a key part of the EU's negotiating strategy.

In pursuit of Kyoto ratification, the EU used economic instruments in support of its climate policy. During 2004 when Russian ratification was absolutely necessary if the Protocol was to enter into force, a deal was concocted whereby the *quid pro quo* for Russian agreement would be EU support for its WTO membership plus some adjustments to the terms on which Russian gas entered the SEM (Bretherton and Vogler 2006: 109). This may not be a particularly strong example, because no great sacrifices were entailed for Russia which enjoys substantial quantities of 'hot air' under the Kyoto arrangements. Elsewhere it is difficult to establish such a clear linkage, but it is probably significant that the

EU has prioritised support for its climate change positions within the trade-dominated bilateral relations that it maintains with the majority of states in the international system. The EU-China relationship, which has been developed very extensively in the last decade, is of critical importance but it is unlikely that direct economic inducements figure here, the emphasis being placed rather more on technological collaboration (see Chapter 17).

Diplomatic co-ordination, sometimes at head of government level, was certainly achieved in the campaign to recruit ratifications of Kyoto. Between COP6 of 2000 and COP6bis, EU diplomatic missions were undertaken to Australia, Canada, Japan, the Russian Federation and Iran. The UK, France and Germany made high-level representations to various other governments and notably Japan (*Earth Negotiations Bulletin* 2001: 176). Similarly, there was a co-ordinated EU diplomatic effort through the G8 under the British and German joint Presidencies in 2005 and 2008 (see also Chapters 8 and 12). High-level activism is evident in advance of the Paris COP at the G7 summit in June 2015 but perhaps most significantly in the detailed diplomatic mediation and entrepreneurial work within the UNFCCC and bilaterally with a wide range of Parties to create and sustain the Durban Platform (Bäckstrand and Elgström 2013).

However, EU foreign policy co-ordination has not always been achieved, given the dispersion of effort between Commission delegations and member states' embassies, part of a more general problem arising from the different Community and intergovernmental 'pillars' of the EU prior to 2010. Subsequently the new European External Action Service (EEAS) was designed to integrate diplomatic staff and trade and other policy resources into a more effective collective effort. It also took over the existing 'Green Diplomacy Network' that since 2002 had linked interested member state diplomats. Understandably events in the EU's turbulent neighbourhood preoccupied the Foreign Affairs Council but in the approach to Paris 2015 climate diplomacy began to appear on its agenda and plans were made to co-ordinate diplomatic action in its aftermath (Council 2015, 2016; see also Chapter 5).

The conditions of actorness – presence and opportunity

Understanding the conditions under which the EU became an actor (and might indeed recede as an international actor) involves not only the possession of relevant capabilities but also the underlying conditions of presence and the opportunity structure within which the Union has to work. The concept of presence has been variously used in discussions of EU external policy (Allen and Smith 1990) but is defined here as:

> …the ability of the EU, by virtue of its existence, to exert influence beyond its borders. An indication of the EU's structural power, presence combines understandings about the fundamental nature, or identity of the EU and the (often unintended) consequences of the Union's internal policies.
>
> (Bretherton and Vogler 2006: 24)

Presence is, thus, essentially a consequence of 'being' rather than denoting purposive action. The most obvious aspect of the EU's presence in the international system is the enormous scale and attractiveness of the SEM for outsiders coupled with the magnetic effects of the Union for states and populations within its immediate and expanding orbit. One might add to this its internal policies and standards which are often simply adopted by external actors, because it makes good commercial sense to do so. Such presence often provides a necessary foundation for the development of international 'actorness' and the exercise of a different form of structural leadership that does not rely upon the direct use of economic instruments.

In terms of climate change the starting point for a discussion of presence must be the scale of economic activity in the Union, its historic, present and future contribution to climate change. Since 1992 very substantial structural change has occurred in the world political economy serving to diminish the relative position of the Union and it is easy, but perhaps excessively simplistic, to make a connection between this loss of standing and events at and after the 2009 Copenhagen climate conference (COP15) including the changing role of the Union. The rise to prominence of the BASIC group (Brazil, South Africa, India and China) reflects a longer process of economic growth following the ending of the Cold War. Chinese growth was most remarkable and Brazilian and Indian GDP both more than trebled in the period 1990–2010. Nonetheless in sheer GDP terms the SEM was still slightly larger than the US and substantially more than that of China (in trillions of dollars the 2014 figures are 18.5, 17.4 and 10.3 respectively) (World Bank 2016)). The scale of economic activity is related to responsibility for GHGE both cumulative, current and projected. For a long period, covering the foundation of the climate regime, the EU was second only to the US as an emitter of GHGE. In 2002 US emissions totalled 6.9 million metric tonnes of CO_2, the EU 4.8, Russia 1.9 and Japan 1.3. By 2005 the second place in the international league of emitters had been taken by a rapidly developing China which was to rival and even overtake the United States. In 2014 the US and China together account for around 45 per cent of total greenhouse gas emissions while the EU's share has fallen to around 10 per cent. The essential set of participants in a comprehensive bargain on mitigation is usually seen to comprise six Parties. Although ranking third among these six emitters responsible for roughly 70 per cent of global emissions (the others are India, Russia and Japan in descending order), the EU's emissions presence has been very much reduced (EU Edgar 2015).

Other important aspects of EU presence are its corpus of internal climate policies, from the ill-fated attempt to devise a carbon tax in the early 1990s through the development of the European Climate Change Progamme (ECCP) and, of course, the ETS. If the latter were to be adopted outside the Union, as is evidently the ambition of its designers, then EU presence would be extended in ways similar to those experienced with the introduction of the euro in the international monetary system – where the Union still cannot be considered to be a monetary actor. With the ETS 'the EU has "first mover" experience in setting up the EU ETS as the world's largest cap and trade system' and, in an example of exemplary leadership it was hoped that this would influence the creation of similar interlocked systems elsewhere (Commission 2009:11). Widespread adoption might provide an instance of polycentric governance in the creation of new carbon markets although continuing difficulties with the ETS carbon price do not bode well.

The conversion of presence into actorness, even where capabilities already exist, requires opportunity. Opportunity refers to the external environment of ideas and events that enable or constrain purposive action. In very general terms the creation of the EU occurred within the opportunity structure of the Cold War involving the strategic protection of the US bloc and benefiting from fixed currency parities pegged to the dollar. During the Cold War the development of European actorness was also constrained by the robust opposition of the Soviet Union to any dealings with the European entity. The first serious foray of the Community into global environmental diplomacy in 1979 was in fact enabled by the willingness of the Soviet Union (as part of the détente process) to treat with it on environmental issues, apparently in the expectation that Council for Mutual Economic Assistance (Comecon) would be granted similar REIO status.

The end of the Cold War also coincided with a marked rise in the salience of environmental issues in world politics, symbolised by the convening of the 1992 Rio Earth Summit. Both factors provided opportunities for the external projection of European environmental policies. However, the major opportunity seized by EU leaders to establish policy leadership was the abdication by the United States of its previously leading role. This occurred in a number of areas, but its attitude to the emergent climate regime from Rio onwards was as important as any. The US virtually invented modern environmental policy and had been a pioneer in, for example, the use of market-based instruments in contrast to the EU's reliance on 'command and control' (see Chapter 16). This lends a touch of irony to the EU's crusading stance on behalf of emissions trading, which it had opposed at Kyoto. On a whole swathe of issues, US governments had led the way and the EU could often be portrayed as a laggard. However by the early 1990s the situation had changed as the US resisted binding commitments under the draft UNFCCC. This pattern tended to continue through subsequent administrations, culminating in the 1997 Byrd-Hagel Resolution and George W. Bush's 2001 denunciation of the Kyoto Protocol. All this provided opportunities to be exploited by Brussels in developing its actorness and burnishing its identity as a climate leader, with the US, often unfairly, characterised as the main obstacle to progress.

In retrospect the 2008 global financial crisis appears to have marked a turning point. It underlined a trend that had been under way at least since the millennium in which, at the WTO and elsewhere, the supremacy of the US, the EU and other large developed nations had been effectively challenged by China and India leading to a reliance upon the G20 formation. The creation of the BASIC group can be interpreted as a particular instance of this broader trend. The Union had, after its period of confrontation with the US, worked hard from the Bali COP onwards to re-engage its Atlantic partner in UNFCCC climate diplomacy. This approach came to fruition with the election of President Obama who increasingly took up the challenge of climate change policy; found ways of circumventing a hostile Congress and re-asserted US leadership, pursuing a new climate understanding as one positive aspect of an otherwise conflictual relationship with China (see Chapter16).

Constructions of actorness

In the final analysis the links between presence, opportunity and capability are socially constructed by subtle processes whereby the Union is treated as an actor and followers are prepared to recognise its leadership. The identity of the Union, it may be argued, is subject to continuous re-definition through its interaction with outsiders. This is of some consequence for the EU as an actor, not only because conceptions of interest logically precede interests, but also because 'understandings about the external context of ideas and events, or the appropriateness or feasibility of alternative courses of action, are shaped by identity constructions that are themselves shifting and contested' (Bretherton and Vogler 2006: 37). In a mundane, but important, sense the EU is constructed as an actor in day-to-day accounts of international climate politics and this is more than convenient shorthand for the Union acting within its competences as an REIO and its member states. On occasion the attribution of 'actorness' to the Union has gone well beyond the strict letter of international law.

A significant dimension of the EU's collective identity is both essentially civilian and normative (Manners 2002) and this may certainly influence the willingness of outsiders to accept cognitive leadership. Alternatively, it has been subject to pejorative comparison with the US (Kagan 2002). Both the Commission and the European Council are at pains to reiterate the leading and normative role of the Union on climate change and, unlike other Union policies, it is one that enjoys consistently high levels of popular support and involved flattering comparisons with the inaction of the United States (Vogler and Bretherton 2006). Alongside the anti-death penalty campaign and support for the International Criminal Court, climate change formed part of the EU's international moral identity, integral to a transformational leadership style. A US government analyst put it like this. Support for the UNFCCC was one example of 'what Robert Kagan has described as the European predilection for establishing a Kantian world order, in which contentious issues are addressed, and potential conflicts resolved, through the establishment of suitably empowered global structures of governance' (Schmidt 2008: 94). This is in part a product of the EU's own history, but also, Schmidt (2008) argues, a consequence of social democratic approaches to policy and a regulatory culture of precaution and risk aversion. In fact this means that the EU is much more of an actor in climate policy than it is in other areas, notably the CFSP, where individual national interests tend to prevail but:

> on issues like global warming, where the risks involved invite precaution and uniquely or readily lend themselves to institutional regulatory solutions, Europeans are more likely to act in an EU context, pursue global remedies and in so doing give expression to their social democratic roots.
>
> (Ibid.: 94)

It goes without saying that this view has not necessarily been shared in the past by India (see Chapter 18) and many other countries critical of EU climate policy. In the views of many of the EU's protagonists the adoption of a high moral tone on the climate obligations of developed countries and on 'common but differentiated responsibilities' also includes an element of hypocrisy, given the rather favourable terms that the EU managed to negotiate for itself at Kyoto. The Copenhagen outcome delivered a severe blow to the previously confident assertion of EU 'actorness' and there have been serious internal differences over new targets, the future of the ETS and the question of aviation emissions. In the background the EU's wider international reputation was being tarnished by the long agony of the Eurozone crisis and by the arguments over continued UK membership, yet the Paris 2015 outcome could be seen as a modest success for the Union, but one that was achieved by 'engagement with many other partners in a broad "High Ambition Coalition"' (Council 2016: 3).

Conclusion

The Westphalian state model cannot serve as a template for an entity such as the EU. Instead this chapter has attempted to consider the attributes of an international actor beyond the framework of statehood in terms of autonomy, volition, negotiation and implementation capacity as well as the significant aspects of third party recognition and construction. In global environmental policy and more specifically in the international politics of climate change, the Union has clearly been an identifiable and purposive actor and possesses many of the associated capabilities. Thus it is possible to refer to the EU itself, rather than its component parts, as a leader. The EU as an environmental policy actor has clearly been a great deal more successful than the EU as a foreign policy actor under CFSP. Although expectations of climate change leadership have been raised, a process encouraged by the Commission and successive European Councils, there is far less of a 'capability expectations gap' than in other domains of external policy. This is not to say that the EU is always an effective actor and some significant problems with its capabilities have been identified. As an actor the EU, as distinct from its individual member states, has been able to exercise some structural leadership and a great deal of cognitive and entrepreneurial leadership. Its particular forte has been to encourage emulation through the promulgation of 'science-based' targets and timetables resting upon its internal policies – in short – exemplary leadership.

It has been argued that EU actorness and by implication its climate leadership role have come about because of favourable circumstances in terms of EU presence and the available opportunity structure. Such circumstances did not persist. The EU's global presence is in relative decline, in economic terms but also in respect of its share of global GHGE. At the same time the very favourable conditions, enabling the EU 'bubble' will not recur. The Union has become larger, but inevitably poorer through enlargement since 2004. With the exit of the UK, a major economy and 'lead state' on climate policy, the EU's external weight will be further reduced (see Chapter 12). Externally the opportunity structure within which the EU had been able to develop its role as 'the' climate leader changed markedly as the BASIC countries ceased to be mere 'bystanders' and the US re-asserted itself. In consequence it appears that the character of EU leadership changed, becoming perhaps less structural, and less transformational, than it had been during the decade of Kyoto Protocol development and ratification, but relying rather more on those cognitive and entrepreneurial capabilities that facilitated the multilateral deals upon which the Durban Platform and Paris Agreement depended.

References

Allen, D. and Smith, M. (1990) 'Western Europe's presence in the contemporary international arena' *Review of International Studies*, 16(1): 19–37.

Bäckstrand, K. and Elgström, O. (2013) 'The EU's role in climate change negotiations: from leader to "leadiator"', *Journal of European Public Policy*, 20(10): 1369–1386.

Bretherton, C. and Vogler, J. (1999 and 2006) *The European Union as a Global Actor*, London: Routledge.

Bretherton, C. and Vogler, J. (2013) 'A global actor past its peak?', *International Relations*, 27(3): 375–390.

Commission (2009) *Towards a comprehensive climate change agreement in Copenhagen*, COM(2009) 39 Final, 28.1.2009, Brussels: Commission of the European Communities.

Commission (2011) *A Roadmap for moving to a competitive low carbon economy in 2050*, COM(2011)112final, Brussels: Commission of the European Communities.

EU EDGAR (2016) *Emissions Database for Global Atmospheric Research*, CO2 Time Series 1990–2014 per region/country' and '...per capita for world countries, Commission of the European Union Joint Research Centre www.edgar.jrc.ec.europa.eu/overview.php?v=CO2ts_pc1990-2014 (accessed 1 January 2016).

Council (2015) *Council Conclusions on Climate Diplomacy*, Press Release 602/15, 20/07, Brussels: Council of the European Union.

Council (2016) *European climate diplomacy after COP21: Council Conclusions 15 February*, Brussels: Council of the European Union.

Delreux, T. and Van den Brande, K. (2013) 'Taking the lead: informal division of labour in the EU's external environmental policy-making' *Journal of European Public Policy*, 20(1): 113–131.

European Council (2005) Presidency Conclusions, 22/23 March, Brussels: European Council.

European Council, (2007) Presidency Conclusions. 8/9 March, Brussels: European Council.

European Council (2008) *Climate change and international security:* S113/08 Brussels: European Council.

European Council, (2014) Presidency Conclusions, 23/24 October Brussels: European Council.

European Council for Foreign Relations (2012) *Foreign Policy Scorecard*, www.ecfr.eu//scorecard/home (accessed 20 June 2013).

Grubb, M. (2010) 'Copenhagen: back to the future?', *Climate Policy*, 10(2): 127–130.

Grubb, M. and Yamin, F. (2001) 'Climate collapse at the Hague. What happened and where do we go from here?', *International Affairs*, 77(2): 261–276.

Hill, C. (1993) 'The capability-expectations gap or conceptualising Europe's international role', *Journal of Common Market Studies*, 31(3): 305–325.

House of Lords (2008) *Select Committee on European Union. Tenth Report*, London.

Kagan, R. (2002) 'Power and weakness', *Policy Review*, June–July, 3–28.

Keating, D. (2014) 'Anatomy of a climate deal' *European Voice*, 20(39): 13.

Manners, I. (2002) 'Normative power Europe: a contradiction in terms?, *Journal of Common Market Studies*, 40(2): 235–258.

Ringius, L. (1997) *Differentiation, Leaders and Fairness: Negotiating Climate Commitments in the European Community*, Oslo: Cicero, Report 1997: 8.

Schmidt, J.R. (2008) 'Why Europe leads on climate change', *Survival*, 50: 4, 83–96.

van Schaik, L. and Egenhofer, C. (2003) *Reform of EU institutions: Implications for the EU's performance in climate change negotiations*, Brussels: CEPS Policy Brief No. 40.

Vogler, J. (1999) 'The European Union as an actor in international environmental politics', *Environmental Politics*, 8(3): 24–48.

Vogler, J. (2009) 'Climate change and EU foreign policy: the negotiation of burden sharing', *International Politics*, 46, 469–490.

Vogler, J. and Bretherton, C. (2006) 'The European Union as a protagonist to the United States on climate change' *International Studies Perspectives* 7, 1–22.

World Bank (2016) 'GDP data by country', http://data.worldbank.org/country (accessed 2 February 2016).

Part II
EU institutions

3 The Commission's shifting climate leadership

From emissions trading to energy union

Jon Birger Skjærseth

Introduction

Ever since the 1990s, the EU has aimed to play a leadership by example role in international climate negotiations, adding increasingly ambitious targets and policy instruments. In October 2014, the EU agreed to cut emissions of greenhouse gases (GHG) by 40 per cent in its member states between 1990 and 2030. The European Commission (henceforth: Commission) has played a shifting role in promoting this development. This chapter argues that the type and style of leadership exercised by the Commission have varied over time.

'Leadership' is here understood as an asymmetrical relationship of influence, where one actor guides or directs the behaviour of others towards a certain goal over a certain period of time (Underdal 1991: 140). Leadership is distinguished from the actions of an agent who actively engages in ordinary policy-making and bargaining (Malnes 1995). To qualify as leader, the Commission should have an independent influence on policymaking exceeding its formal role.

The Commission has the potential to act as an *entrepreneurial* leader. Its resources may be used to formulate new policy ideas, mobilize support and craft consensus. Entrepreneurial leadership can be seen as a matter of finding the means and guiding others toward a common end. In climate policy, entrepreneurial leadership will require conviction, skill, energy and formal status. The scope for entrepreneurial leadership is likely to increase with uncertainties concerning decision makers' preferences and possible solutions.

A *cognitive* leader can shape and influence the preferences of other actors. Access to, and control over, the production and dissemination of relevant information is a key resource for cognitive leadership. A study of Commission expert groups indicates that the potential for cognitive leadership is particularly high in Directorate General (DG) Environment and DG Transport and Energy.[1] Cognitive leadership exercised by the Commission presupposes asymmetrical knowledge vis-à-vis EU decision makers. Knowledge networks within the Commission and between the Commission and other actors can also take the form of an epistemic community with a shared set of beliefs in specific 'diagnoses' and 'cures'. However, entrepreneurial and cognitive leadership are not mutually exclusive: they may prove particularly effective in combination.

The Commission by itself can hardly exercise leadership by example, as it is not an implementing agent. It can, however, promote internal policies that boost the credibility of EU leadership by example internationally. Likewise, the Commission does not possess the resources traditionally needed for *structural* leadership, which is associated with some type of power or force. However, the Commission's formal role and informal initiatives can affect others by holding them accountable to joint commitments. This way of directing the behaviour of others resembles structural leadership.

Whereas types of leadership point to the means and resources applied for guiding the actions of others, leadership *styles* emphasize the vision and timescale of leadership. In the context of climate change, transactional/humdrum leadership will aim at changes that are incremental and short-term; transformational/heroic leadership aims at transformative and long-term energy changes toward a low-carbon economy (see also Chapter 1).

Commission leadership speaks to the EU integration literature on supranationalism and supranational entrepreneurship (e.g. Pollack 1997; Sweet 1997; Hodson 2013). In a multi-level climate governance context, Commission influence depends on various factors, including decision-making procedures and the distribution of interests, norms, positions and behaviour of those making EU climate policy at various levels. These include non-state actors (see Chapters 14 and 15), national governments (see Chapters 6–12) and the European Parliament (EP) (see Chapter 4). Moreover, the Commission's room for manoeuvre is affected by external energy events and international commitments.

The Commission has responsibility for proposing legislation and ensuring implementation by EU member states. The legal basis for these responsibilities in climate policies is provided by successive treaties. The 1987 Single European Act (SEA) introduced climate change as a part of the environmental chapter, and the 2009 Lisbon Treaty referred specifically to climate change, codifying established practice. The Commission President is a member of the European Council – a body composed of the EU leaders who have become increasingly important in forming the direction of the EU's internal and external climate policy (see Chapter 5).

The Commission may take independent initiatives or respond to requests from member states, from other EU institutions, from stakeholders or – since 2012 – from individual EU citizens. It is tasked with facilitating agreement – a role that has expanded significantly over time (Barnes 2010). The Commission is appointed to represent the interests of the EU as a whole, but its 28 Commissioners – one from each member state – also serve as a clearinghouse for the interests of member states vis-à-vis the Commission (Egenhofer *et al.* 2011). Nationality seems to play a minor role for officials, but is one of several components of the Commissioner's role (Egeberg 2012). In the international climate negotiations, the Commission represents the EU, together with the current and incoming Presidency (see also Chapters 2 and 5).[2]

The Commission is divided into 32 DGs, of which a few have key roles in climate policy. These include DG Environment, DG Transport and Energy,[3]

DG Enterprise/Growth and DG Competition, with the first two as the most important. These and other DGs differ in interests and culture, tending to put forward competing policy positions (Hix 2005). Even contracted Commission staff are loyal to the units they formally serve (Murdoch and Trondal 2013). In the following, the Commission is *not* seen as a unitary actor. Moreover, this chapter focuses mainly on the Commission's internal leadership role. EU-internal climate policy development serves as a necessary precondition for the EU's leadership by example ambitions in the international climate negotiations.

Phases of EU climate policy

From the late 1980s, the Commission worked at crafting a climate and energy package of policies to show its leadership by example at the 1992 Rio Conference (Skjærseth 1994). The Commission immediately started to hammer out a joint package of policy instruments for delivering on the EU 1990 stabilization target. Central elements included proposals for a harmonized carbon/energy tax, a system for monitoring emissions, strengthening the existing energy-efficiency programme and establishing a programme on renewables. The tax proposal was the key climate policy instrument and it led to some of the most ferocious lobby activity ever seen in Brussels. Climate change enabled the Union to play an international leadership role while addressing its internal energy security needs at the same time. However, its only success as regards binding measures was the monitoring mechanism, adopted in 1993. The Council was not able to adopt any of the other proposals prior to the Rio Conference, whereupon Environment Commissioner Ripa de Meana resigned in protest (Skjærseth 1994).

The key principles for today's climate policy were established in those early years, but climate policy development stagnated until the 1997 Kyoto Protocol, with failed efforts at getting the carbon/energy tax unanimously adopted. From the Kyoto Protocol, the Commission's initiation of climate policies can be divided roughly into three phases.

In the first phase from 1997, climate policies gained momentum with the adoption of the burden-sharing agreement and the EU Emissions Trading Scheme (ETS). The burden-sharing agreement divided member state responsibilities for meeting the EU's 8 per cent emissions reduction commitment under the Kyoto Protocol. In October 2001, the Commission proposed the ETS, intended as an innovative market-based policy instrument with the world's first-ever international cap-and-trade system, targeting 11,000 industrial installations. The final ETS Directive was formally adopted in July 2003. The associated Linking Directive was adopted in 2004, connecting compliance with the ETS to purchase of credits through the Kyoto Protocol's flexible project mechanisms.

The second phase began with the first Barroso Commission and ratification of the Kyoto Protocol in 2005. In January 2008, the Commission formally proposed the climate and energy package of policies for achieving the targets agreed by the European Council in March 2007: to cut GHGE, to increase the share of renewable energy consumption, and indicatively to increase energy efficiency by

20 per cent by 2020. For the first time, climate and energy targets and policies became linked at the EU level. The package involved two cross-sector climate instruments. The first was a revised EU ETS aimed at reducing emissions in ETS-sectors by 21 per cent below 2005 emission levels, between 2013 and 2020. The proposal included a transition from a decentralized system to an EU-wide cap, to be reduced annually by 1.74 per cent. Allocation procedures were altered, from free allowances to a system based on payment by auctioning as the main principle.

The second instrument was an Effort Sharing Decision (ESD) based on national targets, intended to yield a 10 per cent reduction for sectors not covered by the ETS, like transport, agriculture, waste and buildings. In addition, the core package contained two specific low-carbon directives: one on the promotion of renewable energy consumption (Renewable Energy Directive, RED) based on differentiated national targets, and a proposal for the world's first international legal framework for safe capture and storage of carbon (CCS). Transport-related climate policies developed by the Commission and measures to promote energy efficiency were prepared according to a different time schedule.[4] The main structure of the package remained intact throughout the 2008 EU negotiations, despite several changes to the Commission's original proposal (Skjærseth 2013, 2014a).

In the third phase, the Commission initiated new long-term climate policies based on the October 2009 European Council agreement to support an EU goal of reducing GHGE between 80 and 95 per cent by 2050 against 1990 levels (European Council 2009; see also Chapter 5). In January 2014, the Commission formally proposed the new climate and energy policy framework for 2030, finally adopted in modified form by the European Council in October 2014. New goals include domestic reductions of at least 40 per cent GHGE compared to 1990. Part of the agreement involves strengthening the EU's ETS by a Market Stability Reserve and by lowering the cap from 2021. Emissions from non-ETS sectors are to be cut by 30 per cent below the 2005 level, through continued effort sharing among the member states. The 2030 climate and energy policy framework is to be specified by new legislation yet to be adopted. The Paris Agreement will put pressure on the timing and adequacy of new EU legislation, but it is unlikely to affect the EU's 2030 targets (Council, 2016; see also Chapters 1, 2 and 5).

Type and style of Commission leadership in different phases

From the 1997 Kyoto Protocol as flagged up above, three successive phases can be distinguished with reference to EU climate targets and corresponding policy instruments.

First phase: entrepreneurial and cognitive leadership in making the EU ETS[5]

The main EU climate policy achievement in the first phase was the EU ETS. With the Kyoto Protocol commitments and the voluntary international emissions-trading option, DG Environment underwent changes in staff and in its position on emissions trading. Its head, Jørgen Henningsen, was replaced by Jos Delbeke, who became the leader of the team developing the EU ETS. Delbeke had been responsible for economic instruments, involved *inter alia* in the unsuccessful efforts to get the EU carbon/energy tax adopted. During the first half of 1998, Delbeke persuaded Environment Commissioner Ritt Bjerregaard to support the plans for emissions trading as the key EU climate-policy instrument (Skjærseth and Wettestad 2008).

One of the first challenges for Delbeke's group was to build up expertise on what an EU ETS could look like. The EU lacked experience with emissions trading, which had not been tested as a climate instrument in an international context. The development from vague idea to a specific design proposal can be traced back to deliberate expertise building in the Commission. Experts were commissioned to undertake a study of design options for an EU emissions trading system. Their work and reports discussed all the main design issues subsequently included in the Green Paper issued by the Commission in March 2000. Delbeke's group gained a lead of almost two years over most member states in expertise on how an EU ETS might look. This asymmetrical distribution of knowledge provided the basis for the Commission's cognitive leadership.

The second challenge for Delbeke's group was to muster political support for emissions trading. Few member states were familiar with emissions trading; few were enthusiastic about the emerging plans; and some important states, Germany among them, initially opposed the idea of an EU ETS. Industry was basically more positively disposed to emissions trading than to taxes, but energy-intensive industry generally preferred voluntary agreements. On the other hand, the electric power industry – shielded from competition beyond Europe – was positive, and the oil giants BP and Shell had implemented company-internal emissions trading that inspired the EU ETS (Skjærseth and Skodvin 2003; Chapter 14). The 'green' groups were sceptical or in opposition, known for their slogan 'trading pollution is not a solution' during the Kyoto negotiations (see Chapter 15). Other parts of the Commission, like DG Enterprise/Growth, did not like the idea of introducing potential competitive disadvantages to European energy-intensive industry; and early statements from the Environment Committee of the European Parliament indicated limited knowledge and diverging views.

Parallel with an inclusive consultation process based on the 2000 Green Paper, the Commission initiated ten exclusive stakeholder meetings under the umbrella of the European Climate Change Programme (ECCP). Participants were selected by Delbeke and his team to include representatives of various green groups and industrial interests positive to emissions trading, and those with substantial interests in the matter. In addition to careful selection of participants, stability and continuity in personal representation were encouraged. A situation in which the same representatives met almost every month for a year served to promote interpersonal relationships based on mutual confidence and understanding. It was Delbeke and his team who chaired the meetings, presented various background documents and wrote the proceedings – thereby framing the developing consensus within the group to some extent (Skjærseth and Wettestad

2010). This deliberate consensus-building effort provided the basis for entrepreneurial leadership.

In these meetings and in general, the Commission emphasized emissions trading as an instrument that had something for everybody. To industry, it was framed as a cost-effective tool that could even provide economic opportunities for those with shrinking emissions to sell allowances. To green groups (and the EP), it was presented as environmentally effective: appropriately designed, it would automatically lead to the cap that had been set. To governments, these arguments were combined and linked to implementing the burden-sharing agreement and the Kyoto Protocol targets. Throughout the meetings it was emphasized that the Kyoto targets were fixed; ratification was expected; and emissions trading would represent a cost-effective way for governments and industry to achieve the targets. This framing proved effective in reducing resistance and building support within the Commission, member states, industry and green groups. In its final report, the working group unanimously recommended that emissions trading should start as soon as practicable. However, disagreement on important design issues remained.

The US exit from the Kyoto Protocol in March 2001 (see Chapter 16) led the Commission to advance its agenda on the ETS Directive proposal significantly. Bush's withdrawal served to unite the EU in an extraordinary way. It made the entry into force of the Kyoto Protocol uncertain – and the EU was determined to take the lead in winning support from the other states needed for the Protocol to enter into force (Skjærseth and Wettestad 2008; see also Chapters 1, 2 and 5). The EU ETS became important in the EU's efforts to show the world that it was indeed taking action on climate change. In the end, a rather weak and decentralized system based on free allowances was unanimously agreed by the EU-15 and endorsed by the European Parliament. The 'shadow' of qualified majority voting proved effective in getting still sceptical countries like Germany on board. The Commission leadership style was more transactional than transformational. The ETS was not intended to facilitate a carbon price sufficiently high to spur transformative long-term energy system changes unless comparable actions were taken by other major emitters.

Second phase: entrepreneurial leadership in linking climate and energy policies for 2020

A clash between DG Transport and Energy and DG Environment surfaced in the wake of the adoption of the EU ETS. Energy Commissioner de Palacio challenged Environment Commissioner Wallström over the economic costs of her climate leadership by example strategy (Skjærseth *et al.* 2016). Commission President Romano Prodi publicly criticized de Palacio, and stressed the importance of keeping the Commission unified in support of the EU's leadership role in international climate policies.

This incident shows that GHGE mitigation had significantly lower priority in DG Transport and Energy than in DG Environment. In 2005, internal disagreement prevented the Commission from proposing a new climate target. But one hurdle for linking climate and energy policies was to disappear: during its autumn 2005 EU Presidency, the UK shifted from resisting to supporting EU-level energy policies (see Chapter 13).

By 2006, oil prices were rising and climate concern was sweeping Europe. Energy security climbed the political agenda also because of the Ukraine–Russia energy dispute that threatened European gas supplies. In March 2006, the Commission issued a Green Paper, prepared by DG Transport and Energy, on a European strategy for sustainable, competitive and secure energy (Commission 2006). It painted a bleak picture of energy challenges for the world's largest energy importer and stressed the need for a new energy policy as an *integrated* part of EU climate policy.

By autumn 2006, favourable external and internal conditions for linking climate and energy policies at EU level had placed the issue firmly on the Commission's agenda. Three additional factors were decisive for preparing the climate and energy package. First, the package initiative came from DG Environment, which needed DG Transport and Energy to strengthen the case for more ambitious climate-policy targets and a stronger and more harmonized ETS within the Commission. DG Enterprise/Growth was, unsurprisingly, sceptical to unilateral climate policy, seeking to protect the interests of Europe's energy-intensive industries. DG Transport and Energy needed DG Environment and the increasing political saliency of climate change to develop a common EU energy policy.

Second, the new Barroso Commission introduced new commissioners: Stavros Dimas as Commissioner for the Environment and Andris Piebalks as Commissioner for Transport and Energy. They managed to put an end to the conflict between their predecessors, although Barroso himself initially prioritized economic policy before climate policy. Various officials within the Commission were also important. Jos Delbeke, the EU ETS entrepreneur, requested his staff in DG Environment to develop early sketches of a possible package based on the RED, CCS and revision of the ETS with auctioning that could be used to subsidize lower-income member states. This linking of policies showed entrepreneurial leadership by crafting policies that gave something to all pivotal policymakers – including fossil-fuel interests (CCS), the new Central and Eastern European states (ETS revenues) and those prioritizing renewable energy. The three biggest member states were also on board, and Germany announced that climate and energy policy would be key priorities during the German 2007 EU Presidency (Skjærseth *et al.* 2016).

With 2007 came a turning point for EU climate and energy policy. In January, the Commission issued two key communications on energy and climate policy strategies for 2020 and beyond. These communications indicated a turn towards a transformational leadership style: 'Strong scientific evidence shows that urgent action to tackle climate change is imperative' (Commission 2007a: 3). The response was to set about transforming Europe, catalysing a new industrial revolution by accelerating the shift to low-carbon growth: 'The EU would have set the pace for a new global industrial revolution' (Commission 2007b: 5, 21).

These communications proposed the 20–20–20 targets. Moreover, the target of reducing GHGEs by at least 20 per cent would be upped to 30 per cent if an adequate international climate treaty were agreed. The twin targets represented an effort to demonstrate international leadership on a post-2012 climate treaty to be negotiated in Copenhagen 2009 (COP15).

The communications underlying these commitments were prepared by DG Transport and Energy and DG Environment respectively and were published jointly on the same day by the Commission, showing the close collaboration between the two Commissioners. Synergies between climate and energy were underscored: an ambitious climate policy would contribute to the achievement of energy goals; an ambitious energy policy would contribute to the achievement of climate policy goals. CCS would reduce resistance from the oil industry and serve as 'political glue' to alleviate the trade-off between mitigation of climate change and security-of-supply for countries highly dependent on indigenous coal, like Poland. A Strategic European Energy Technology Plan (SET Plan) was proposed, to lower the cost of clean energy and put the EU at the forefront of the low-carbon technology sector. Largely swept under the carpet were potential trade-offs and risks like uncertainties about the costs and prospects of CCS technologies, downward pressures in carbon prices caused by renewables and energy-efficiency measures in the ETS sectors, and the incompatibility between national renewable energy policies and the internal energy market.

The proposals also combined interests and values between DG Transport and Energy and DG Environment and among member states. DG Environment favoured a stringent climate policy; DG Energy was more concerned with energy security. The 10 Central and Eastern European States (CEES) that joined the EU between 2004 and 2007 were more concerned about energy security, whereas the EU-15 generally favoured a more stringent climate policy, as expressed by the adoption of the ET Directive.

In March 2007, the European Council agreed on the key elements of the new climate and energy policy. The Commission had received backing from the highest political level before the package was proposed in January 2008. This climate and energy package was based on a thorough assessment of how the ETS, Effort Sharing Decision, and RED proposals would work together, and was framed as fair. To make the package politically acceptable, significant member-state differences in costs by 2020 were to be levelled out to compensate poorer members: by setting differentiated national targets in the non-ETS sectors (Effort Sharing Decision) based on GDP/capita; by setting different national targets for the share of EU energy consumption to be achieved by renewable energy (RED) based on a combination of GDP and flat-rate increase in the share of renewable energy; and by using auctioning revenues (ETS) to compensate lower-income member states.

The integrative and flexible nature of the package facilitated agreement in December 2008, despite high complexity and the financial crises unfolding from autumn 2008. Member states unanimously adopted a compromise package, and the Parliament endorsed it – including the reduction targets for ETS and non-ETS sectors and different national targets in the Effort Sharing Decision and RED. French leadership was instrumental in forging the compromise (see Chapter 7).

Third phase: decreasing leadership in the 2030 framework and beyond?

The climate and energy package aimed at strengthening the credibility of EU climate policy leadership internationally. However, the international conditions set by the EU for upping to a 30 per cent GHGE reduction target were not met in Copenhagen. Moreover, changes in circumstances led to new policy initiatives from the Commission. The economic crises unfolding from 2008 brought emissions down in the ETS and non-ETS sectors, reducing the carbon price to just above €5 in spring 2014 and thereby incentives for investing in measures that could ensure a low-carbon economy.

The Commission responded in 2010 by presenting an analysis of options for unilaterally moving beyond the 20 per cent GHGE reduction target (Commission 2010). Throughout 2010 a 30 per cent unilateral target was discussed internally within the Commission, pushed by Climate Commissioner Connie Hedegaard and DG Climate Action. However, a new unilateral target had few supporters among member states. Poland in particular had been a long-standing opponent of tougher emissions cuts (Skjærseth 2014b). Vigorous opposition came also from the European Alliance of Energy Intensive Industries (see also Chapter 14).

Lack of broad-based support stalled the Commission's plans for moving unilaterally towards 30 per cent reduction by 2020. The Commission then used the 2009 European Council agreement on decarbonizing the EU by 2050 as a foundation for stepping up policies (see Chapter 5). In March 2011, the Commission issued a roadmap prepared by DG Climate Action for moving towards a competitive low-carbon economy by 2050 (Commission 2011a). In December 2011, it followed up with the publication of an Energy Roadmap 2050 prepared by DG Energy (Commission 2011b). However, Poland vetoed both roadmaps despite the Commission's efforts to hammer out a compromise (see Chapter 10).

Failure to facilitate agreement on a more ambitious 2020 climate target or stepwise targets toward 2050 reduced the Commission's room for manoeuvre. But something would have to be done to rescue the EU's climate policy flagship – the EU ETS. The surplus of allowances was expected to build up in 2012 and 2013, to reach over two billion allowances that would continue beyond 2020. In July 2012, the Commission initiated legal changes to merely postponing or 'backloading' auctioning of 900 million allowances from the beginning to the end of the 2013–2020 period. Even so, the proposal proved extremely controversial, and was not adopted until January 2014. The Commission's unsuccessful efforts to facilitate agreement on responses to changing circumstances led it to follow a more cautious, member state-sensitive strategy. Illustrative here is the Market Stability Reserve (MSR) proposed by the Commission in 2014 for dealing with the problem of surpluses. In 2015, the EU adopted by qualified majority a more ambitious MSR than originally proposed by the Commission, including prevention of the release of the 900 million backloaded allowances into the over-flooded carbon market.

Caution was also evident in the deliberations on the new 2030 climate and energy framework. In January 2014, the Commission proposed a revised framework on climate and energy policies for 2030. The proposal, which built on extensive consultations with stakeholders, reflected the economic crisis, lack of international progress and differing member state positions. The tone had changed markedly from 2007: gone was the rhetoric of a new 'green' industrial revolution, synergies and energy-technological innovation. This indicated a shift back from a transformational to a more transactional leadership style. With Poland in the lead, Bulgaria, the Czech Republic, Hungary, Romania and Slovakia demanded full national control over their energy mixes, compensation from the EU, and burden sharing. Poland also demanded that new EU climate and energy policies be made conditional on the achievement of a global climate deal in Paris 2015 (Skjærseth 2014b; see also Chapter 10)).

In October 2014, the European Council adopted the framework for EU climate and energy policies for 2030, including a new 40 per cent domestic reduction goal compared to 1990 levels, a 27 per cent renewable energy consumption goal at EU-level, and an indicative target of a 27 per cent increase in energy efficiency as against the business-as-usual scenario (see Chapter 5). The first paragraph in the conclusions reflects a more active role in internal climate policies from the heads of states and governments: the European Council will keep all the elements of the framework under review and will continue to provide strategic orientations as appropriate, notably with respect to *consensus* on ETS, non-ETS, interconnections and energy efficiency (European Council 2014). Concessions given to Poland and its allies further constrain the Commission's room for manoeuvre in promoting a low-carbon economy. The framework recognizes that energy security can be bolstered by having recourse to indigenous resources, such as coal and shale gas. Full respect for the freedom of member states to determine their energy mix is also explicitly included – as are a non-binding energy-efficiency goal and a new renewables goal that will not be transformed into binding national goals. Instead, a watered-down energy governance mechanism was agreed.

The Energy Union has been made a central priority of the Juncker Commission, with a new Vice-President tasked with improving horizontal coordination for realizing such a union. A common Commissioner has been appointed for climate *and* energy policy execution. It is too early to say how this will play out in terms of Commission leadership towards a low-carbon economy. However, the first communication on the Energy Union Package – framed as a *coherent* climate and energy package – does not appear very promising (Commission 2015). The Energy Union initiative was triggered by the Russia–Ukraine crisis and picked up by Donald Tusk, who had set energy security as top priority in new climate and energy policy development during his time as Polish Prime Minister. Tusk later became President of the European Council. This background is reflected in the communication, which reflects the deep divisions among member states.

The Energy Union initiative and the 2030 climate and energy policy framework reflect a more diversified, re-nationalized landscape compared to the 2020 targets. This diversification has constrained the Commission's room for manoeuvre as an entrepreneurial and cognitive leader. The European Council, where the individual member states meet at top level, gained a more prominent role in overseeing policy development; member state preferences became more asymmetrical; and the positions of the 'least ambitious actors' became increasingly firm and vocal. Greater certainty as to pivotal actors' preferences and possible solutions resulted from experiences with the making and implementation of policies for 2020 (Skjærseth *et al.* 2016).

Leadership in a multi-level governance context

The climate policy of the EU is shaped by non-state actors, member states, EU institutions, the international climate negotiations, and external events. The Commission has played a crucial but shifting role in setting the agenda, drawing attention, inventing policy options, framing solutions, mobilizing support and linking policies in ways that can promote agreement.

The initiation of the EU ETS shows the Commission's potential as the 'engine' of climate policy development. With the assistance of the Kyoto Protocol, the Commission took the initiative, built up independent knowledge and crafted support among stakeholders. In essence, it acted as an entrepreneurial and cognitive leader in making the EU ETS. Delbeke's team within the Commission linked to other trading positive actors served as an 'epistemic community' with a shared belief in emissions trading as a cost-effective means for reducing GHGE. The influence of the Commission's entrepreneurial and cognitive leadership was accordingly high – but the style transactional, as the EU ETS aimed at incremental changes.

The entrepreneurial role played by the Commission in crafting the climate and energy package was markedly different from the initiation of the EU ETS. The Prodi Commission had initially been paralysed by internal conflicts, but the first Barroso Commission managed to rally around a proposal aimed at combining interests within and outside the Commission. The proposed package did not include the same extent of asymmetrical knowledge and opportunities for cognitive leadership as the ETS. Still, the role played by the Commission qualifies as entrepreneurial leadership because of its creative framing and combination of policies so as to give something to most pivotal policymakers. The Commission's leadership style was bolder and more transformational, expressed by its vision of speeding up the shift to low-carbon growth to set the pace for a new global industrial revolution. This process was also more complex and challenging as regards issues, policy instruments, number and types of actors involved and the unanimity requirement. External events, changes in member state positions, governmental leadership and the international climate negotiations all contribute in explaining the 2020 package.

Prior to the 2030 climate and energy framework negotiations, the Commission's room for entrepreneurial and cognitive leadership was gradually reduced as a result

Table 3.1 Commission leadership type and style over time

Phases	*Type*	*Style*
First phase: Making the EU ETS	Entrepreneurial and cognitive by shaping positions, mobilizing support and crafting agreement	Transactional
Second phase: Crafting the 2020 package	Entrepreneurial by linking climate and energy policies to facilitate agreement	Transformational
Third phase: Shaping the 2030 framework	Structural leadership efforts by holding member states accountable	Initially transformational, increasingly transactional

of increasing opposition from the 'least ambitious' member states and the unanimity requirement. With the exception of the MSR, which was intended to maintain previous achievements, the Commission has apparently not floated any new creative or bold ideas, nor managed to facilitate a common response to new challenges. Instead, it has sought to respond to new challenges by 'old' means, holding the member states accountable to previous commitments by pushing for stronger policies. This resembles structural leadership efforts. The initial leadership style can be characterized as transformational: the Commission pressed for long-term targets and policies aimed at 2050 decarbonization. Hampered by setbacks and increasing opposition, its leadership style became increasingly transactional.

Conclusions

This chapter has shown the shifting entrepreneurial and cognitive leadership role of the Commission in promoting EU climate policies in three phases after the Kyoto Protocol, from the initiation of the EU ETS to the Energy Union. Leadership style has shifted from increasingly transformational to more transactional. The most important conditions for shifting leadership include internal coherency, external events and commitments, decision making rules and the distribution of preferences among other EU policy makers, notably the member states. The basic constitutional features empowering the Commission to develop climate policy have remained roughly stable throughout the three phases.

Overall, the Commission – rather, DG Environment/Climate Action – has been perhaps the single most important actor in promoting ambitious, long-term climate policies within the EU. Targets and policies have also strengthened the credibility of the EU as a leader by example in the international climate negotiations. The behaviour of the Commission has been based on its unique formal status as agenda setter and policy initiator, the conviction that action for dealing with climate change is imperative, and its skilful bureaucrats. Commission Presidents have responded supportively to lower-level initiatives, also viewing climate policy as an area for legitimizing European integration and helping the EU to play an active role in the international arena.

While compartmentalization may obstruct policy initiation, the distinctive strength of the Commission lies in its ability to shape and frame (climate) policies for the longer term. The non-partisan Commission, led by Commissioners from each member state, is collectively responsible for its decisions, and is appointed to serve EU interests without being directly accountable to the electorate or responsible for the financial resources needed. This enables it to think and act in a more long-term manner than is possible for most individual member states.

Institutionalized cooperation can gather momentum through a 'snowball' effect generating positive feedback from implementation, facilitating further steps. However, experiences from implementing the 2020 package has been mixed, and CCS has failed to give fossil fuel interests a stake in decarbonization (Skjærseth *et al.* 2016). The 2030 negotiations apparently echo the economist's law of diminishing returns. The first steps are likely to be the easy ones, with benefits clearly exceeding costs. Then, when attempts are made to tighten up ambitions, abatements costs tend to increase while benefits decrease. This development may be partly counterbalanced by the 2015 Paris Agreement and change the Commission's role as a climate policy leader towards overseeing implementation. As EU long-term plans stand now, the decarbonization challenge lies increasingly with implementation and transformation in the member states.

Notes

1 In 2007, DG Environment ranked second and DG Transport and Energy ranked fifth among the 22 DGs, with 127 and 94 expert groups respectively (Dreger 2014).
2 In the international climate negotiations, the Commission has the advantage of continuous involvement since the Rio Conference, but acts formally on a mandate from the member states.
3 From February 2010: DG Energy and DG Climate Action.
4 These included a regulation covering emissions from new cars, primarily applicable to member states with car manufacturers; and a directive on fuel quality, including a required reduction of the carbon footprint of road fuels from well-to-wheel, primarily applicable to oil companies.
5 This section draws heavily on Skjærseth and Wettestad, 2010.

References

Barnes, P.M. (2010) 'The role of the Commission of the European Union: creating external coherence from internal diversity', in R. Wurzel and J. Connelly (eds), *The European Union as a Leader in International Climate Politics*, London: Routledge.

Commission (2006) *Green Paper on a European strategy for sustainable, competitive and secure energy*, COM (2006) 105 final, 8 March, Brussels: European Commission.

Commission (2007a) *Limiting global climate change to 2 degrees Celsius: the way ahead for 2020 and beyond*, COM (2007) 2 final, 10 January, Brussels: European Commission.

Commission (2007b) *An energy policy for Europe*, COM (2007) 1 final, 10 January, Brussels: European Commission.

Commission (2010) *Analysis of options to move beyond 20% greenhouse gas emission reductions and assessing the risk of carbon leakage*, COM (2010) 265 final, 26 May, Brussels: European Commission.

Commission (2011a) *A roadmap for moving to a competitive low carbon economy in 2050*, COM (2011) 112 final, 8 March, Brussels: European Commission.

Commission (2011b) *Energy roadmap 2050*, COM (2011) 885 final, 15 December, Brussels: European Commission.

Commission (2014) *A policy framework for climate and energy in the period from 2020 to 2030*, COM (2014) 015 final, 22 January, Brussels: European Commission.

Commission (2015) *Energy Union Package: a framework strategy for a resilient Energy Union with a forward-looking climate change policy*, COM (2015) 80 final, 25 February, Brussels: European Commission.

Council (2016) *Outcome of the Environment Council Meeting 4 March. 3452nd Council meeting, 4 March 2016*, Brussels: Council of the European Union.

Dreger, J. (2014) *The European Commission's Energy and Climate Policy: A Climate for Expertise?*, London: Palgrave Macmillan.

Egeberg, M. (2012) 'Experiments in supranational institution-building: the European Commission as a laboratory', *Journal of Public Policy*, 19(6): 939–59.

Egenhofer, C., S. Kurpas, P.M. Kaczynski and L.G. van Schaik (2011), *The Ever-Changing Union: An Introduction to the History, Institutions and Decision-making Processes of the European Union* (2nd edn), Brussels: Centre for European Policy Studies.

European Council (2014) *Presidency conclusions from European Council, 23 and 24 October 2014*, Brussels: European Council.

Hix, S. (2005) *The Political System of the European Union*, New York: Palgrave.

Hodson, D. 2013 'The little engine that wouldn't: supranational entrepreneurship and the Barroso Commission', *Journal of European Integration*, 35(3): 301–14.

Malnes, R. (1995) '"Leader" and "entrepreneur" in international negotiations: a conceptual analysis', *European Journal of International Relations*, 1(1): 87–112.

Murdoch, Z. and J. Trondal (2013) 'Contracted government: unveiling the European Commission's contracted staff', *West European Politics*, 36(1): 1–21.

Pollack, M.A. (1997) 'Delegation, agency and agenda setting in the European Community', *International Organization*, 51(1): 99–134.

Skjærseth, J.B. (1994) 'The climate policy of the EC: too hot to handle', *Journal of Common Market Studies*, 32(1): 25–45.

Skjærseth, J.B. (2013) *Unpacking the EU climate and energy package: causes, content, consequences, FNI Report 2/2013*, Lysaker: Fridtjof Nansen Institute.

Skjærseth, J.B. (2014a) 'Linking EU climate and energy policies: policy-making, implementation and reform', *International Environmental Agreements: Politics, Law and Economics*, Published online 02.10.2014, 17 p. DOI: 10.1007/s10784-014-9262-5.

Skjærseth, J.B. (2014b) *Implementing EU Climate and Energy Policies in Poland: From Europeanization to Polonization?*, FNI Report 8/2014, Lysaker: Fridtjof Nansen Institute.

Skjærseth, J.B. and T. Skodvin (2003) *Climate Change and the Oil Industry: Common Problem, Varying Strategies*, Manchester: Manchester University Press.

Skjærseth, J.B. and J. Wettestad (2008) *EU Emissions Trading: Initiation, Decision-making and Implementation*, Aldershot: Ashgate.

Skjærseth, J.B. and J. Wettestad (2010) 'Making the EU Emissions Trading System: the European Commission as an entrepreneurial epistemic leader', *Global Environmental Change*, 20(2): 314–21.

Skjærseth, J.B, P.O. Eikeland, L.H. Gulbrandsen and T. Jevnaker (2016) *Linking EU Climate and Energy Policies: Decision Making, Implementation and Reform*, Cheltenham: Edward Elgar.

Sweet, A.S. (1997) 'European integration and supranational governance', *Journal of European Public Policy*, 4(3): 297–317.

Underdal, A. (1991) 'Solving collective problems: notes on three modes of leadership', in *Challenges of a Changing World, Festschrift to Willy Østreng*, Lysaker: Fridtjof Nansen Institute, 139–53.

4 The European Parliament and climate change

A constrained leader?[1]

Charlotte Burns

Introduction

The European Parliament (EP) has historically had limited scope to shape the EU's external climate change policy as the lead has traditionally been taken by the Commission and Council. The Parliament has a circumscribed *de jure* role in international environmental politics: under Article 218 of the Treaty on the Functioning of the European Union it has a right to be consulted by the Council and in some circumstances may offer its formal consent to international agreements to which the EU is party. But, while the Commission and Council are obliged to keep the EP informed about the conduct of negotiations, the Parliament has limited ability to shape international negotiations. Consequently, the EP has limited external policy-making power, such that while the EP has sent delegations to international events such as the regular Conference of Parties (COP) meetings on climate change and issued reports and resolutions for the attention of the other institutions, its policy impact has been limited. This fact is reflected in the literature on the EU and climate change where the EP is barely mentioned and then only in passing (e.g. Schreurs and Tiberghien 2007; Oberthür and Roche-Kelly 2008; Wettestad *et al.* 2012; see Burns and Carter 2011 and Biedenkopf 2015 for notable exceptions).

Thus, while the EP is keen to present itself as a leader within EU environmental politics more generally, it has struggled to extend that leadership to the field of climate change. The main way in which it has been able to shape the climate agenda has been through its internal powers as a co-legislator, which have enabled it to amend policies brought forward by the Commission to give effect to the EU's wider external climate ambitions. Here the Parliament is constrained by the need to achieve internal cohesion as well as keep the Council on side.

The EP has sought to exercise cognitive and entrepreneurial leadership but its approach has typically been symbolic (Wurzel and Connelly 2011): it tries to shape policy through its debates and non-legislative resolutions, which allow it to air ideas but do little to alter the substance of policy or to shape the structure of incentives faced by the key policy actors. Attempts by the Parliament to use its legislative powers to move beyond symbolic leadership have been constrained. Below, first we outline the EP's role as an environmental actor, before evaluating its attempts to exercise leadership over time, and then assessing its approach to the green economy.

The EP as an environmental actor

The EP has developed a reputation over the years of being the EU's 'environmental champion' (Weale *et al.* 2000; Burns and Carter 2010; Burns 2013), it 'often sees itself, and is seen by others, as the defender of environmental interests' (Weale *et al.* 2000: 91), and has in relation to certain environment policies performed an 'important leadership role' (Zito 2000: 4). The EP's scope to offer a traditional leadership role has been limited by its institutional location: it is essentially a reactive chamber that amends policy proposals put forward by the Commission, and must negotiate with the Council to see its preferences realised in the final legislative text (Zito 2000; Judge and Earnshaw 2008). Nevertheless, the Parliament has consistently endeavoured to shape and strengthen EU environmental policy, deploying entrepreneurial leadership approaches to achieve its political ambitions (Zito 2000). Over the years the EP's Environment Committee has built close relations with officials in the Commission's DG Environment, in the Council and with the wider policy community of business and NGOs in order to shape the policy agenda (Judge 1992; Hubschmid and Moser 1997). Thus it has been a key entry point to the legislative process for the many external actors that may otherwise be excluded from policy-making. The Parliament has also sought to shape the wider environmental policy agenda through the adoption of own-initiative resolutions even when its legislative powers were more limited (see Judge 1992).

With the advent of codecision in 1993, the EP became better placed to shape legislation. Under codecision (or, as it is now known, the ordinary legislative procedure [OLP]) the EP is able to act as a co-legislator with the Council – taking responsibility for negotiating joint texts with the Member State representatives and having an ultimate power of veto if the negotiations do not produce an acceptable compromise (Shackleton and Raunio 2003). The Environment Committee has consistently been the 'largest customer' of codecision legislation within the Parliament (EP 1999, 2004, 2010a, 2015a). As such the personnel on the Committee – which has the reputation of being more ambitious than the EP's plenary (EP 2006a: 4–5) have been able to secure important compromises on a range of environmental measures. Indeed, the entrepreneurship of key individuals and the relative radicalism of the Environment Committee seem to have been critical in cementing the EP's reputation as an environmental leader. However, more recent empirical studies have suggested that as the Parliament has become more powerful its green ambition has waned (Burns and Carter 2010; Burns *et al.* 2013; Gravey 2015) and certainly post-2009 there is evidence that the Parliament has weakened Commission proposals on economic grounds (see below).

The EP's evolving leadership in climate change politics

The EP has used its internal powers to support, strengthen and contribute to the EU's wider external leadership ambitions, especially in pushing for cognitive commitments to be backed up with concrete policy commitments thereby enhancing the EU's claims to lead by example. The EP's attempts to exercise leadership within the field of climate change fall into three distinct phases. First, the 1990s to 2001, when the EP attempted to offer cognitive leadership by articulating the case for more stringent standards and the use of a mix of instruments but was constrained by its lack of internal legislative and external policy-making powers. Second, from 2001 to 2009: the extension of codecision following the entry in force of the Amsterdam Treaty coincided with a range of climate proposals, thereby opening a window of opportunity for the EP to offer cognitive leadership on core issues relating to the use of particular policy instruments to deliver climate change emission reduction targets, and entrepreneurial leadership by establishing a new institutional forum (the Temporary Committee on Climate Change) to address the issue of climate change. Third, 2009 to the present where the leadership ambition exhibited by the EP in the second phase has been tempered by a range of factors: the onset of the economic crisis; the growing assertiveness of Central and East European states (CEES) in the Council; and the challenges associated with securing internal cohesion within the Parliament. This limit to EP leadership reflects the wider shift to a more pragmatic approach to climate leadership by EU negotiators in the face of a changed and challenging external environment (e.g. Bäckstrand and Elgström 2013; see also Chapter 3).

Phase 1 – rhetorical reinforcement

In the 1990s the EP's lack of internal legislative powers in this field limited its scope to offer external leadership. The EP issued statements and resolutions in the run-up to COP meetings and statements, afterwards delivering its verdict on the outcome of negotiations, but it was unable to influence international negotiations in any substantive way. Indeed, in its resolutions relating to international conferences, the EP consistently complained about its delegations being excluded from key meetings and not being consulted (see, for example, EP 2000, 2002a, 2002b, 2006b). During this period the EP's approach was therefore based upon calling for the ratification of Kyoto, for the Commission and Council to offer leadership on the international stage, for the Commission to bring forward legislation to implement the Kyoto targets and chiding the United States for withdrawing from the Protocol (EP 1998a, 1998b, 2002a, 2002b). In essence, the EP pursued a rhetorical attempt to reinforce the EU's burgeoning external climate leadership (Schreurs and Tiberghien 2007), thereby offering cognitive leadership though offering support for more stringent measures and wider policy instruments.

Thus the EP supported the Commission's attempts to promote the climate change agenda through the use of new instruments such as the carbon dioxide

(CO_2)/energy tax, (EP 1998b), although the Parliament was much more sceptical about the use of voluntary agreements, such as that between the Commission and car manufacturers to reduce carbon emissions (ibid.). The EP has also consistently questioned the use of the Clean Development Mechanism (CDM) and Joint Implementation (JI) mechanism of the Kyoto Protocol in its resolutions on climate negotiations. Like several member states, the EP's position was not based upon denying the potential utility of such instruments, but upon pointing out that these approaches should be used in addition to, rather than instead of, other more traditional means of cutting carbon emissions, such as regulation (e.g. see EP 2002b).

In addition to these statements of principle the EP also used its legislative powers under codecision consistently to tighten legislative proposals that limited emissions of greenhouse gas, such as the auto-oil programme which sought to combine regulation of emissions from cars with new rules on fuel quality standards. However, it should be noted that between 1993 and 1999 the EP could only offer opinions under codecision on proposals brought forward under the single market provisions of the treaty, which limited its scope to shape proposals falling under the environmental provisions of the Treaty. The revisions of the Amsterdam Treaty in 1999 extended codecision to most environmental policy (excluding land use planning, water management and fiscal measures), thereby widening the scope of the EP's competence. Thus, once the Commission started bringing forward proposals to implement the EU's Kyoto targets a window of leadership opportunity opened for the Parliament.

Phase 2 – the window of leadership opportunity

Eight key pieces of legislation adopted between 2001 and 2009 provided the foundation for the EU's putative leadership in international climate change negotiations: the initial Emisisons Trading Scheme (ETS) directive; the extension of ETS to aviation; and six pieces of legislation that comprised the climate change and energy package, encompassing five directives covering the extension of ETS, effort-sharing, Carbon Capture and Storage (CCS), renewable energy, and fuel quality, and a regulation on limiting CO_2 emissions from cars. As all these proposals were subject to codecision, the EP had the opportunity to amend their content and thereby help to strengthen standards within the EU as well as to consolidate the EU's wider leadership role on the international stage (see Wettestad *et al.* 2012). Thus the EP had the opportunity to use its internal powers to bolster the EU's external climate ambition, which it did by adopting a consistent stance based upon strengthening standards and, where appropriate, introducing elements of auctioning into the ETS.

On emissions trading and aviation it is clear that all three legislative institutions (the Commission, Council and EP) were committed to the ETS project and to establishing the EU as a climate change leader. The Parliament had long been sceptical of the use of emissions trading as a policy instrument for achieving the EU's Kyoto targets, and had stated its preference for regulation and taxation as a means to bring down CO_2 emissions. However, the EP has also consistently called for the Commission and Council to take a leadership role at the international level. Once the US had withdrawn from the Kyoto process, and the rules on how to pursue emissions cuts had been tightened in the Marrakesh Accords, the Commission's proposal for an emissions trading scheme offered a genuine opportunity for the EU to be seen to be taking a transformational 'leadership-by-example' role (see Chapter 1). Thus, notwithstanding the Parliament's initial scepticism and in view of the shifting consensus that emissions trading was a viable option for delivering emissions cuts, the EP engaged in a strategy of reinforcing the EU's external leadership ambition by seeking to strengthen legislative proposals, i.e. it used its internal powers to bolster the EU's external climate ambitions. Thus, while the EP was supportive of the Commission's proposals, it still adopted amendments that tightened limits and introduced some auctioning requirements (Wettestad 2005; Burns and Carter 2011).

The climate change and energy package

On the one hand, the run up to the launch of the climate change and energy package was auspicious in witnessing the emergence of a sense of public urgency about the need for a more radical response to climate change. In particular, the Stern Review (Stern 2007) published in the autumn of 2006 made a compelling economic case for radical action on climate change and the Intergovernmental Panel on Climate Change's fourth assessment report early in 2007 made clear that climate change is advancing faster than anticipated. In March 2007, in the wake of these events and in preparation for Copenhagen 2009, the European Council launched its 20/20 by 2020 strategy. It committed the EU by 2020 to a 20 per cent reduction of 1990 CO_2 emissions levels – increasing to a 30 per cent reduction if other leaders took the same approach in a new post-Kyoto international agreement – and to a 20 per cent share of energy supply from renewable sources. Six pieces of legislation, to be adopted under codecision, were brought forward to deliver these aims.

Yet the subsequent global economic and financial crisis pushed climate change down the policy agenda and strengthened the case against measures that might harm economic competitiveness. The industrial lobby – particularly the well-resourced car manufacturing and energy industries – was increasingly vocal and persuasive in resisting the adoption of stringent limits and many member states, particularly new entrants, expressed their opposition to potentially costly legislation claiming that they had already made emissions cuts in the 1990s and that their contribution should now be recognised (*Guardian*, 16 October 2008). Thus, the EP found itself constrained by events: what may have been possible before the crisis struck was no longer so and the risk of failure – the prospect of no legislation – loomed over all the institutions. Within the Parliament, the Environment Committee, which had argued in favour of a low-carbon economy through a green economy approach now found it much more difficult to convince more sceptical EP committees (such as the Industry Committee) that ambitious EU climate change and energy policy measures would lead to a 'double dividend' that could deliver environmental protection and economic growth through the creation of new jobs in 'green' industries.

An important consequence of the changed economic context was that the key actors now advancing climate change leadership – namely the EP, the Commission and environmental non-governmental organisations (ENGOs) – found themselves in a defensive position of simply trying to prevent the Commission's original climate change and energy proposals from being significantly weakened or vetoed, rather than pushing for a more radical agenda. As one ENGO representative put it 'the Parliament was forced into trying to hold the line' (Interview, March 2009).

The EP adopted a consistent approach on all the dossiers based around trying to tighten or give legal force to reduction targets, to introduce sanctions for states that fail to meet the targets, to reduce the use of CDM credits and to make sustainability criteria transparent and meaningful. Unsurprisingly, the EP was more successful on amendments and legislation that addressed industry concerns and implied lower costs. For example, on the ETS proposal the EP supported several measures designed to reassure industrial interests concerned about the negative impact of inclusion in the ETS on their competitiveness. The EP therefore proposed that those sectors at significant risk of 'carbon leakage' would be entitled to receive up to 100 per cent of permits free of charge, although the qualifying sectors would not be named until an international agreement had been reached, and allocation would be at the level of the benchmark of the best available technology. The EP also raised the threshold for installations affected by the ETS from 10,000 to 25,000 tonnes of annual CO_2 emissions.

Overall, despite the Parliament's best efforts its ability to achieve its preferred outcome was limited. There are several reasons for the apparent failure of the EP to provide leadership and to achieve its goals: the changed economic context; the presence of hostile member states in Council from both East and West; a vigorous industrial lobbying campaign; and the fact that this was a complicated, highly technical package of legislation that was negotiated at speed. Given these prevailing conditions it is unsurprising that the EP's ability to offer entrepreneurial and cognitive leadership was limited. However, the amendment of the climate change and energy package was not the only platform for leadership available: in April 2007 the EP established its Temporary Committee on Climate Change (CLIM) and while this body had limited legislative power it provided an alternative venue for the EP to express its preferences on climate change, thereby giving it a platform for cognitive leadership.

CLIM was liberated from the constraints facing the legislative committees whose *rapporteurs* had to tailor their proposals to attract cross-group support within the EP and sufficient support within the Council to stand a chance of being adopted (see Burns and Carter 2011 for detail). While the CLIM *rapporteur*, Karl-Heinz Florenz, also had to secure support from the different political groups within the EP, as CLIM's final report had no legislative force, the EP could use it as an opportunity to make stronger statements about the actions necessary to combat climate change. Thus the Florenz report called for a review of whether the EU's commitment to cut temperatures by 2°C would be sufficient to avoid 'dangerous climate change' (European Parliament 2009a: 16) and demanded a 25–40 per cent reduction of greenhouse gas emissions (GHGE) by 2020 and 80 per cent by 2050. These positions were considerably more ambitious than anything member states were prepared to see included in legislation. The CLIM report also reminded parties to the Kyoto Protocol and Marrakesh Accords that they were expected to reduce domestic GHGE before taking advantage of external flexible mechanisms such as the CDM and stated 'that excessive CDM/JI use undermines the credibility of the European Union in the international UN negotiations and thus its leadership role in fighting climate change' (ibid.: 17). These statements are a clear indication of the Parliament's frustration at the member states' desire to using the CDM to reach their targets rather than reducing domestic GHGE. Thus, when freed from the limitations imposed by the legislative process the EP returned to its usual position of offering clear cognitive leadership on the issue of climate change, but its efforts were symbolic rather than transformational.

Phase 3 – pragmatism prevails

The Copenhagen summit was seen as a low point for EU climate diplomacy; however, there has been a concerted effort since 2009 on the part of the Commission to re-establish the EU's reputation as a climate leader but by pursuing a more consensual and negotiated approach with potential partners (Van Schaik and Schunz 2012; Bäckstrand and Elgström 2013; see also Chapter 3). Central to this strategy was the creation of a 'High Ambition Coalition' comprising the EU and 79 'poor and vulnerable nations' (Shankelman and Murray 2015), which played a key role in securing agreement at the COP21 in Paris. Broadly speaking the EU emerged from Paris with its reputation somewhat restored, largely thanks to able French chairing of the event, and notwithstanding the reluctance of some EU states to contemplate the inclusion of decarbonisation in the agreement text (Tobin and Burns 2016).

The EP has continued to broadly support the Commission in its external leadership ambitions in this post-2009 period. Thus, it has adopted amendments to Commission proposals, resolutions on the EU's negotiating position and continued to insist upon its inclusion in negotiating fora at the COP meetings (e.g. see EP 2012a, 2014a, 2015b). The Parliament sent a delegation to Paris in 2015 and supported the EU's calls for an 80–95 per cent cut in GHG emissions by 2050 and reinforced calls for regular reviews of the intended nationally determined contributions (INDCs) in order to improve the chances of the Paris agreement achieving its stated goals of limiting temperature increases to between 1.5 and 2°C (EP 2015b). The EP also supported the Commission's diplomatic work in Paris, bolstering efforts to see the 'High Ambition Coalition' secure key demands that benefitted poorer countries, for example on climate financing.

The EP has also used its internal legislative powers to strengthen the Commission's Market Stability Reserve (MSR) by bringing forward the deadline for the back-loading of carbon permits under the MSR that seeks to bolster the struggling emissions trading scheme (EP 2015c). However, there have also been examples of the EP weakening Commission proposals, on the extension of ETS to aviation and the setting of emission limits for light commercial vehicles (LCVs). The extension of ETS to aviation bitterly divided the Parliament. The Environment Committee *rapporteur* broadly supported and tried to strengthen the Commission's proposal that sought to extend emissions trading to international flights entering its territory despite the on-going failure to reach agreement with non-EU states about emissions from aviation via the International Civil Aviation Organization (ICAO) (European Commission 2013; EP 2014b). Members of the Industry Committee wanted to maintain the *status quo* (EP 2014b). A compromise was negotiated with the Council that weakened the Commission proposal by continuing to limit the application of the regulation on aviation emissions to domestic traffic until 2016. The centre right European People's Party (EPP) were fully behind this weaker proposal but the Group of the Progressive Alliance of Socialists and Democrats (S&D) and Liberals both split over the issue with those opposed claiming that the EU had been bullied by large states and industrial pressure (EP 2014c).

On LCVs the EP weakened Commission proposals by increasing emission limits from the 135 g CO_2/km proposed by the Commission to 147 g CO_2/km (EP 2011a). However, here, unlike the aviation case the EP was fairly united across political groups. The primary reason for weakening the Commission proposal was concern for the implications of the proposals for the competitiveness of the car industry. As noted by the spokesperson for the Committee on Transport and Tourism,

> In our rush to cut emissions... we should not undermine the competitiveness of European car manufacturers. This requirement is particularly pressing in view of the ongoing financial crisis, which has had serious repercussions on the European car industry so far.
>
> (Oldřich Vlasák, EP 2011b)

So what we see in the post-Copenhagen era is a more cautious Parliament that has shown itself prepared upon some occasions to weaken Commission proposals to secure internal parliamentary support as well as agreement with the Council, with wider economic concerns being used as a justification for this less radical approach. The EP is consequently sending mixed messages. On the one hand it has offered rhetorical leadership with regard to the EU's external climate ambition, but on the other, that leadership position is undermined by its actions on internal climate legislation. This ambiguity suggests that despite enjoying significant internal powers, the EP has continued to struggle to shed its reputation as a merely symbolic climate leader.

The Parliament and the green economy

The EP has long supported the use of new environmental policy instruments and the development of the green economy as a way of getting business to support the environmental policy agenda. The institution's approach to the use of new policy instruments and of innovation within the field of climate change legislation has been characterised by remarkable consistency. While the Parliament has always endorsed a mix of policy instruments it has traditionally favoured those new environmental policy instruments that have as part of their repertoire an element of regulation or taxation. It has consistently called for limited use of offsetting via the CDM and JI mechanisms of the Kyoto Protocol. Its amendments to the climate change and energy package continued its long held policy of tightening limits and introducing sanctions where appropriate. This continuity reflects the fact that the EP operates through coalition and consensus with voting behaviour dividing along ideological lines (Hix *et al.* 2007). Indeed, what is interesting about the passage of the climate change and energy package was the fact that the most contested divisions emerged within the EPP, rather than between the EPP and S&D. These splits reflect the controversial nature of the proposals under discussion and the strongly held national preferences that saw governments and industry fiercely lobbying their MEPs (see Burns 2013). The behaviour of those MEPs also reflects the unique nature of the European Parliament situated as it is at the interface between national and European levels of governance. MEPs may legislate at the European level but they are accountable to their national voters and parties, consequently when matters of national interest are at stake, particularly close to an election, such splits on national line are unsurprising.

Challenges

The EU's drive to play an international leadership role on climate change has presented an opportunity for the EP to use its internal legislative powers under codecision in combination with its symbolic agenda-setting capacity though CLIM to advance the EU's external leadership on the world stage. The EP achieved some success in strengthening the original ETS legislation, and its extension to aviation, notably by increasing the auctioning requirements. Significantly, this legislation was dealt with in a context when concern about climate change was steadily rising and all three institutions agreed on the need for action. However, the passage of the climate change and energy package and the adoption of subsequent legislation designed to implement that package has taken place in very different circumstances against the backdrop of a deepening international economic crisis, which had reduced the appetite for a radical climate change package among the EU's member states. In addition, there has been a shift within the Council that saw previously committed heads of government leaving office or, in the case of German Chancellor Angela Merkel, shifting position (see Chapter 8). We have also seen the emergence of a bloc of accession states prepared to wield their power in order to gain concessions. This development reflects a new confidence among those states – for the first few years after joining the EU the new states kept a relatively low profile in Council (Burns *et al.* 2012) – but the climate change and energy package suggests that where there are key economic interests at stake they will voice concerns. The fact that in most of those states the environment has low political salience (see Chapter 10) may not bode well for future EU environmental policy.

For the Parliament the passage of the climate change and energy package served to underline the limitations of its powers under codecision when the Council is intransigent. Rather than providing an opportunity for the EP to realise its long held ambition to offer entrepreneurial leadership, the climate change and energy package provided an important reminder that the agreement of both co-legislators was required for the package to be adopted. The COPs held since 2009 further underline the EP's limited power in this particular field of environmental policy. The resolutions adopted by the Parliament prior to the COPs consistently call for the EP to be included in meetings of the EU negotiating team but the Parliament's delegation continues to be excluded (e.g. EP 2009b, 2010b, 2011c, 2012b, 2013, 2014d, 2015b).

A further challenge facing the Parliament is the balance of power within the chamber: in 2014 the European elections saw the emergence of a far-right group dominated by the French *Front National*. However, on environmental matters this ideological shift looks to have had limited impacts, as the centre-left and centre-right groupings continue to cooperate to secure majorities. More of a concern for the Parliament is the shift in emphasis of the Juncker Commission appointed in 2014 in which climate and environmental policy appear to have slipped down the EU policy agenda (Čavošcki, 2015) resulting in fewer policy proposals leaving MEPs wondering what they are supposed to do with their time (Interview, March 2015).

Conclusions

This analysis of the EP's role in climate change politics allows us to draw several conclusions relating to the core themes of this book: leadership, new environmental policy instruments and the green economy First, the shifting fortunes of the EP in the field of climate change very clearly demonstrate that the EP's ability to exercise influence and therefore leadership is contingent upon a range of factors (Judge *et al.* 1994) – not least, the prevailing exogenous conditions. The EP was forced to adopt a defensive position when dealing with the 2008 climate change and energy package to prevent the package being weakened or shelved altogether, and economic considerations have also been used subsequently by MEPs to justify weakening Commission proposals.

The Parliament has adopted a fairly consistent position over the years towards the use of new environmental policy instruments associated with a green economy approach. While sceptical about the use of some instruments it has been prepared to engage in a policy of constructive engagement. However, the passage of the climate change and energy package demonstrates that despite the 'win-win' discourse of the green economy that the Commission championed during Barroso's second term (2009–2014) there remains a clear trade-off between economic growth and environmental protection. In uncertain and difficult economic times heads of state and government in the EU and elsewhere will shrink from offering leadership where leadership implies costs, and MEPs will be subject to pressure from national interests and governments. Their unique position at the interface of national and European levels of governance allows them, for much of the time, to be insulated from national pressures, but where the costs are high and key national interests are at stake they are as vulnerable as domestic politicians to pressure from their parties and governments.

Finally, it is clear that the enlargement of the European Union has changed the dynamics of environmental decision-making in the EU; it has shifted the political centre of gravity of the Union and its institutions ideologically to the right and geographically to the East. With 28 members, many of which are poorly prepared to withstand a deep global downturn and view environmental policy-making as an expensive luxury, it is difficult for the EU to develop and maintain a coherent leadership position in international environmental politics. Governance in this diverse multi-level system of governance has become even more complex and this increased heterogeneity is having a clear impact upon preference formation within the institutions. These shifts have certainly constrained the EP's capacity to assume an internal leadership role as both the changing intra and inter-institutional balance of power effect the Parliament's ability to shape the EU's external leadership ambition.

Note

1 This chapter draws upon evidence from interviews with actors based in the EU's institutions and NGOs, conducted in 2009, 2013 and 2015. All interviewees wished to remain anonymous and emphasised that their views were personal and did not represent those of the institutions and organisations for which they work.

References

Bäckstrand, K. and Elgström, O. (2013) 'The EU's role in climate change negotiations: from leader to leadiator', *Journal of European Public Policy*, 20(10): 1369–1386.

Biedenkopf, K. (2015) 'The European Parliament in EU external climate change governance', in S. Stravridis and D. Irrera (eds) *The European Parliament and its International Relations*, London: Routledge, 92–106.

Burns, C. (2013) 'The European Parliament', in A. Jordan and C. Adelle (eds), *Environmental Policy in the European Union*, 3rd edn, London: Earthscan, 132–151.

Burns, C. and Carter, N. (2010) 'Is codecision good for the environment?', *Political Studies*, 58: 123–142.

Burns, C. and Carter, N. (2011) 'The European Parliament and climate change: from symbolism to heroism and back again' in R.K.W. Wurzel and J. Connelly (eds) *The European Union as a Climate Change Leader*, London: Routledge, 58–73.

Burns, C., Carter, N. and Worsfold, N. (2012) 'Enlargement and the environment: the changing behaviour of the European Parliament', *Journal of Common Market Studies*, 50(1): 54–70.

Burns, C., Carter, N., Davies, G.A.M. and Worsfold, N. (2013) 'Still saving the Earth?: The European Parliament's environmental record', *Environmental Politics*, 22(6): 935–954.

Čavoški, A. (2015) 'A post-austerity European Commission: no role for environmental policy?', *Environmental Politics*, 24(3): 501–505.

European Commission (2013) *Proposal for a directive of the European Parliament and of the Council amending Directive 2003/87/EC establishing a scheme for greenhouse gas emission allowance trading within the Community*, COM/2013/0722 final, Brussels: European Commission.

EP (1998a) *Resolution on Environmental Policy and Climate Change following the Kyoto Summit*, B4–0142, 0143, 0144, 0145, 0151, 0164 and 0165/98, OJC 80, 16.03.1998, 227–231, Brussels: European Parliament.

EP (1998b) *Resolution on Climate Change in the Run-Up to Buenos Aires (November 1998)*, B4–0802/98, OJC 313, 12.10.1998,169–172, Brussels: European Parliament.

EP (1999) *Activity Report 1 November 1993–30 April 1999 From Entry into Force of the Treaty of Maastricht to Entry into Force of the Treaty of Amsterdam of the Delegations to the Conciliation Committee*, PE 230.998, Brussels: European Parliament.

EP (2000) *Resolution on Climate Change; Preparing for the Implementation of the Kyoto Protocol*, B5–0118/1999, OJC 107, 13/04/00, 112–115, Brussels: European Parliament.

EP (2002a) *European Parliament Resolution on the Outcome of the Bonn Conference on Climate Change*, B5–0539, 0540, 0541, 0543, 0551 and 0552/2001, OJC 72E, 21.03.2002, 321–323, Brussels: European Parliament.

EP (2002b) *European Parliament Resolution on the Follow-Up to Parliament's Opinion on the European Union's Strategy for the Marrakesh Conference on Climate Change*, B5–0686/2001 OJC 212E, 09.05.2002, 299–301, Brussels: European Parliament.

EP (2004) *Activity Report 1 May 1999 to 30 April 2004 (5th Parliamentary Term) of the Delegations to the Conciliation Committee*, PE 287.644, Brussels: European Parliament.

EP (2006a) *Conciliations and Codecision Activity Report: July 2004 to December 2006 (6th Parliamentary Term. First Half-Term)*, Brussels: European Parliament.

EP (2006b) *Resolution on the European Union Strategy for the Nairobi Conference on Climate Change*, B6–0543/2006, OJC313E, 20/12/2006, 439–442, Brussels: European Parliament.

EP (2009a) *2050: The Future Begins Today – Recommendations For the EU's Future Integrated Policy on Climate Change*, T6–0042/2009 (04/02/2009), Brussels: European Parliament.

EP (2009b) *European Parliament Resolution of 25th November 2009 on the EU Strategy for the Copenhagen Conference on Climate Change (COP15)*, B7–0141/2009, (25.11.2009), Brussels: European Parliament.

EP (2010a) *Activity Report on Codecison and Conciliation 1 May 2004 to 13 July 2009 (6th Parliamentary Term)*, URL: www.europarl.europa.eu/code/information/activity_reports/activity_report_2004_2009_en.pdf (accessed 18 September 2015).

EP (2010b) *European Parliament resolution of 25 November 2010 on the climate change conference in Cancun (COP16)*, URL: www.europarl.europa.eu/sides/getDoc.do?type=TA&language=EN&reference=P7-TA-2010-442 (accessed 18 September 2015).

EP (2011a) *European Parliament legislative resolution of 15 February 2011 on the proposal for a regulation of the European Parliament and of the Council setting emission performance standards for new light commercial vehicles as part of the Community's integrated approach to reduce CO 2 emissions from light-duty vehicles*, URL: www.europarl.europa.eu/sides/getDoc.do?type=TA&language=EN&reference=P7-TA-2011-53, (accessed 18 September 2015).

EP (2011b) *PE Debates Tuesday 15/02/11*, URL: www.europarl.europa.eu/sides/getDoc.do?pubRef=-//EP//TEXT+CRE+20110215+ITEM-003+DOC+XML+V0//EN (accessed 18 September 2015).

EP (2011c) *European Parliament resolution of 16 November 2011 on the climate change conference in Durban (COP17)*, URL: www.europarl.europa.eu/sides/getDoc.do?type=TA&language=EN&reference=P7-TA-2011-504 (accessed 18 September 2015).

EP (2012a) *European Parliament Resolution of 15 March 2012 on a Roadmap for Moving to a Competitive Low Carbon Economy in 2050*, URL: www.europarl.europa.eu/sides/getDoc.do?type=TA&language=EN&reference=P7-TA-2012-86 (accessed 18 September 2015).

EP (2012b) *European Parliament resolution of 22 November 2012 on the Climate Change Conference in Doha, Qatar* (COP18) URL: www.europarl.europa.eu/sides/getDoc.do?type=TA&language=EN&reference=P7-TA-2012-452 (accessed 18 September 2015).

EP (2013) *European Parliament resolution of 23 October 2013 on the climate change conference in Warsaw, Poland (COP19)*, URL: www.europarl.europa.eu/sides/getDoc.do?type=TA&language=EN&reference=P7-TA-2013-443 (accessed 18 September 2015).

EP (2014a) *European Parliament Resolution of 5 February 2014 on a 2030 Framework for Climate and Energy Policies*, (2013/2135(INI)), URL: www.europarl.europa.eu/sides/getDoc.do?type=TA&language=EN&reference=P7-TA-2014-0094 (accessed 18 September 2015).

EP (2014b) *Report on the proposal for a directive of the European Parliament and of the Council amending Directive 2003/87/EC establishing a scheme for greenhouse gas emission allowance trading within the Community, in view of the implementation by 2020 of an international agreement applying a single global market-based measure to international aviation emissions*, Committee on the Environment, Public Health and Food Safety, Brussels: European Parliament, URL: www.europarl.europa.eu/sides/getDoc.do?type=REPORT&mode=XML&reference=A7-2014-0079&language=EN (accessed 18 September 2015). </TitreType>.

EP (2014c) *PE Debates Wednesday 02/04/1*4, URL: www.europarl.europa.eu/sides/getDoc.do?pubRef=-//EP//TEXT+CRE+20140402+ITEM-025+DOC+XML+V0//EN&language=EN (accessed 18 September 2015).

EP (2104d) *Resolution on the 2014 UN Climate Change Conference – COP20 in Lima, Peru (1–12 December 2014)* URL: www.europarl.europa.eu/oeil/popups/ficheprocedure.do?reference=2014/2777(RSP)&l=en, (accessed 18 September 2015).

EP (2015a) *Activity Report on Codecision and Conciliation 14 July 2009–30 June 2014 (7th Parliamentary Term)*, URL: www.europarl.europa.eu/code/information/activity_reports/activity_report_2009_2014_en.pdf (accessed 18 September 2015).

EP (2015b) *Report Towards a new international climate agreement in Paris*, P8_TA-PROV(2015)0359, Committee on the Environment, Public Health and Food Safety, Brussels: European Parliament.

EP (2015c) *European Parliament legislative resolution of 8 July 2015 on the proposal for a decision of the European Parliament and of the Council concerning the establishment and operation of a market stability reserve for the Union greenhouse gas emission trading scheme and amending Directive 2003/87/EC* URL: www.europarl.europa.eu/sides/getDoc.do?type=TA&language=EN&reference=P8-TA-2015-0258 (accessed 18 September 2015).

Gravey, V. (2015) *Stronger, not Greener? The European Parliament in the 2013 Reform of the EU's Common Agricultural Policy*, Paper prepared for the 16th UACES Student Forum Conference, 29–30 June 2015, Belfast: Queen's University Belfast.

Hix, S., Noury, A.G. and Roland, G. (2007) *Democratic Politics in the European Parliament*, Cambridge: Cambridge University Press.

Hubschmid, C. and Moser, P. (1997) 'The co-operation procedure in the EU: why was the European Parliament influential in the decision on car emission standards?', *Journal of Common Market Studies*, 35: 225–242.

Judge, D. (1992) 'Predestined to save the Earth: the Environment Committee of the European Parliament', *Environmental Politics*, 1: 186–212.

Judge, D. and Earnshaw, D. (2008) 'The European Parliament: leadership and 'followership', in J. Hayward (ed.), *Leaderless Europe*, Oxford: Oxford University Press, 245–268.

Judge, D., Earnshaw, D. and Cowan, N. (1994), 'Ripples or waves: the European Parliament in the European Community policy process', *Journal of European Public Policy*, 1: 27–52.

Oberthür, S. and Roche-Kelly, C. (2008) 'EU leadership in international climate policy: achievements and challenges', *The International Spectator*, 43: 35–50.

Shackleton, M. and Raunio, T. (2003) 'Codecision since Amsterdam: a laboratory for institutional innovation and change', *Journal of European Public Policy*, 10: 171–188.

Shankelman, J. and Murray, J. (2015) *COP21: EU teams up with Africa and small islands as battle lines drawn with China and India*, businessGreen, 08/12/15, URL: www.businessgreen.com/bg/news/2438364/cop21-eu-teams-up-with-africa-and-small-islands-as-battle-lines-drawn-with-china-and-india (accessed 21 February 2016).

Schreurs, M.A. and Tiberghien, Y. (2007) 'Multi-level reinforcement: explaining European Union leadership in climate change mitigation', *Global Environmental Politics*, 7: 19–46.

Stern, N. (2007) *The Economics of Climate Change*, Cambridge: Cambridge University Press.

Tobin, P. and Burns, C. (2016) 'EU facilitates Paris Agreement, but economic concerns linger', *Environmental Europe, Ideas on Europe* 13/01/16, URL: http://environmentaleurope.ideasoneurope.eu/2016/01/13/eu-cop21-climate-paris/ (accessed 21 February 2016).

Van Schaik, L. and Schunz, S. (2012) 'Explaining EU activism and impact in global climate politics: is the Union a norm- or interest-driven actor?', *Journal of Common Market Studies*, 50(1): 169–186.

Weale, A., Pridham, G. Cini, M., Konstadakopulos, D., Porter, M. and Flynn, B. (2000) *Environmental Governance in Europe*, Oxford: Oxford University Press.

Wettestad, J. (2005) 'The making of the 2003 EU emissions trading directive: an ultra-quick process due to entrepreneurial proficiency?', *Global Environmental Politics*, 5(1): 1–24.

Wettestad, J., Eikeland, P. and Nilsson. M. (2012) 'EU climate and energy policy: a hesitant supranational turn?', *Global Environmental Politics*, 12(2): 67–86.

Wurzel, R.K.W. and Connelly, J. (eds) (2011) *The European Union as a Leader in International Climate Change Politics*, London: Routledge.

Zito, A. (2000) *Creating Environmental Policy in the European Union*, Basingstoke: Palgrave.

5 The Council and the European Council

Stuck on the road to transformational leadership

Claire Dupont and Sebastian Oberthür

Introduction

In this chapter, we analyse the role of the Council of the European Union (the Council) and the European Council of heads of state and government in the evolution of EU leadership in international climate policy. We focus on the development of domestic EU climate policy within the context of international climate negotiations, because credible international leadership requires appropriate domestic measures ('leadership by example') (Gupta and Grubb 2000; Bretherton and Vogler 2006). We argue that the Council and the European Council have remained crucial in developing the leadership record of the EU on climate change, but that this record, having moved from a rhetorical leadership style in the 1990s towards the beginnings of a transformational leadership style in the 2000s, has stagnated on the way to such transformational leadership. Divergences among member states have impeded further progress, while the evolving external politics of climate change have constituted a major challenge.

Nevertheless, climate change is firmly established in the realm of 'high politics', and the European Council has moved into an increasingly central position in climate policy. The European Council engages more in climate policymaking by providing clear directions for the course and ambition of internal policy measures. Since the peak of climate leadership in the EU, a certain degree of 'new intergovernmentalism' (Bickerton *et al.* 2015) has become evident in the development of internal and external climate policy – with the Council, and especially the European Council being the key decision-makers.

In this chapter, we first discuss the role of the Council and European Council in internal and external EU climate policy in general. We then trace their leadership role, highlighting the move towards a transformational leadership style over the course of the 2000s, leading to stagnation in advances in leadership since the late 2000s. Next, we discuss the Council's and European Council's narrative on the green economy, and their role in the multi-level governance of climate change.

Council and European Council in internal and external EU policymaking on climate change

The Council and the European Council are examples of multi-level governance in action. They are simultaneously meeting places for member states (at ministerial level in the Council and at the level of heads of state and government in the European Council), and institutions in their own right, which decide on internal policy and external negotiating positions. The development of internal climate policy is particularly important for the EU to be a credible international leader on climate change: without ambitious domestic policies, there is little credibility to claims of leadership from the EU. By providing the example of how to act on climate change, it becomes a leading actor that can attract followers (Gupta and Grubb 2000).

The European Council holds no formal competence in the legislative procedures of the EU, but its impact derives from its importance as the highest political gathering of the EU member states. Relevant statements on the direction of the Union's development are released after each European Council meeting as 'Conclusions'. Internally, the conclusions can provide the political impetus for the development of domestic climate and energy policies. Externally, the conclusions of the European Council may give signals on the political commitment and leadership of the EU to international partners (McCormick 2001; Oberthür and Pallemaerts 2010; Oberthür and Wyns 2014).

The Council shares its power with the European Parliament and the European Commission for internal policymaking (see Chapters 3 and 4). In the applicable ordinary legislative decision procedure for the adoption of EU legislation, the Council can block climate and energy policy proposals, but it cannot easily move domestic EU climate policy forward on its own. The adoption of most relevant policy proposals requires a qualified majority within the Council, although the Council frequently decides without recourse to a formal vote. The Council of Environment Ministers is one of 10 Council formations (as of 2016) and has traditionally taken the lead in developing EU internal and external climate policy, but the Council of Energy Ministers has become increasingly involved in domestic policy development. Climate policy has become more intertwined with energy policy over the years, with several energy policies adopted to combat climate change (such as measures to promote renewable energy and energy efficiency). The involvement of energy policymakers in internal policy development is especially clear since the elaboration of the first climate and energy package towards 2020 in 2007–2009 (see below). Since 2009, the EU has specific competence on energy policymaking, including on the promotion of 'energy efficiency and energy saving and the development of new and renewable forms of energy' (Art. 194.1(c) of the Treaty on the Functioning of the European Union, TFEU). However, the Council has to decide unanimously on measures 'primarily of a fiscal nature' (e.g. environmental or energy taxation; see Arts. 192.2 and 194.3 TFEU), as well as 'measures significantly affecting a Member State's choice between different energy sources and the general structure of its energy supply' (Art. 192.2 TFEU; see also Art. 194.2 TFEU). Individual member states thus possess veto power on parts of climate and energy policy (Wettestad 2000).

In contrast, the Council is the most important actor shaping EU *external* climate policy. Climate policy is an area of 'shared competence' between the EU and its member states (see Chapters 3 and 4). Both the EU (represented by the European Commission) and its member states are parties to the UN Framework Convention on Climate Change (UNFCCC) and participate in the international climate negotiations. In practical terms, the Council determines the international negotiation position of the EU with active participation and input from the Commission. A working group of the Council develops the positions, which are then usually reflected in Council conclusions. These Council conclusions provide the basis for the member states and the Commission to coordinate their strategy on a daily basis at the international negotiations (Article 218 TFEU) (van Schaik 2010; Oberthür 2011). The European External Action Service (EEAS), launched in 2011, now supports the EU member states in carrying out climate diplomacy with international partners, but is less involved in internal policymaking (see Chapter 3). The Parliament has also gained more relevance for the international negotiations, as since the entry into force of the Lisbon Treaty its consent is required for the EU to conclude international agreements (Art. 218.6 TFEU; see also Chapter 4).

Rotating among the member states every six months, the Presidency of the Council plays slightly different roles in internal and external EU climate (and energy) policymaking. Internally, as the chair of Council meetings, it shapes the agenda and the conduct of meetings. It also plays a significant role in brokering compromise solutions and taking political initiatives. Externally, the Presidency acts as the main contact point and spokesperson for the EU, although the Commission has taken on an enhanced role over the years.

Stagnating leadership: stuck halfway towards transformational leadership?

The leadership style of the Council and the European Council seemed to be moving towards transformational leadership in the 2000s (Oberthür and Dupont 2011). Changing internal and external dynamics since the late 2000s have challenged the EU's leadership ambitions and halted the march towards truly transformational leadership. The Council and European Council nevertheless remain important.

Throughout the 1990s, EU leadership on climate change was rhetorical because relatively ambitious positions were rarely underpinned by effective domestic climate policy. This record improved in the 2000s when the EU passed and implemented an increasing number of climate policies, including the EU Emissions Trading Scheme (ETS). Although the Council and the European Council never became major 'green' forces, they dealt more regularly with climate policy. By the mid-2000s, the EU had arguably established the most advanced climate policy framework of the major economies worldwide and had to some extent narrowed the previously existing credibility gap. At the same time, EU coherence and coordination toward the international negotiations also improved (Dupont and Oberthür 2012; 2015c). At this time, climate change was also firmly entrenched in high politics and regularly discussed at the European Council (see Figure 5.1). Both internal policy measures and external positions were part of the Council and European Council's negotiations.

In March 2007, the European Council provided a major impetus to the development of EU climate policy, by making an 'independent commitment' to reduce GHGE in the EU by 20 per cent from the 1990 level by 2020. In addition, the European Council agreed to increase the share of renewable energy sources in EU final energy consumption to 20 per cent and to save 20 per cent on the EU's projected energy consumption for 2020 (European Council 2007).

The Council and European Council played leading, but ambiguous, roles in the elaboration of the new climate policies to achieve the 2020 goals, including the adoption of a renewable energy directive (Directive 2009/28/EC) and a revised ETS Directive (2009/29/EC). They clearly supported such strengthened measures, but member states also engaged in haggling and trade-offs that ultimately weakened a number of the policy measures (Dupont 2016). Unusually, the European Council took the lead in agreeing the policy measures in December 2008 – formally a deal between the Council and the European Parliament. Nevertheless, the targets to 2020 are just a small step on the road to meet the EU's objective of keeping the increase in global mean temperature below two degrees Celsius or to reduce GHGE by 80–95 per cent by 2050 compared to 1990 (European Council 2009b; Dupont and Oberthür 2015b).

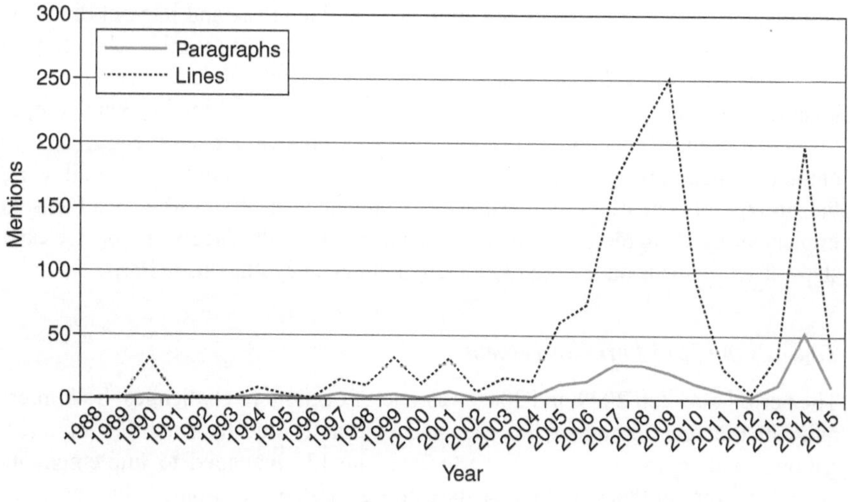

Figure 5.1 Mentions of climate change in the conclusions of the European Council 1988–2015.

Source: own counting of European Council conclusions.

Over the years, climate policy has become increasingly integrated with energy policy measures – moving from a European Climate Change Programme (ECCP) in the early 2000s to a 'climate and energy policy framework' since the mid-2000s. This growing link between climate and energy policy has had an ambivalent effect, sometimes pushing for more ambitious climate action and sometimes weakening climate leadership. It also adds the complexity of energy policy decision-making to the development of climate policy. Member states continue to guard their sovereignty over the shape of their energy structure. The three pillars of EU energy policy – sustainability, competitiveness and security – are rarely given equal weight at all times by each member state. External shocks or events that affect energy prices or supply (e.g. the Ukraine crisis or gas supply crises in the 2000s) may change energy priorities in member states, either raising or lowering the importance of adopting climate-friendly energy policies. Internal divergences in national interest on energy policy play out in the negotiations and trade-offs in the Council, leading to more or less ambition on climate policy, and hence more or less of a transformational leadership style.

Just as the Council and European Council were at the peak of their leadership role in the second half of the 2000s, the 2009 Copenhagen Climate Conference raised questions about the viability of the EU's external climate policies for maintaining international leadership. The 2020 climate and energy package served as the EU's input for the 2009 Copenhagen climate conference, where the EU acted – unsuccessfully – as the major pusher for a global, ambitious and comprehensive post-2012 agreement. However, the EU's ambitious internal policies had little influence on the negotiations, as the EU was side-lined between the US and the BASIC countries (Brazil, South Africa, India and China) (Oberthür 2011; van Schaik and Schunz 2012; Groen and Niemann 2013). Furthermore, the economic and financial crises from 2008 onwards, the Ukraine crisis, and internal divisions between the member states dampened possibilities for far-reaching climate policies. Jobs and competitiveness became the predominant political discourses, with countries such as Spain (see Chapter 11) and Greece, for example, making competiveness a primary concern. This competition for political attention contributed to climate change moving down or off the agenda after 2010 (Figure 5.1). Following the disappointment of the Copenhagen conference, the EU shifted its focus to implementing its agreed policies and beginning difficult discussions on the next steps. It set its focus on the road to the 2015 climate negotiations in Paris.

Towards the 2015 Paris Agreement

The period between 2009 and 2015 was one of even more challenges, both internally in the functioning of the EU, and externally due to a number of international and regional crises. Nevertheless, the EU managed to implement its agreed internal policies and adapt its external strategy to promote international action on climate change.

The EU adopted policy measures on CO_2 emissions and cars, on backloading (removing allowances from the carbon market for a limited period of time to contain oversupply of emission allowances), on energy efficiency, and on further measures to fix the problem of oversupply of allowances in the ETS (a Market Stability Reserve, MSR). Throughout this period, the Council was the least ambitious of the EU deciding institutions. It watered down the 2012 Energy Efficiency Directive (Boasson and Dupont 2015), and negotiations on the MSR were symptomatic of the deepening divisions among member states. Poland, Bulgaria, Romania, Croatia and Hungary all voted against the MSR (Wettestad and Jevnaker 2016). These divisions were entrenched in the growing unity of the Visegrad group (see also Chapter 10) of Central European Countries (the Czech Republic, Poland, Hungary and Slovakia) speaking with one voice in opposition to the mainly Western European 'Green Growth Group' of member states (Belgium, Denmark, Estonia, Finland, France, Germany, Italy, the Netherlands, Portugal, Slovenia, Spain, Sweden, the UK). Furthermore, growing EU scepticism has also restrained climate policy ambition in the EU, as such scepticism often comes hand-in-hand with climate scepticism and opposition to climate policies, especially in the discourse of right-wing political ideologies.

In 2014, the European Council nevertheless provided its important political backing to develop the future climate and energy policy framework in the EU. In October 2014, the European Council agreed on three headline targets to (1) reduce GHGE by at least 40 per cent; (2) increase the share of renewable energy to at least 27 per cent; and (3) improve energy efficiency by at least 27 per cent by 2030 with an option to increase this target to 30 per cent before 2020 (European Council 2014). While the first target is to be binding as in the past, the renewable energy target is intended to be binding at EU level (i.e. there will be no binding targets for member states) and the energy efficiency target is non-binding. This agreement was the result of intense discussions throughout 2014, with the European Council showing lower ambition than other European institutions. The Commission had proposed a formula of '40–27–30' in January and July 2014 (40 per cent GHGE reduction, 27 per cent share of renewable energy and 30 per cent improvement in energy efficiency by 2030), whereas the European Parliament had advocated '40–30–40' (all binding) in February 2014. The European Council's agreement served as the EU's input to the international Paris Agreement, adopted in 2015 (Dupont and Oberthür 2015c).

Importantly, the European Council also agreed to establish two funding schemes. First, building on a similar scheme in place for 2013–2020, a facility fed from the sale of 400 million ETS allowances will support carbon capture and storage, renewables, and low-carbon innovation in industrial sectors. Second, another 2 per cent of ETS allowances will be used to fund energy efficiency improvements and the modernisation of energy systems in low-income member states (especially in Central and Eastern Europe). Depending on the details to be worked out, both could be significant instruments for advancing the structural changes required for decarbonisation (Dupont and Oberthür 2015c).

Importantly, the EU has already surpassed its 2020 target of reducing GHGE by 20 per cent (see Figure 5.2). Preliminary figures show that the EU had achieved a GHGE reduction of more than 24 per cent in 2014, facilitated by a

Figure 5.2 GHGE of the EU-28, 1990–2013, with preliminary figures for 2014.

Source: European Environment Agency data (EEA 2015).

relatively mild winter. GHGE are expected to decline to at least 24–25 per cent below 1990 levels by 2020, which may raise questions about the level of ambition of a 40 per cent GHGE reduction target by 2030 (EEA 2015).

The European Council's 2030 climate and energy policy framework agreed in 2014 also provided the basis for the EU's involvement in the international negotiations towards the Paris Agreement adopted in December 2015. The EU submitted the binding target of a GHGE reduction of at least 40 per cent by 2030 as its contribution to the international efforts, the highest reduction figure of the major players. It did so as among the first major parties (second only to Switzerland), thereby trying to 'lead by example'. It furthermore demanded fair, ambitious and binding GHGE limitation commitments from all countries, a regular five-yearly review to enhance global ambition, and robust rules on transparency and accountability (Oberthür and Groen 2016).

These policy objectives were in line with, and a result of, the revision of the EU's leadership strategy towards 'leadiating' (i.e. leading and mediating; see Bäckstrand and Elgström 2013), after the failure of the Copenhagen conference in 2009. Hence, the EU not only agreed to enshrine its 20 per cent reduction target in a second commitment period of the Kyoto Protocol (2013–2020), which was adopted in 2012, but it also moderated its international demands and reoriented its international strategy towards coalition and bridge building. It also reached out to ambitious developing countries such as small island states, least developed countries and several Latin-American countries, including through the Cartagena Dialogue for Progressive Action established after Copenhagen, and made compromise proposals trying to build bridges. This adapted strategy responded to an evolving international constellation that was characterised by the reengagement of the US in international climate policy (see Chapter 16) and the rise of emerging powers, including the BASIC countries (Bäckstrand and Elgström 2013; see also Oberthür 2011).

Led by the Council and the European Council (together with the European Commission), this recalibration of the EU strategy proved relatively successful. While the Commission continued to play a prominent role, the leading role of the Council in external climate policy was not challenged, as the broader EU crises tempered the desire of the Commission to push for change. As a matter of practice, the EU negotiations were led by 'lead negotiators' appointed by the Council Presidency, which primarily came from the big member states and the Commission. A first important fruit of the EU's adapted strategy was the 2011 Durban mandate to elaborate 'a protocol, another legal instrument or an agreed outcome with legal force' to be adopted in 2015 that was brought about by a 'Durban coalition' of ambitious developed and developing countries, including the EU (Bäckstrand and Elgström 2013).

The Paris Agreement adopted in December 2015 has been widely heralded as exceeding expectations and, hence, as a success for the EU (although this view is not shared by all environmental NGOs; see Chapter 15). The Agreement establishes GHGE neutrality in the second half of the twenty-first century as a central objective, commits all parties to submitting successive climate action plans every five years, and establishes a unitary and ambitious transparency and accountability framework for all parties (Obergassel *et al.* 2016; Bodle *et al.* 2016). It thereby goes a long way to satisfying the core EU demands, that the EU actively promoted on the basis of its recalibrated strategy. This included intensified climate diplomacy efforts with the help of the EEAS, which implemented a climate diplomacy action plan in 2015 (Oberthür and Groen 2016).

Looking back at developments since Copenhagen in 2009, it would thus seem that the EU has successfully reviewed and revised its international climate policy in view of a changing international environment. For the time being, the 'leadiator' role it has carved out for itself seems to provide a new model. Whether the EU can further sustain that model will not least hinge on its ability to continue to lead by example through the development of its domestic climate and energy policies as well as on the further evolution of the international climate politics landscape.

Green economy

In the early years of international climate policy development, climate change was largely considered a threat to modern society. As it became a more prominent policy issue and entered the realm of high politics, the economic opportunities presented by moving to a low-carbon society grew more prominent in political discourse. The narrative of moving towards a 'green' or 'low-carbon' economy with multiple benefits for the climate, economy and competitiveness has become established in internal policymaking and has played a growing role in external climate diplomacy, and the EU's role in altering the discourse demonstrates a degree of cognitive leadership. Both the Council and the European Council have regularly highlighted the win-win opportunities for the EU in moving towards a 'green economy' or a 'low-carbon economy' and as a motivation for external partners to develop similar policies. The 'green economy' discourse has much broader relevance than to climate policy alone, and – especially in the Environment Council – is often deployed to increase motivation for policy development on a range of issues.

Internally, the narrative on the 'green economy' seems to have become increasingly entrenched in the discussions on EU climate and energy policy development. Since the leadership stance of the Council and European Council began to level off in the late 2000s, these arguments have been used to bolster the motivation to agree on climate policy measures – even in the face of economic and financial crises. References to the green economy can be seen as a rationale for policy development across a number of portfolios, and as a justification for ensuring environmental and climate objectives are integrated into other policy sectors. For example, in 2009, the European Council stated that the 'challenges posed by both climate change and the economic and financial crisis will open up new opportunities and make it possible to move to a safe and sustainable low-carbon economy capable of generating growth and creating new jobs' (European Council 2009a: 11). In 2013, the Environment Council called for 'more climate resilient investment … to contribute to a transition to a green economy and the creation of new job opportunities' (Council of the European Union 2013: 2). In 2015, obtaining 'green growth' was underlined as a reason to push forward policies on the protection of biodiversity; on the circular economy; and to push for higher degrees of environmental integration into economic instruments such as the Europe 2020 growth strategy and the annual growth surveys associated with it and the European semester (an annual analysis of the implementation of EU economic rules). In international climate policy, the Council regularly emphasises the potential of moving to a 'climate-friendly economy' to stimulate climate policy action in other countries.

In general, the language of the 'green economy' has become more prominent over time. Moving to a low-carbon economy is an overarching goal of the EU, as inherent in the European Council's objective to decarbonise the economy by 2050. In the Environment Council, such a narrative is put forward as a means to achieve policy integration and 'win–win' policy measures. The 'green economy' discourse has therefore played a growing role in pushing internal policy developments on climate change, especially in the context of adopting policies that have mainly long-term benefits (and that may have short-term costs). This narrative has also found broad support among EU citizens. In 2014 and 2015 about 80 per cent of citizens surveyed believed that fighting climate change and using energy efficiently can boost the economy and jobs (Eurobarometer 2014 and 2015). The narrative can aid the leadership stance of the EU more generally as the adoption of credible internal policies that lead to a low-carbon economy demonstrates that economic prosperity and climate protection are synergistic objectives.

Multi-level governance

The Council and the European Council are supranational institutions of the EU, while also being meeting places for member state representatives. There is thus a clear interaction among levels of governance in their leadership roles on climate change. Internally, the political dynamics among member states and between the Council and the European Council have changed over the years. The Council and European Council also operate within a broader context of international challenges, which compete for attention with the climate issue. This multi-level interaction has been both an accelerator of EU policy and a brake on progress.

Internally, since the EU enlarged to 28 member states with distinct national interests, national energy priorities and external relations, climate policy development has proved more challenging. Opposition from some member states (e.g. the Visegrad group, see above and Chapter 10) to strong climate policies, as well as internal economic and debt crises, EU scepticism, international conflict, migration and energy crises have all added to the challenging context for climate leadership. Even states like the UK, Germany and Denmark (see Chapter 12, 8 and 6 respectively), which had previously passed strong domestic policies on climate change, renewable energy and energy efficiency, have increasingly faced competing priorities at national level leading to less urgency in advancing EU-level climate and energy policy. Such internal dynamics can be linked to the structural distinctions among member states, including industrial competence and structure, government changes and economic strength (Boasson and Dupont 2015; Dupont and Oberthür 2015c).

At the same time, EU citizens have continued to support action on climate change through phases of EU enlargement, and through internal and external economic and financial crises. Eurobarometer surveys have shown sympathy for the seriousness of the climate problem, and also the responsibility of national governments and the EU to act. Climate change has regularly featured in the top four most serious problems since 2008 and EU citizens see clear broader benefits in taking a leading role on climate change (Eurobarometer 2008, 2011, 2014 and 2015). While the Council and European Council have also reflected this win-win opportunity, this narrative has been insufficient, in the light of the internal and external challenges, to push the Council and European Council further on the road towards true leadership by example and a transformational leadership style.

As a result of several trade-offs, a number of member states are taking on a greater share of the effort (and reaping the modernisation benefits that come with the transformation to a low-carbon economy). Where some member states are eager and motivated to move forward on policy – such as in the negotiations on the Market Stability Reserve to fix the ETS – such trade-offs can ensure winning coalitions in the Council. Where there are not enough member states willing to take on ambitious policy measures, policies are weakened considerably in the negotiations (the case of energy efficiency policy, for example), even with pressure from Parliament (see Chapter 4). Internal policy developments vacillate between long-term goals and objectives, usually established and agreed at

European Council level, and short-term concerns of particular member states. Where short-term concerns take precedence, climate policies may be weakened, leaving much of the work of policy development and implementation to later decades (Dupont and Oberthür 2012). The need to agree on compromises and trade-offs is compounded by the culture of adopting policies by consensus in the Council, even when qualified majority voting is possible (Häge 2012). These dynamics heighten the challenge in sustaining leadership on climate.

Finally, the institutional traditions and working relationship between the Council and the European Council have changed over time. In an examination of economic governance, Uwe Puetter described the shift in division of work between the Council and European Council to informal settings where more deliberation takes place. He furthermore underlined the heightened role in EU governance taken on by the European Council, which now 'intervenes on a regular basis and constantly deals with matters of day-to-day politics' (Puetter 2012, 168). This finding can also be evidenced in the role played by the European Council in the adoption of the 2020 climate and energy framework, and since then in the delineation of ambitions and policy directions for 2030 and 2050.

On an international level, the Council and European Council interact in the context of the multilateral setting of the UNFCCC negotiations. These negotiations have created critical windows of opportunity and decision moments for EU policy development, as the EU has attempted to demonstrate international leadership (Oberthür and Pallemaerts 2010). For example, the preparation for the Copenhagen Summit in 2009 motivated and drove, to a significant extent, the 2009 EU climate and energy package. Agreement on the cornerstones of the policy framework for 2030 in the European Council in October 2014 was prompted in part by the shadow of the 2015 Paris conference, which adopted a new international climate agreement for beyond 2020. Other international and transnational venues and forums, such as the G7 or the G20, have also acquired importance (Bulkeley and Newell 2010; Wurzel and Connelly 2011).

While slow progress after 2009 has made the international negotiations less of a driving force for EU climate policy and leadership, the 2015 Paris Agreement may reverse that trend to some extent. Concern over lack of progress internationally and among international partners rose after the Copenhagen conference and undermined the EU's rationale for engaging in 'leadership by example', in view of its own declining share in global GHGE (to around 10 per cent in the 2010s). As a result, international climate politics were increasingly questioned as a point of reference for EU policy development after the disappointment of the Copenhagen Summit (Groen and Niemann 2013). However, with the successful conclusion of the Paris Agreement in 2015, international climate policy may enable (rather than constrain) EU leadership by example again in the future, although possibly at a lower level than pre-2009. It may also enhance the role of effective climate policies and sustainable technologies as a source of international influence of the EU – a field that the EU can develop through its own means (Oberthür 2016).

Conclusions

After an improving leadership record from the 1990s to the 2000s, moving from symbolic leadership to a more transformational style of leadership by example, the 2010s have seen stagnation in EU leadership on climate change. The Council and the European Council have remained key players in EU internal and external climate policy development, with the European Council even growing in importance over the years. In 2014, for example, the European Council adopted more conclusions on climate policy than the Environment Council. The European Council has provided ever more detailed guidance on internal and external climate policy directions and ambitions. This reality may reflect the 'new intergovernmentalist' turn in EU integration (Bickerton *et al.* 2015). Changing political dynamics among the 28 member states, and changes in the working relationship between the Council and the European Council have led to a challenging environment within which to adopt a truly transformational style of leadership. The external political context has also resulted in much competition for attention on the agenda of the Council and European Council, although climate change remains firmly fixed as an issue of high politics.

Since the height of EU climate leadership in 2008/2009, internal policy development has moved forward only slowly, following a pattern of 'catch-up governance', where policy measures are strengthened too little too late to be truly transformational (Dupont and Oberthür 2015a). Externally, the EU has successfully adapted its climate diplomacy strategy since 2009 to respond to difficult international framework conditions. As a result, the EU largely achieved its aims in the 2015 Paris negotiations, also because it had moderated the ambition of these aims after the Copenhagen negotiations in 2009. The discourse on the 'green economy', the interconnections with energy policy and the reality of functioning within a complex multi-level system have added to the difficulty in enacting strong climate leadership.

If 'new intergovernmentalism' is the new normal of EU internal and external climate policy development, we can expect the roles of the Council and European Council to remain crucial for climate leadership into the future. Internally, the development of the 2030 climate and energy policy framework and further policy measures to 2050 will rely on the level of ambition and engagement of member states within the Council and European Council. Overcoming the pattern of 'catch-up governance' may require further impetus, perhaps through external shocks, other international partners moving beyond the level of ambition of the EU, or a coalition of advanced member states moving forward in great leaps. Externally, the international context may play a greater role in enabling and pushing EU leadership after the successful Paris negotiations. While awaiting further implementation, the 2015 Paris Agreement paves the way for the implementation of the EU's 2050 decarbonisation objective. Achieving this objective will require a considerable effort from the EU to transform its own economy and demonstrate the potential for low-carbon development. Under the circumstances, the Council and the European Council are the EU institutions that are crucial for the materialisation of the political will that is required to achieve decarbonisation in the EU.

References

Bäckstrand, K. and Elgström, O. (2013) 'The EU's Role in Climate Change Negotiations: From Leader to "Leadiator"', *Journal of European Public Policy*, 20(10): 1369–1386.

Bickerton, C.J., Hodson, D. and Puetter, U. (2015) 'The New Intergovernmentalism: European Integration in the Post-Maastricht Era', *Journal of Common Market Studies*, 53(4): 703–722.

Boasson, E.L. and Dupont, C. (2015) 'Buildings: Good Intentions Unfulfilled', in C. Dupont and S. Oberthür (eds), *Decarbonization in the European Union: Internal Policies and External Strategies*, Houndmills: Palgrave Macmillan, 137–158.

Bodle, R., Donat, L., and Duwe, M. (2016) 'The Paris Agreement: Analysis, Assessment and Outlook', *Climate and Carbon Law Review*, 10(1): 5–22.

Bretherton, C. and Vogler, J. (2006) *The European Union as a Global Actor*, London: Routledge.

Bulkeley, H. and Newell, P. (2010) *Governing Climate Change*, London: Routledge.

Council of the European Union (2013), *Environment Council Conclusions on an EU Strategy on Adaptation to Climate Change, 18 June*, Brussels: Council of the European Union.

Dupont, C. (2016) *Climate Policy Integration into EU Energy Policy: Progress and Prospects*, London: Routledge.

Dupont, C. and Oberthür, S. (2012) 'Insufficient Climate Policy Integration in EU Energy Policy: the Importance of the Long-Term Perspective', *Journal of Contemporary European Research*, 8(2): 228–247.

Dupont, C. and Oberthür, S. (2015a) 'Conclusions: Lessons Learned', in C. Dupont and S. Oberthür (eds), *Decarbonization in the European Union: Internal Policies and External Strategies*, Houndmills: Palgrave Macmillan, 244–265.

Dupont, C. and Oberthür, S. (eds) (2015b) *Decarbonization in the European Union: Internal Policies and External Strategies*, London: Palgrave Macmillan.

Dupont, C. and Oberthür, S. (2015c) 'The European Union', in E. Lövbrand and K. Bäckstrand (eds), *Research Handbook on Climate Governance*, London: Edward Elgar, 224–236.

EEA (2015) *Approximated EU GHG Inventory: Proxy GHG Estimates for 2014*, Copenhagen: European Environment Agency.

Eurobarometer (2008) *Europeans' Attitudes Towards Climate Change. Special Eurobarometer 300*, Brussels.

Eurobarometer (2011) *Climate Change. Special Eurobarometer 372*, Brussels.

Eurobarometer (2014) *Climate Change. Special Eurobarometer 409*, Brussels.

Eurobarometer (2015) *Climate Change. Special Eurobarometer 435*, Brussels.

European Council (2007) *Presidency Conclusions, March 2007*, Brussels: Council of the European Union.

European Council (2009a) *Presidency Conclusions, 18/19 June 2009*, Brussels: Council of the European Union.

European Council (2009b) *Presidency Conclusions, 29/30 October 2009*, Brussels: Council of the European Union.

European Council (2014) *Conclusions, Document EUCO 169/14, 23 and 24 October*, Brussels: European Council.

Groen, L. and Niemann, A. (2013) 'The European Union at the Copenhagen Climate Negotiations: A Case of Contested EU Actorness and Effectiveness', *International Relations*, 27(3): 308–324.

Gupta, J. and Grubb, M. (eds) (2000) *Climate Change and European Leadership*, Dordrecht: Kluwer Academic Publishers.

Häge, F.M. (2012) 'Coalition Building and Consensus in the Council of the European Union'. *British Journal of Political Science*, 43(3): 481–504.

McCormick, J., (2001) *Environmental Policy in the European Union*, Hampshire: Palgrave Macmillan.

Obergassel, W., Arens, C., Hermwille, L., Kreibich, N., Mersmann, F., Ott, H.E. and Wang-Helmreich, H. (2016) *Phoenix from the Ashes – An Analysis of the Paris Agreement to the United Nations Framework Convention on Climate Change*, Wuppertal: Wuppertal Institute for Climate, Environment and Energy.

Oberthür, S. (2011) 'The European Union's Performance in the International Climate Change Regime', *Journal of European Integration*, 33(6): 667–682.

Oberthür, S. (2016) 'Where to Go from Paris? The European Union in Climate Geopolitics', *Global Affairs*, 2(2): 119–130.

Oberthür, S. and Dupont, C. (2011) 'The Council, the European Council and International Climate Policy: From Symbolic Leadership to Leadership by Example', in R.K.W. Wurzel and J. Connelly (eds), *The European Union as a Leader in International Climate Change Politics*, London and New York: Routledge, 74–91.

Oberthür, S. and Groen, L. (2016) 'The European Union and the Paris Agreement: Leader, Mediator or Bystander?' *WIREs Climate Change* (forthcoming).

Oberthür, S. and Pallemaerts, M. (eds) (2010) *The New Climate Policies of the European Union: Internal Legislation and Climate Diplomacy*, Brussels: VUB Press.

Oberthür, S. and Wyns, T. (2014) Paris Climate Agreement 2015 – EU Needs to Ensure "Signal" and "Direction", *IES Policy Brief*, (8).

Puetter, U. (2012) 'Europe's Deliberative Intergovernmentalism: The Role of the Council and European Council in EU Economic Governance', *Journal of European Public Policy*, 19(2): 161–178.

van Schaik, L.G. (2010) 'The Sustainability of the EU's Model for Climate Diplomacy', in S. Oberthür and M. Pallemaerts (eds), *The New Climate Policies of the European Union: Internal Legislation and Climate Diplomacy*, Brussels: VUB Press, 251–280.

van Schaik, L.G. and Schunz, S. (2012) 'Explaining EU Activism and Impact in Global Climate Politics: Is the Union a Norm- or Interest-driven Actor?', *Journal of Common Market Studies*, 50(1): 169–186.

Wettestad, J. (2000) 'The Complicated Development of EU Climate Policy', in J. Gupta and M. Grubb (eds), *Climate Change and European Leadership: A Sustainable Role for Europe?*, Dordrecht: Kluwer Academic Publishers, 25–45.

Wettestad, J. and Jevnaker, T. (2016) *Rescuing EU Emissions Trading*, London: Palgrave Macmillan.

Wurzel, R.K.W. and Connelly, J. (eds) (2011) *The European Union as a Leader in International Climate Change Politics*, London: Routledge.

Part III

Member states and neighbouring European states

6 Denmark

Small state with a big voice and bigger dilemmas

Mikael Skou Andersen and Helle Ørsted Nielsen

Introduction

Denmark is a relatively small country by any standard and its influence in international climate negotiations remains limited. Nevertheless, the pioneering of certain climate-related policies (including renewables, energy efficiency and taxes on carbon dioxide (CO_2)) has, over the previous decades, been attracting international interest and has provided Denmark with some leverage to act as instigator and self-declared leader in international climate diplomacy. The independent Climate Change Performance Index, for instance, over three consecutive years ranked Denmark as its top performer in acknowledgement of the will and skill to transform into a low-carbon economy (Burck *et al.*, 2014). Not least a strong policy dimension, including the propensity to speak with 'a loud voice', seems decisive for the top ranking. While Francis Fukuyama has proclaimed the 'getting to Denmark' a desirable approach (cf. Meaney, 2011), there are numerous dilemmas both at the substantive and diplomatic level when it comes to climate policy. Denmark's actual greenhouse gas emissions (GHGE) per capita remain above the EU average (OECD, 2012: 73), reflecting a high level of individual consumption as well as sectoral interests related to resource extraction, intensive livestock farming and international shipping, that occasionally spill over into the policy-making processes.

National attitudes to climate change

Danes have long been convinced that climate change is occurring and a solid majority believes that man-made emissions of greenhouse gases must be curbed (Mandag Morgen, 2007), but the salience on the public agenda and the vigour with which this recognition has been pursued has waxed and waned over the years.

In surveys Danes consistently rank at the top among Europeans concerned about climate change. For instance, in both 2011 and 2013, about 30 per cent of Danes ranked climate change as 'the single most serious problem facing the world as a whole', while 73 per cent of the Danish respondents identified climate change as one of the most serious global problems (Eurobarometer, 2014). This general concern only partially translates into political pressure, however. In a 2013 poll conducted by the think tank CONCITO, 44 per cent of the respondents said that climate change would be one of the issues determining their vote in the next election (Concito, 2014). When asked about the top electoral issues, significantly fewer voters ranked climate and environment as their top policy priorities, i.e. around 10 per cent in 2014 and 2015 (TV2, 2014, 2015). Hence, climate policy did not feature as a salient issue in the election campaigns of 2011 and 2015.

As regards the political agenda, climate and energy policies have repeatedly received parliamentary attention, typically in connection with the passing of energy policies in the 1990s and again leading up to COP15 in 2009 in Copenhagen (Reenberg, 2014).

Attitudes in the business community have evolved from a strong resistance to energy taxes in the 1990s to a more varied response with some parts of the Danish business sector, including energy producers, arguing that Denmark should act as a pioneer in climate and energy policies, while more heavy energy users emphasize that competitiveness considerations should trump climate objectives. There are several environmental NGOs (ENGOs) which have been lobbying in favour of unilateral mitigation policies; apart from the more conventional ones (e.g. Greenpeace and NOAH/Friends of the Earth) the more expertise-based 'Ecological Council' as well as the climate think-tank CONCITO have been quite active. A particular feature of the Danish NGO scene is the influential Association of Windturbine Owners with a broad-based membership of both farmers and city dwellers.

Phases of domestic climate change policy

While Denmark's climate policy has varied in political salience and level of ambition, it has steadily revolved around an active energy policy, aiming for energy efficiency, transition towards cleaner energy sources in the energy system and support for development and use of renewable energy sources. This has been paralleled by step-by-step increases in targets for reducing GHGE and ever higher shares of renewable energy. In the last 10 years, climate mitigation efforts have been accompanied by national strategies for adaptation to climate changes.

1988–1993: from energy security to pioneering climate change awareness

Denmark initiated its active energy policy in the latter half of the 1970s, when the policy was framed in terms of energy security rather than climate change.

Following the 1987 Brundtland Commission's report and the 1988 International Conference on the Changing Atmosphere in Toronto, energy policies were linked to climate change. Thus, in 1990 Denmark outlined its first ever comprehensive energy policy embedded in a climate change frame *Energy 2000: Action plan for sustainable development* (ME, 1990). The plan constituted an effort at cognitive and exemplary leadership, stating that Denmark as one of the largest per capita CO_2 emitters carried a special responsibility for contributing to efforts towards an international agreement on preventing the greenhouse effect, and that this contribution would depend on the ability and determination to implement sustainability principles in Denmark's own energy system. The Danish Parliament adopted a 20 per cent reduction target for CO_2 emissions by 2005 (1988 baseline) and a 15 per cent reduction in energy consumption. This was to be achieved through conversion to cleaner energy sources for supply and increased integration of wind power in electricity production, underpinned by tax signals. The Conservative–Liberal government at the time had entrusted the portfolio of energy to Jens Bilgrav from the Social-Liberals, a party traditionally in a pivotal position between left and right in Danish politics. As it left the coalition and joined forces again with the Social Democrats a new era of more profound activist climate policy began.

1993–2001: taking the lead on the climate agenda

Throughout the 1990s Svend Auken (Social Democrat), a high-profile Minister of Environment and Energy, provided significant cognitive and entrepreneurial leadership whereby Denmark proclaimed its intention to serve as a pioneer for sustainable development and climate policy (MoE, 1996, 2000). Thus, in order to help stabilize global emissions of greenhouse gases at the 1990 level, as agreed under the UNFCCC in 1992, Denmark reconfirmed its commitment to a unilateral reduction of its CO_2 emissions. Perhaps more importantly, Minister Auken pushed for an ambitious EU position leading up to the Kyoto negotiations, thereby casting Denmark in an entrepreneurial leadership role at the EU level (Ringius, 1999). During the ensuing negotiations on 'burden-sharing', Denmark also acted as an exemplary leader, committing to a 21 per cent reduction of its GHGE (MoE, 2000). Underpinning these international obligations were policy initiatives at home that pioneered a green tax reform, particularly CO_2 taxes on fossil based sources of energy and electricity, support for renewable energy, both onshore and offshore windpower, and subsidies for energy efficiency investments in households and industry (MoE 1996; MoE 2000). Thus, in part driven by the transformational leadership of Svend Auken, during this period Denmark served both as an entrepreneurial and exemplary leader in international negotiations, but also as a pioneer through its ambitious energy policies.

2001–2007: environment worth the money

At the end of 2001 a Liberal–Conservative government took office. Following scepticism voiced by Prime Minister (PM) Anders Fogh Rasmussen (Liberals) about the causes of climate change, the government forged a novel discourse by emphasizing that climate and environmental policies had to be economically sound. 'Environment worth the money' became the policy mantra. In line with this, the government provided generous research funding for climate sceptics who found mitigation policies too costly or suggested that climate change could be caused by solar activity rather than by excessive CO_2 emissions (Politiken, 2007a, 2007b). Institutionally, energy policy was subsumed under the Ministry of Economics and Business Affairs, signalling a new approach.

Denmark's international commitments to some extent tied the hands of the new government, and the Kyoto protocol was grudgingly ratified. Still, the 2003 action plan *Green Market Economy* shifted emphasis towards greater use of flexible mechanisms of the Kyoto Protocol, indicating that more reductions should be undertaken outside Denmark (Regeringen, 2003: 9). This stance was mirrored in a halt in the development of Danish wind power as financial support for renewable energy was reduced and off-shore windpower cancelled (IEA, 2006; Ryland, 2010: 81; Toke and Nielsen, 2015). The political majority during this period seemed more preoccupied with reigning in what was seen as excessive – or excessively expensive – environmental and climate policies than with pioneering or leading in the policy field.

2007–2011: pioneering and brokering climate policy

A softening of the course on energy and climate policy surprisingly emerged following the Intergovernmental Panel on Climate Change's (IPCC) *4th Assessment Report* of 2007. The same Liberal–Conservative government proposed a long term goal of a fossil-free energy system, based entirely on renewable energy with a medium-term goal of 30 per cent renewables in the energy supply by 2025 (MTE, 2007). Concurrently, Connie Hedegaard was appointed Denmark's first Minister of Climate and Energy. A Climate Commission of experts was set up to analyse how the long-term vision of a fossil-free energy system could be realized. But for the short-term, an energy agreement signed the following year by most of the political parties in parliament set a target that 20 per cent of the energy use in 2011 should come from renewable sources. The agreement included a redesign of feed-in tariffs for wind power and biogas as well as increased and new energy-related taxes (Agreement, 2008). Moreover, PM Fogh Rasmussen undertook, in 2008, a noteworthy U-turn on his hard-line stance on climate policy when he fully embraced a green growth agenda in a speech at the annual assembly of the Liberal Party. These policy changes resonated with the perceived long-term risks of being 'addicted to' energy supplies from Muslim countries, surfacing in the wake of Denmark's 'cartoon crisis' (Carle, 2006; Meilstrup, 2010b).

In the international arena Climate Minister Hedegaard also gradually pulled Denmark into an entrepreneurial leadership role through the Greenland dialogue and eventually by offering to host the COP15 negotiations (see below).

2011–2015: climate policy constrained by the financial crisis?

In 2011 a Centre-Left coalition government of Social Democrats, Social Liberals and the Socialist People's Party took office – three parties which in recent history had been high-profile on environmental and climate policy. While economic recovery became by far the most salient priority of the government, its inaugural programme linked the solutions to the financial crisis and the climate/environmental crisis (Regeringen, 2011). The Danish strategy would be to strive for a 'green transition' that would create jobs while also reducing climate gas emissions and other environmental pressures, serving as a cognitive and exemplary pioneer. The government soon proceeded to forge a broad political compromise committing all political parties (but one) to a new energy plan covering the period until 2020. The initiatives included in the plan were to achieve a 34 per cent reduction of Denmark's GHGE by 2020 over the 1990 baseline, and further reductions were to be found in the transport and agricultural sectors to accomplish the overall target of a 40 per cent reduction in GHGE – in line with the upper bound of the reduction commitment interval for annex 1 countries suggested by the IPCC (ENDS, 2012b; MoC, 2012). The plan included a renewed emphasis on wind power, which was to supply 50 per cent of electricity consumption by 2020 through new onshore and offshore facilities. The most contentious issue was the cost of the plan to be passed on to energy users through taxes and tariffs. Two years later, under an agreement regarding economic growth, certain levies on electricity and gas were reduced in order to improve the competitiveness of Danish businesses.

Institutional responses, policy instruments and programmes

In 2011 the International Energy Agency (IEA) called Denmark 'a leader among OECD member countries in terms of renewable energy, energy efficiency and climate change policies' and noted that Denmark had managed to decouple economic growth, energy consumption and GHGE (IEA, 2011: 7). The IEA pointed out, however, that policy efforts are still lagging in some sectors, notably transport and agriculture.

Increased energy efficiency in industry and households and transition to cleaner and renewable energy sources in energy and electricity production constitute the two primary strategies in the regime of successive energy plans and climate policies that has evolved. These strategies have been implemented through a broad range of policy interventions.

The policy instruments implemented to achieve higher energy efficiency have included taxes on energy, subsidies for investments in energy saving equipment, regulation through building codes that set standards for equipment and insulation and requirements for energy savings from energy companies, and information-based tools such as labelling schemes for energy efficient appliances, cars and buildings (Nielsen and Pedersen, 2013). Market-based instruments have taken a more prominent role, but merely complementing standards, planning and regulations.

To promote substitution towards cleaner energy sources, a CO_2 tax scheme for fossil fuels has been in place since 1993 (Andersen and Ekins, 2009). Direct investments into natural gas conversion of central energy systems as well as government R&D funding has also been important for the conversion of energy supply (IEA, 2011). Support for renewable energy constitutes a key policy focus. Renewable energy has been promoted through planning policies, feed-in tariffs, purchasing obligations and support for technology development. The 2008 Energy Agreement improved framework conditions for renewables, including biomass, biogas, waste, solar and wind power. For instance, the agreement raised the feed-in premiums on power from wind, biomass and biogas (IEA, 2011) and set out plans for the construction of 400 MW new offshore wind farms. In order to foster acceptance the agreement introduced compensation schemes and a scheme for those living near turbines. The 2012 agreement included plans for significant expansion by 2020 of off-shore and onshore wind power and further improved framework conditions for biogas. These initiatives were financed by levies on electricity consumption.

Climate and energy policies and the links between them have been formulated in successive political agreements, typically involving a broad majority of the Danish political parties (IEA, 2011). This systematic approach to climate policy-making has been further institutionalized with the adoption of a Climate Law in 2014. The law aims to provide a strategic framework to reach Denmark's 2050 target of a low-carbon society. It involves three activities: (1) a process for the setting of national climate policy objectives, (2) an annual climate policy report by the government accounting for GHGE development and Denmark's progress towards meeting its international obligations, and (3) the establishment of an independent Climate Council, consisting of experts from a wide range of fields, who are to provide independent assessment of Danish climate policy and recommendations on how to move towards a low-carbon society (Climate Act, 2014).

Institutionally, the energy and climate portfolios have been shifted among different ministries, typically with the change of governments, suggesting also different framings and foci. Hence, the first significant shift in institutional anchoring occurred in 1994 when energy policy after a short-lived period as a separate ministry was moved into the Ministry of the Environment, reflecting a strong linkage between climate, environment and energy policy. This was reversed with the government change in 2001, when energy was subsumed under the Ministry of Economics and Business Affairs. To signal a stronger environmental course climate and energy policy was merged into an independent ministry in 2007, allowing for a focused effort. Despite recurrent changes in ministry portfolios, a stable institutional actor has been the Danish Energy Agency, the operational unit in charge of overseeing and preparing all regulations and policies in the field since 1976.[1]

Multi-level and polycentric climate governance

As a broker of the internal burden-sharing agreement in the EU prior to the 1997 Kyoto conference, Denmark had accepted a domestic reduction commitment of 21 per cent that went well beyond what a more proportional allocation of reduction targets among member states would have suggested. By doing so Denmark provided support for Germany to commit to a similar reduction target, which was of significance for the overall EU reduction potential due to the crucial share of Germany's emissions delivering 80 per cent of the net EU reduction (Ringius, 1999; see also Chapter 8). In Kyoto, Denmark's Minister Auken engaged considerably in the negotiations too, notably in obtaining the initial USA approval of the protocol, which won him much acclaim in the climate policy community.

Unfortunately the emissions that Denmark had been reporting to UNFCCC included a correction for electricity trade (high imports in the baseline year) which soon triggered a domestic backlash. Critics argued that the actual reduction commitment was 28–29 per cent, since UNFCCC rules do not allow for electricity trade corrections. With Germany benefitting from free reductions of 'hot air' in its eastern *Länder* (former German Democratic Republic) (see also Chapter 8), it was implied that Denmark had accepted a target way above that of any other EU Member State, and the critics expressed concerns for the costs involved, which soon triggered a long and emotional debate over the Kyoto Protocol.

Eventually, following the 2001 parliamentary elections the critics formed the government, which led to a U-turn also in the strategy pursued in EU and international negotiations. Denmark held the EU Presidency in the second half of 2002, when the 10-year Rio follow-up UN conference gathered. Widespread confusion about priorities emerged as Denmark's new PM found it 'refreshing' to hear statements about scrapping the Kyoto Protocol and that securing clean drinking water was of higher importance than climate policy – in some disregard of his role as Chair of the European Council who ought to represent the EU (DR, 2002).

Growing tensions within the Liberal–Conservative government over the hard-nosed climate policy led to the appointment of Connie Hedegaard (Conservative) as new Minister of Environment only two years later. Not only did she present a revival of a more concerned attitude about climate change, she also reinstated Denmark's previous role as broker and entrepreneurial leader in international climate negotiations through the ambitious 'Greenland dialogue' which proved a useful venue for promoting consensus in international climate change negotiations.

In 2005 Hedegaard initiated a series of informal meetings where ministers and senior negotiators from countries which were key players in global reduction efforts could exchange views (OECD, 2007: 195). The meetings were launched in Greenland, where the participants could see glaciers melting with their own eyes. The Greenland dialogue involved the key partners including US and China and continued in meetings in South Africa (2006), Sweden (2007) and Argentina (2008) returning to Greenland in 2009. Although the expressed purpose of the Greenland dialogue had been to build trust, not to make decisions as such, it contributed to an emerging consensus on the 2 degree target.

In tandem with these meetings senior US senators and policymakers (e.g. John McCain) had been invited to Greenland to try to soften their hard stance on climate change. This was made possible with Denmark being a close ally of the US and an active participant in the Iraq war. A congressional hearing on clean energy in Washington, DC featuring Connie Hedegaard was one of the outcomes of the bilateral process during the George W. Bush administration (see also Chapter 16).

It was during the Greenland dialogue that Denmark offered to host the 15th Conference of the Parties (COP15) to the UNFCCC, creating expectations that it would be feasible to reach a binding agreement for a successor to the Kyoto Protocol, effectively committing countries to a long-term effort for reducing greenhouse gases. With the offer Denmark shifted its role from a cognitive pioneer to providing entrepreneurial leadership, but in reality COP15 would prove to be one of the biggest disappointments in international climate diplomacy for years and a huge blow to Danish ambitions.

> Copenhagen failed to live up to even the lowest expectations. What is more, the summit produced diplomatic chaos on a scale the world has seldom seen. When US President Barack Obama arrived at the Bella Center on the last day of the negotiations, Hillary Clinton welcomed him by saying, 'Mr. President, this is the worst meeting I've been to since the eighth-grade student council.' The outcome of the meeting, the Copenhagen Accord, was heavily criticized for being inadequate and only 'a letter of intent'. The conflicts between developed and developing nations were monumental. The US and China were not able to settle well-known, deep disputes, and the Danish Prime Minister was humiliated on the UN podium while the world was watching. The most significant effort ever by a Danish government to position itself as a global, political leader turned into the biggest international, diplomatic defeat for decades.
>
> (Meilstrup, 2010a: 114)

A deep internal division in Denmark between supporters and critics of the Kyoto Protocol had penetrated preparations for COP15, reflected in an enduring rivalry between the PM's office and Hedegaard's ministry. The PM's office at several occasions intervened and insisted to lead the drafting of a set of key conference conclusions known as the gold-paper for the meetings in the PM's golden cabinet room. Hedegaard and her skilled negotiators were well acquainted with the UNFCCC framework, unlike the PM, who, distrusting the UN system, wanted to provide the conference with a firm steer and in doing so managed to alienate the emerging economies by appeasing Denmark's close ally, the US. Negotiations at COP15 collapsed exactly when a version of the gold-paper was leaked and in addition to the well-known differences, objections could be raised on the formal point that the Danish paper did not stem from the UNFCCC led process, thus lacking legitimacy (Meilstrup, 2010a, 2010b).

As the PM had stepped down to become NATO's General Secretary his inexperienced substitute – also named Rasmussen – added to the diplomatic chaos that resulted. While chairing the COP15 Assembly he made the unfortunate remark

'I don't know your rules', giving the impression of neglect – and to some even of schizophrenia, as the Danish government had continued to fund climate sceptics generously while preparing and hosting COP15 (Kristeligt Dagblad, 2010).

The new Obama administration had not had time to prepare domestic legislation that would allow the US to enter into a binding agreement in Copenhagen, nor did it control the required 67 seats in the Senate for a subsequent ratification, and it was under these circumstances Denmark had begun navigating towards a political deal rather than a legal one. China, for its part, had been involved in the Greenland dialogue through NDRC, the National Development and Reform Commission, the highest planning body and a stronghold of technocrats, whereas its diplomats from the Ministry of Foreign Affairs came to lead China at the conference (see also Chapter 17). With the latter being one of the few remaining strongholds of ideology and Marxism in Beijing, its diplomats were fairly distrustful about the negotiations (Andersen, 2010).

In response to the quest for a political rather than a legal agreement, an alliance was forged in Beijing with India, Brazil and South Africa a few weeks before the conference against the host country Denmark and a number of rich countries, but the alliance went unnoticed in Copenhagen's ministries. The final COP15 negotiations proved disastrous not only to Denmark but to the EU as a whole, whose Heads of State were left behind in flimsy conference backrooms as the Copenhagen Accord was drawn up in direct negotiations between the US and China, flanked by its allies.

Perhaps not surprisingly the COP15 process altogether demonstrated critical deficiencies in Denmark's capacity to take the lead in high-level negotiations with the largest powers in the world. Denmark as a small country did not have a large corps of influential diplomats that could be mobilized and it made little use of the Nordic Council with its regional allies. Traditional Danish virtues such as consensus seeking, talking directly about the issues and maintaining an informal approach fell short of the requirements and expectations for international negotiations. An illustrative major blunder was the neglect to invite China's Premier formally for the final nightly negotiations, leading to convoluted procedures when only a vice-minister arrived to the meeting of Heads of States (Meilstrup, 2010b: 223).

At a deeper level Denmark's climate policy and leadership ambitions were hampered by internal rivalry and differing ideologies as to the proper fora and methods for international cooperation, providing a novel domestic dimension to the concept of polycentricity in climate change governance.

Climate change as a threat or opportunity?

The string of energy and climate plans that have been at the core of Denmark's climate policies repeatedly stress that climate change offers opportunities for sustainable or green growth. Thus, Energy 2000 (ME, 1990), which was the first plan to link energy and climate policy, highlighted the opportunities for export of green technologies, particularly renewables and energy efficiency equipment, and the follow-up plan Energy 21 (MoE, 2007) also highlighted the economic opportunities flowing from an active energy policy.

A turning point was the first Fogh Rasmussen government (2001–2005) which with its emphasis on cost-effective solutions and cost-benefit valuations claimed a trade-off between climate policy and economic benefits. In contrast the recent Centre-Left government, despite making economic recovery and growth its strong priority, positioned its climate policy in a green transition framework (Regeringen, 2011). By pursuing ambitious climate and environment policies Denmark would serve as a green laboratory and reap economic first-mover benefits. As such, apart from the first half of the 2000s and across political ideology, Denmark has acted fairly consistently as a pioneer, setting an example via its energy policies.

But the strength of commitment to the transition principle appears to have diminished in 2013 when CO_2/energy taxes paid by businesses were lowered in an effort to spur economic growth, and in 2014 when the government gave in to arguments that the renewable levies on electricity were hurting the competitiveness of Danish companies. Nevertheless Denmark continued to support EU efforts to step up emission reduction targets to 30 per cent, arguing for a green transition (Skovgaard, 2014).

These different approaches suggest an ambiguity underlying the Danish approach to seeing climate change as an opportunity. First, active green growth strategies have been pursued most actively by environment and climate ministries, while the Ministry of Finance has been known to be outright sceptical about the realism of promised co-benefits (Skovgaard, 2012). And as an increasingly strong player the attitudes of the Ministry of Finance may be decisive in decisions on climate policy.

Second, the debate about energy policy also reveals a further split: some businesses in Denmark which have indeed been international first movers on renewable energy and green technologies, strongly pushed for and embodied the green growth strategy, while traditional industries were sensitive to changes in energy prices in order to remain competitive in the world market. In other words, the win-win discourse may hold for some businesses, who can benefit from Danish energy policies to develop leading positions in the global market. But not all sectors are equally well positioned to exploit green growth opportunities. Regulation of agricultural GHGE, for instance, has not been given nearly the same level of attention as policies on the energy sector.

The domestic implementation of EU and international commitments

The domestic climate mitigation policies Denmark had pioneered during the 1990s, prior to the Kyoto Protocol, have been delivering reductions of carbon emissions in specific sectors of the economy. Expansion of wind power limited emissions from the power sector (Jensen, 2002), while the CO_2 taxation scheme, targeted at business and industry, spurred both energy efficiency and fuel shifts

(Enevoldsen *et al.*, 2007). As a result the overall domestic CO_2 emissions had been reduced by about 13–14 per cent by the time the Kyoto Protocol entered into force (Enevoldsen, 2005). Unfortunately, due to the base-year controversy (see above) these reductions were hardly noticeable within the official UNFCCC emissions accounting framework.

The preoccupation with the possible costs of the Kyoto Protocol triggered a two-track policy for compliance, where the emphasis was on making the best use of flexible mechanisms for cheap reductions outside Denmark and on lobbying hard within the EU to obtain a correction for the base-year problem. The European Commission acknowledged within less than a year Denmark's base-year issue, but it took six years before a partial correction could be approved in the Council, providing Denmark with a relief of only two million tonnes of CO_2 to the baseline. By this time Denmark had put considerable effort into acquiring emission allowances from the Clean Development Mechanism (CDM) and Joint Implementation (JI) projects in a range of countries (Malaysia, Thailand, South Africa, etc.), where experiences and contacts from previous environmental assistance projects could be put to use. A third policy track initiated in a later stage consisted in making use of the complicated accounting rules for LULUCF (Land Use, Land-Use Changes and Forestry). Meanwhile Denmark was named and shamed with only two other footdragging countries (Spain, Italy) in the annual compliance checks over several years (EEA, 2007, 2010), pointing out that Denmark would not meet its target within the first Kyoto commitment period. The fiscal crisis affected the Danish economy severely via the bursting of a housing bubble, which fundamentally altered the trends and helped reduce emissions, and according to EEA's (2014) final progress report Denmark in the end managed to meet the target for the Kyoto commitment period as a whole. Still, about half the reductions have been achieved with flexibility, LULUCF and base-year correction, while the remaining reductions are due both to the fiscal crisis and actual measures, mainly from the 1990s (EEA, 2013: 67).

The outlook for Denmark's compliance with EU's 2020 targets on emissions reductions, energy efficiency and renewables are much better, not least due to the ambitious Energy Agreement approved by parliament in 2012 (EEA, 2013). The agreement spurred a rapid expansion of renewable based electricity, with wind power alone accounting for 41 per cent of total electricity consumption in 2014. In fact, it seems likely that Denmark will overshoot on the EU obligation to have 30 per cent renewable energy by 2020, provided that the planned offshore wind capacity will be implemented.

Despite the promising outlook the official Danish emissions targets omit two significant sources of greenhouse gases: international shipping and overseas territories. While they contribute about 15 per cent to Denmark's GDP, international shipping emissions are not accounted for with the official UNFCCC inventories. This is a serious omission because Denmark's largest shipping conglomerate AP Møller-Mærsk controls a fleet of more than 1,000 vessels and is estimated to emit between 20 and 25 million tonnes of CO_2 annually, adding about 40 per cent to Denmark's gross GHGE (Ingeniøren, 2008; Politiken, 2015).

Danish overseas territories in the North-Atlantic are not part of the EU. While the Faroe Islands are not even a party to the Kyoto Protocol, Greenland has applied for territorial exclusion from the second commitment period. Greenland is not interested in being subdued by emission restrictions as part of the Danish Realm, as it is aiming to boost its GDP with extraction of minerals and oils. The self-ruling territory argues much in line with developing nations and has made the case for a tenfold increase of its emissions to 10 million tonnes annually. Prior to COP15 the Danish government accepted in principle the demand for an emissions increase in Greenland.

Conclusion: political leadership in and by Denmark?

Denmark has provided leadership in climate change politics, consistently pioneering by example through its energy policies, but at times also taking on an active leadership role in climate policy making, exhibiting both cognitive and entrepreneurial leadership.

Both pioneering energy and climate policies and occasional leadership in EU and international climate policy making by Denmark owes a great deal to key individuals, who have ventured beyond simple transactional approaches, notably Ministers Auken and Bilgrav, as well as Minister and EU Climate Commissioner Hedegaard. Without their decisions to make the most of the mandates they held, rather than simply clinging on to their seats, there would have been much less to showcase, and Denmark would have been a far more timid player in international negotiations.

Bilgrav prepared the strategic lines with the drafting of 'Energi-2000',[2] while Auken issued the national planning decree that requested all local authorities to prepare for the siting of wind turbines and pushed for CO_2 taxation of business and industry. Hedegaard initiated the Greenland dialogue and persuaded an unwilling Danish government to embark on a gradual green transition. Coming from different positions in the political landscape and being tied to different groupings in Danish society they complemented each other in identifying the measures and approaches that resulted in a relatively broad-based climate policy to emerge. While many were skeptical about the Kyoto Protocol, the policies that gradually emerged created a domestic tailwind (Putnam, 1988; Andersen and Liefferink, 1997), where many actors and interests could sense that they might gain some opportunity from a low-carbon transition. It led eventually to a tipping point even within the power sector, a traditional enemy of climate change policy, with utilities opening new markets in offshore wind power at home and abroad.

Their relatively ambitious policies would not have been feasible without a resonance and foundation in Danish society at large. The energy efficiency policies emerged from the cold shower that Denmark experienced during the first oil crisis, when OPEC (Organization of the Petroleum Exporting Countries) countries embargoed Denmark specifically in response to a favourable statement on Israel from the Danish PM. The diversification of energy supply took place via a long and intense debate over nuclear power, which involved many citizens who took active part in developing viable alternatives, extending a rural shareholder tradition into the future of energy supply. Wind technology emerged due to local initiatives and guilds and on the basis of a pool of technological know-how dating back more than a century to Denmark's Edison, inventor Poul la Cour, and having survived in embryo with the national test turbine in Gedser town (Christensen, 2009). Export of energy technology has grown at high rates and has sidelined pork and butter as the number one Danish export commodity.

'If there is no struggle, there can be no progress' said Frederick Douglass. It is hardly surprising that the changes required by climate change mitigation policies create anxiety and opposition from many parts of society, with short-term losses being far more visible than the long-term risks of dangerous climate change. The struggle in Denmark with climate change deniers has been no less than in other countries, with deniers occasionally rising to prominence. Conflicts over climate change policy have been at times draconian, as there are large dilemmas involved in Denmark's domestic climate change policies, notably related to the prominent role of intensive agriculture, international shipping and resource extraction. The loud voice with which Denmark's negotiators occasionally have been speaking in international fora is not only there to make a small country heard – it is sometimes probably also helpful in overcoming and disciplining critical elements of a domestic constituency.

Notes

1 Within the Danish administrative system large, semi-autonomous agencies exist in several sectors.

2 The plan was also an inspiration to Schleswig-Holstein (Krawinkel, 1991) and may have nurtured Germany's later *Energiewende*.

References

Agreement 21 February 2008 between the Government, the Social Democrats, Danish People's Party, Socialist People's Party, the Social-Liberals and New Alliance about Danish energy policy 2008–2011.

Andersen, M.S. (2010) 'Brokenhagen: Elementer til en forståelse af COP-15', *Økonomi og Politik* 83(3): 3–17.

Andersen, M.S. (2015) 'Reflections on the Scandinavian model: some insights into energy-related taxes in Denmark and Sweden', *European Taxation*, 55(6); 235–244.

Andersen, M.S. and Ekins, P. (2009) *Carbon-energy taxation: Lessons from Europe*, Oxford: Oxford University Press.

Andersen, M.S. and Liefferink, D. (1997) 'The impact of the pioneers on EU environmental policy', in M.S. Andersen and D. Liefferink (eds), *European environmental policy: the pioneers*, Manchester: Manchester University Press, 1–39.

Burck, J., Marten, F. and Bals, C. (2014) *The Climate Change Performance Index: results 2014*, Bonn: Germanwatch.

Carle, R. (2006) 'Cartoon crisis: Islam and Danish liberalism', *Society Abroad*, 44(1): 80–88.

Christensen, B. (2009) *Wind power the Danish way*, Askov: The Poul la Cour Foundation.

Climate Act no. 716 of 25 June 2014. Act on the Climate Council, climate policy report and national climate policy targets. Concito (2014), Klimabarometeret 2014, Copenhagen. http://concito.dk/klimabarometeret2014 (accessed on 11.04.2016).

DR (Danmarks Radio) 9.9.2002 www.dr.dk/nyheder/politik/de-kristelige-vil-af-med-lomborg (accessed on 11 April 2016).

EEA (2013) *Trends and projections in Europe 2013. Tracking progress towards Europe's climate and energy targets until 2020.* EEA Report 10/2013, Copenhagen: European Environment Agency.

EEA (2014) *Progress towards 2008–2012 Kyoto targets in Europe*. EEA Technical Report 18/2014, Copenhagen: EEA.

ENDS Europe (2012a) Danish Presidency cautious on climate aims, 24 January 2012.

ENDS Europe (2012b) Denmark agrees 'historic' green energy plan, 26 March 2012.

Enevoldsen, M.K. (2005) *The theory of environmental agreements and taxes*, Cheltenham: Edward Elgar.

Enevoldsen, M.K., Ryelund, A.V. and Andersen, M.S. (2007) 'Decoupling of industrial energy consumption and CO_2-emissions in energy-intensive industries in Scandinavia', *Energy Economics* 29(4): 665–692.

Eurobarometer (2014) *Special Eurobarometer 409 Climate Change*, TNS Opinion & Social.

IEA (2011) *Energy policies of IEA countries: Denmark*, Paris: International Energy Agency.

IEA (2006) *Energy policies of IEA countries: Denmark*, Paris: IEA.

Ingeniøren (2008) Mærsk forurener lige så meget som hele Danmark, 7 February 2008.

Jensen, D.B. (2002) 'Vindenergiens udbredelse i Holland og Danmark', *Samfundsøkonomen*, 3: 32–39.

Krawinkel, H. (1991) *Für eine Neue Energiepolitik. Was die Bundesrepublik Deutschland von Dänemark lernen kann*, Frankfurt: Fischer.

Kristeligt Dagblad (2010) Lomborg har fået 138 millioner over 10 år, 9 February 2010.

Mandag Morgen (2007) 1 ton mindre. De holdningsmæssige forudsætninger for klimasagens folkebevægelse, 29 March 2007.

Meaney, T. (2011) Getting to Denmark, *The Nation*, 33–37.

Meilstrup, P. (2010a) 'The Runaway Summit: The Background Story of the Danish Presidency of COP15, the UN Climate Change Conference', *Foreign Policy Yearbook 2010*, Copenhagen: Danish Institute for International Studies.

Meilstrup, P. (2010b) Kampen om klimaet – historien om et topmøde der løb løbsk, Copenhagen: People's Press.

MoC (2012) *Accelerating green energy towards 2020 (Energy Agreement)*, Copenhagen: Ministry of Climate, Energy and Building.

MoE (1990) *Energy 2000 – Action plan for sustainable development*, Copenhagen: Ministry of Energy.

MoE (1996) *Energy 21. The government's energy action plan*, Copenhagen: Ministry of Environment.

MoE (2000) *Climate 2012. Status and perspectives on Danish climate policy*, Copenhagen: Ministry of Environment.

MTE (2007) A visionary Danish energy policy 2025, Copenhagen: Ministry of Transport and Energy.

Nielsen, H.Ø. and Pedersen, A.B. (2013) 'Hvordan kan staten fremme innovation, der fører til bæredygtige energisystemer?', *Politica*, 45(3): 323–343.

OECD (2007) Environmental performance reviews: Denmark, Paris: Organisation for Economic Co-operation and Development.

OECD (2012) *Economic surveys: Denmark*, Paris: OECD.

Politiken (2007a) Dommedag aflyst, 18.10.2007.

Politiken (2007b) Klimakamp: Anders Fogh Rasmussen afviste først: Foghs forvandling, 20.10.2007.

Politiken (2015) Maersk vil sejle længere på literen, 06.04.2015.

Putnam, R. (1988) 'Diplomacy and domestic politics: the logic of two-level games', *International Organization*, 42, 42760.

Reenberg, F. (2014) *Dansk Klimapolitik. Klimaet på dagsordenen*, Roskilde Universitet, ISG–Forvaltning. Speciale.

Regeringen (2011) *Et Danmark der står sammen*, København: Regeringsgrundlag.

Regeringen (2003) *Grøn markedsøkonomi – mere miljø for pengene*, København www.mst.dk (accessed on 11 April 2016).

Ringius, L. (1999) 'Differentiation, leaders and fairness: Negotiating climate commitments in the European Community', *International Negotiation*, 4, 133–166.

Ryland, E. (2010) 'Danish wind power policy', *Environmental Politics*, 19(1): 80–85.

Skovgaard, J. (2014) 'EU climate policy after the crisis', *Environmental Politics*, 23(1): 1–17.

Toke, D. and Nielsen, H.Ø. (2015) 'Policy consultation and political styles: Renewable energy consultations in the UK and Denmark' *British Politics*, 10, 454–474.

TV2 (2014) Her er danskernes vigtigste valgemner, 14 September 2014.

TV2 (2015) Danskernes valg. Sundhed og hospitaler det vigtigste valgtema, 27 May 2015.

7 French climate policy

Diplomacy in the service of symbolic leadership

Pierre Bocquillon and Aurélien Evrard

Introduction

As the host of the 21st Conference of the Parties (COP21) in December 2015, France has been acclaimed for its leadership in brokering the first universally binding climate agreement (e.g. Stothard and Chassany 2015). French climate diplomacy was set in motion long before the start of the 2015 Paris climate conference, mobilizing important administrative and political resources. Prime Minister Manuel Valls declared the fight against climate change a 'major national cause' for 2015,[1] whereas the Ecology and Energy Minister, Segolène Royal, considered the adoption of the French Law on Energy Transition in July 2015 as a new step towards becoming a 'nation of environmental excellence'.

France can boast of its relatively low levels of greenhouse gas emissions (GHGE) per capita and carbon intensity – in most part due to its electricity sector dominated by nuclear and hydroelectricity – which have contributed to position it as an 'inadvertent climate pioneer' (Szarka 2011). On this basis, the French government and administration have often claimed to assume '*leadership by example*'. France's bid for leadership has also been driven by ambitions for diplomatic prestige and international '*grandeur*'. It has opportunely assumed the role of an entrepreneurial and, at times, heroic foreign policy leader, notably during the EU climate and energy package negotiations in 2008 as well as during COP21 (see also Chapters 1, 2 and 5). However, France has tended to follow, rather than anticipate or trigger, European and international climate developments. Moreover, ambitious rhetoric has not always been matched by sustained political commitments and implementation. French climate policy developments remain characterized by acute controversies – for example on environmental taxation – and by a humdrum process of policy change.

Applied to the French case, the distinction between pioneers and leaders (see Chapter 1) raises the following question: Is it possible to pretend to be a leader without being a pioneer? It is indeed this paradoxical approach that seems to characterize most accurately the French strategy. This raises two further questions: How has the gap between French ambitions and achievements been managed? And how have French international pledges influenced domestic developments? This chapter discusses these questions and argues that France should be characterized mainly as a symbolic leader. It aims to explain the convoluted French approach to climate leadership, as well as the country's attempts at closing the gap between its international leadership stance and its reactive national policies, often developed in fits and starts.

National attitudes to climate change

In France, climate change has been politically and socially constructed as a consensual policy problem. During the 1990s, few prominent climate voices had access to the mass media and the issue was mainly framed through scientific arguments. In addition, media coverage was largely determined by international conferences. In the second half of the 2000s, due to international factors (e.g. ongoing international climate negotiations on a post-Kyoto agreement and the release of the Intergovernmental Panel on Climate Change's (IPCC) 4th assessment report), and internal factors (e.g. the importance of environmental issues in the 2007 Presidential campaign) media coverage has increased slowly but steadily, with major peaks of attention around international conferences (Aykut *et al.* 2012). Climate issues have been picked up by journalists (in general media outlets) who have tended to depoliticize the issue, framing it primarily in terms of individual behaviour and 'eco-citizenship' (Comby 2015).

According to a yearly survey published by the French environmental agency, ADEME (*Agence de l'Environnement et de la Maitrise de l'Energie*), in French public opinion climate change has been consistently ranked as one of the three most 'concerning' environmental issues (together with air and water pollution) since the mid-2000s. However, the survey also shows that there have been significant variations (Figure 7.1). The most striking is a peak in 2007–09 in the run up to the 2009 Copenhagen UN climate conference when climate change became the most salient issue for more than 30 per cent of the respondents. In 2010 it dropped to about 17 per cent and since then has not regained its 2009 peak.

This downward shift resulted from a combination of domestic and international factors. The post-2009 period has been characterized by several governmental renouncements in environmental policy (Halpern 2012), while the failure of the 2009 Copenhagen climate conference, combined with the financial and economic crisis, have also contributed to a shift of French public attention away from climate issues. Yet, the attention to climate issues was building up again in preparation for the 2015 Paris climate conference.

Although conducted using other methods compared to the above-mentioned French opinion polls, Eurobarometer surveys allow for comparative analysis. With 14 per cent of respondents considering climate change as the most serious problem in 2013, France lies below the European average (16 per cent), and even more markedly below EU climate pioneers such as Sweden (39 per cent) and Germany (27 per cent) (Eurobarometer 2014). Between 2011–13, the decline (–6 per cent) in climate awareness was stronger in France than in the EU-28 and placed the country alongside Southern and Eastern European countries. Most

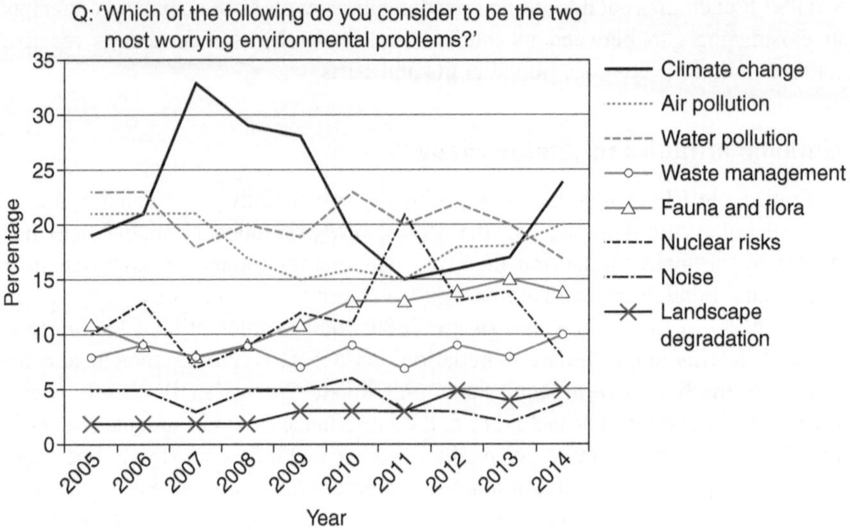

Figure 7.1 Public perception of environmental problems (2005–14).
Source: ADEME (2014).

recently, the 2015 Paris climate conference has helped to reverse this trend although it remains to be seen how long this will last in the face of enduring economic uncertainty and security concerns fueled by the November 2015 terrorist attack in Paris.

Phases of domestic climate change policy: institutional responses, policy instruments and programmes

French climate policy has been so tightly embedded in European and international contexts that it is difficult to separate domestic and foreign policy dimensions. Indeed, domestic policies have been either directly (or indirectly) a consequence of European and international agreements; or have been formulated to establish a foreign policy stance.

From the 1980s to the 1997 Kyoto Protocol: reactive policy, defensive strategy

In France until the end of the 1980s climate change was almost nonexistent as a policy problem, and attention was mainly driven by international climate conferences. Although initially a latecomer and follower, from the early 1990s onwards France developed the ambition of displaying international climate leadership. A working group, the *Groupe Interministériel sur l'Effet de Serre* (GIES), was set up in 1989, and upgraded three years later to the status of an inter-ministerial mission (MIES). On the occasion of its creation, Prime Minister Rocard stated: 'France has actively contributed to international action in this area. It must set an example through an efficient domestic policy' (Virlouvet 2015: 82). Yet, despite such ambitious rhetoric, French climate policy remained mostly reactive and defensive, the country appearing to 'rest on its laurel with little climate innovation' domestically (Szarka 2011: 115).

This ambiguous attitude reflected the specificities of the country's energy system (Giraud *et al.* 1997). France is heavily reliant on nuclear electricity, a low-carbon source of energy that accounted for three quarters of the country's electricity mix in 1990. Large hydro represented another 15 per cent. Due to the development of the electronuclear programme from 1974 onwards, French emissions fell by 23 per cent in the 1980s, making the country's economy one of the least carbon intensive in Europe.[2] This position of 'inadvertent pioneer' (Szarka 2006) did not foster a proactive attitude however. One of the main goals of French climate diplomacy was even to prevent the adoption of precise emission reduction targets. According to public authorities, there was little room for further improvements of GHGE reductions.

The Kyoto Protocol's mixed impact: higher public attention, uneven implementation (1997–2007)

In the context of the Kyoto negotiations, it was impossible for France to uphold the above-mentioned strategy. The 1997 Kyoto climate conference contributed to a rise in public attention to climate change. Moreover, the participation of the Green Party in a left-wing coalition government – the government of *Gauche plurielle* (1997–2002) which included the Green Minister for the Environment Dominique Voynet – influenced the country's position (Evrard 2012). France endorsed the EU burden sharing mechanism in 1998, but it agreed to a mere stabilization of its CO_2 emissions at 1990s level for 2008–12. To implement this modest commitment, a first Climate Plan was adopted in 2000: the *Plan National de Lutte Contre le Changement Climatique* (PNLCC). It set general orientations – developing renewable energy, increasing energy efficiency – and promoted new market based policy instruments – including environmental taxation, the setting up of a carbon market and creation of feed-in-tariffs for renewable energy (Szarka 2006: 630; Evrard 2013; Bocquillon and Evrard 2016).

However, as with other environmental issues, the changes fostered by this new political context were far from radical due to the heavy weight of institutional legacies (Evrard 2012; Aykut and Dahan 2015: 551). For instance, a carbon tax adopted by Parliament in 2000 was eventually rejected by the Constitutional Court in view of its incompatibility with the principle of equality of taxation (due to exemptions for large companies) (Deroubaix and Lévêque 2006). As for the feed-in-tariffs, they were set too low to trigger large-scale renewable energy developments. Humdrum domestic policies remained at odds with the flamboyant public speeches given at the international level. At the World Summit on Sustainable Development in Johannesburg (2002) for instance, President Chirac started his voluntarist speech with the famous quote: '*Our house is burning and we look elsewhere*'.

The consolidation of French climate policies in the 2000s was a direct consequence of the increasing discrepancy between the Kyoto target and rising emissions due to economic growth (see Figure 7.2). A Climate Plan, adopted in 2004 for the period 2004–12,[3] was formulated with a view to meet the target established as part of the EU burden sharing agreement. It explicitly referred to the 'Factor Four' trajectory – a decrease of CO_2 emissions by a factor of four by 2050 – which both Prime Minister Raffarin and President Chirac had pledged to achieve at the international level. It also proposed an array of policy instruments to reach this objective with the right-wing government favouring fiscal incentives and informational instruments over taxation – including incentives for low-carbon vehicles, energy labels, efficiency certificates for buildings, tax credits for efficient appliances and biofuels. Many of the sectoral targets adopted transposed European commitments – e.g. on biofuels, renewable electricity promotion or energy labeling (Bocquillon and Evrard 2016). This period was characterized by 'bounded innovation dynamics' (Szarka 2003), most changes consisting in improving existing policy instruments rather than proposing ambitious new measures.

From Grenelle to COP21: merging energy and climate issues, yearning for exemplarity (2007–15)

In the second half of the 2000s, the French position evolved from foot-dragging to circumstantial leadership. This shift has been accompanied by a reframing of energy and climate policies – two areas traditionally dealt with separately by both the government and administration – as two faces of the same coin. This new approach was first introduced in the 2005 Energy Law (*Programme d'Orientation de la Politique Energétique Française* – or POPE law), which established the fight against climate change as a priority of French energy policy. The bill also reaffirmed the 'Factor Four' target (75 per cent GHGE reduction by 2050) to be reached at a pace of 3 per cent per year on average.

This new orientation was further consolidated in a context of high public attention to environmental and climate issues throughout the second half of the 2000s while reaching its peak with the election of President Sarkozy in May 2007. During his election campaign, under pressure from ENGOs and activist TV presenter Nicolat Hulot, Nicolas Sarkozy signed an 'Ecological Pact' and made strong environmental commitments. On his accession to power, he created a super-Ministry of Ecology, Energy, Sustainable Development and Planning (MEEDDAT) – a merger of four existing ministries – so as to rebalance inter-ministerial relations in favor of environmental issues (Lascoumes *et al.* 2014). The powerful Directorate General for Energy and Raw Materials (DGEMP), which used to be part of the Ministry of Industry, was also merged with other departments into a new Directorate General for Energy and Climate (DGEC). This administrative reorganization aimed to reinforce climate change policies and institutionalized the new energy/climate framing, without fundamentally altering administrative and political practices. A key commitment of President Sarkozy was to set in motion an innovative national consultation process on environmental issues, the '*Grenelle de l'environnement*' (Whiteside *et al.* 2010). Confirming the priority granted to climate change and the new framing of energy, the first of the eight working groups was entitled: '*Fighting against climate change and managing energy*'. The *Grenelle* process clearly contributed to increased public attention to climate change and led to the adoption of a wide range of policy commitments and targets in various sectors, from housing to energy, transport and research.

However, the legislative phase that was supposed to materialize and implement these objectives proved uneven, confirming the long-standing ambiguous attitude of the government (Boy *et al.* 2012). The first law (*Grenelle 1*), which contained the main principles and goals of the *Grenelle*, was adopted almost unanimously by Parliament in August 2009. As for the implementation law (*Grenelle 2*), the legislative process was more controversial, giving birth to a text lacking global coherence and marked by several renouncements. Key measures were postponed, awaiting the adoption of implementation decrees, and sometimes *sine die*. It was most noticeably the case of the *Contribution climat-énergie* (Carbon Tax) and the Heavy Vehicle Transit Tax (*Ecotaxe*). Following a banal blame shifting strategy, the *Ecotaxe* was postponed until the Presidential and Parliamentary elections in 2012. When the incoming President François Hollande and his government tried to implement it, the measure triggered a strong controversy and was finally abandoned in Autumn 2014. The change of political majority did not alter the energy and climate framing however. A carbon tax on fossil fuel use (*Contribution Climat Energie* – CCE) was adopted discreetly in late 2013 but initially set at low levels.

More importantly, a Law on Energy Transition, replacing the Climate Plan, was passed in July 2015, after a round of public consultations and a lengthy legislative process. The law sets a 40 per cent GHGE reduction target by 2030 and confirms the 'Factor Four' objective for 2050. It also sets a national target of 32 per cent of renewables in energy consumption (40 per cent in electricity) and proposes to halve final energy consumption by 2030. The emphasis is put on energy efficiency in buildings and clean transport, most objectives and measures now awaiting implementation decrees. After opposing it, the government finally accepted a last minute amendment, proposed by the Green Party and its own majority to raise the carbon tax to €56 per ton in 2020 (a four-fold increase) and €100 in 2030. As the host of COP21, the French Government has pursued the same discursive strategy, invoking the Law on Energy Transition as a symbol of its exemplarity and leadership at the international level.

Multi-level and polycentric climate governance in France

National centralization and the belated empowerment of local authorities

Since the ground-breaking laws of decentralization of 1982–84 and as a consequence of various other reforms, local authorities have been progressively entrusted with new powers in a variety of areas. Yet, these changes did not initially affect the governance of energy and climate change which remained largely centralized. French environmental and climate policy-making was characterized as a form of meso-corporatism in which powerful sectoral interests were entrusted with policy stewardship in collaboration and under the supervision of a specialized central administration (Szarka 2006). This centralized approach also characterized the energy sector which was dominated by two national public monopolies created after World War II: *Electricité de France* (EDF) and *Gaz de France* (GDF, now Engie). Despite a process of liberalization and partial privatization after 2000, the two energy giants have maintained close ties with the state administration – especially the powerful DGEMP/DGEC – and preserved their hold on productive and political structures (Poupeau 2014).

Since the 1990s, in the face of a lack of formal competences for local authorities over energy and environmental issues, the ADEME and NGO networks tried to promote local energy and climate initiatives on a voluntary basis but only with limited success. The situation started to change with the adoption of the first national Climate Action Plan (CAP) in 2004, which invited local authorities to establish voluntary plans with their local partners (Yalçin and Lefèvre 2012). The 2005 Energy Law also encouraged local authorities to develop renewable energy and energy management. But it is only since the *Grenelle de l'environnement* that local authorities – mainly the regions but also large urban areas – have really started to seize on climate issues (Nadai *et al.* 2015: 282). The two *Grenelle* laws made local CAPs compulsory for 400 authorities of over 50,000 inhabitants. In addition, they imposed the creation of Regional Schemes for Climate Air and Energy (SRCAE) which are non-prescriptive plans elaborated in collaboration between the state and the regions that integrate various planning documents related to energy and climate at the regional level.

The empowerment of local authorities has been further reinforced by European dynamics, notably the adoption of the 20–20–20 targets and EU climate and energy package. The European Commission launched the 'Covenant of Mayors' in 2009 to support local authorities' efforts in developing energy efficiency, renewable energy and climate measures. Local authorities have increasingly used direct references to EU objectives in their local sustainable development plans. This movement towards regionalization has also stimulated the development of local networks and initiatives promoted by NGOs. For instance, the TEPOS network (*Territories à Energies Positives* – Positive Energy Territories) was set up in 2011 by the CLER, a French renewable NGO, to support the energy transition in rural areas (Nadai *et al.* 2015).

The new decentralization law adopted in 2015 endows regions with competences on environment and energy. It replaces the SRCAE, as well as various other planning documents by a new planning framework for regional and sustainable development – the *Schéma Régional d'Aménagement, de Développement Durable et d'Egalité des Territoires* (SRADDET) – which is now made binding (Roussel 2015). Along with the new Law on Energy Transition (2015), it confirms the increasing role of local levels in energy and climate change.

Nevertheless, as Poupeau argues (2014), this recent 'activism' at the local level has not overturned traditionally centralized patterns of policy-making, as key policy instruments remain out of the hands of local authorities (e.g. in terms infrastructure planning, finance, etc.). The state has combined the mobilization of local authorities for energy and climate action – e.g. to promote energy efficiency measures and behaviours, or sustainable urban planning – with containment to preserve its legitimacy and control of a strategic sector. To date, local climate plans have helped raise awareness and bring stakeholders together, but have faced difficulties in setting ambitious strategies and in implementation (Yalçin and Lefèvre 2012; Virlouvet 2015).

French activism in European and international fora: a mixed record

From the mid-2000s onwards, France has increasingly demonstrated entrepreneurial leadership at the EU and international level (Schreurs and Tiberghien 2007: 39). If its initial efforts yielded limited results, the French government has been able to claim some successes.

Following the 1992 UN Rio Conference, the French governments promoted, in 1993, an international distribution of climate efforts based on GHGE per capita for both industrialized and developing countries (Szarka 2008: 126). This was in large part motivated by the relatively low GHGE per capita of France, and to a lesser extent by the will to demonstrate cognitive leadership. This approach met with little success in view of the scepticism of other industrialized countries and was eventually abandoned. The internationalist and developmentalist ambitions of the French government – be they real or rhetorical – have not disappeared. In the run-up to the 2009 Copenhagen climate conference, Prime Minister Borloo presented a 'Climate Justice Plan' which proposed a financial mechanism to support developing countries in their climate effort with a view to building a climate coalition for Copenhagen (Szarka 2011: 119–21). Although promoted at the highest level through multilateral diplomacy, this new attempt at exercising cognitive leadership came too late in the negotiation process and carried little weight in the face of the structural power exercised by the US and China; France and the EU were sidelined in Copenhagen.

At the EU level, the French government has also become an ardent promoter of a collective approach. In the early 1990s, it supported the project of an EU-wide carbon/energy tax on condition that it would apply only to the carbon content of fuels – in order not to penalize nuclear energy – and, to defend France's competitiveness, include safeguards should other industrialized countries fail to adopt equivalent measures. After the abandonment of the EU and national carbon tax projects (in 1994 and 2000 respectively), the French government waited circumspectly for the Emission Trading Scheme (ETS) which was finally adopted in 2003. Although it officially supported this new policy instrument, the government did not abandon the idea of a European carbon tax. In 2006, Prime Minister De Villepin proposed a carbon tax at EU borders to reduce the impact of the ETS on European competitiveness. Again, President Sarkozy floated this idea following the rejection of the French carbon tax in 2010, although without more success. A constant in the French approach has been to privilege EU and international solutions over go-it-alone policies, in the name of national competitiveness.

One of the French government's greatest international achievements has been the adoption of the EU Climate and Energy Package during the French EU Presidency in the second half of 2008. In Autumn 2007, it became clear that this package would be a priority and flagship project for the newly elected President Sarkozy. The package embodied domestic environmental and climate commitments and represented an opportunity for the government to position itself at the vanguard of EU climate leadership. The French EU Presidency committed to its adoption by the end of its term and drove the negotiations at an astonishingly fast pace, combining top level diplomacy in the European Council and national capitals – including Berlin and Warsaw – as well as intense negotiations at lower levels of the EU Council, to conclude a deal in time for Copenhagen, be it at the expense of the environmental integrity of the package (Bocquillon 2016). In this context, France supported a binding GHGE reduction target of 20 per cent by 2020, accepted to reduce its emissions by 14 per cent in sectors not covered by the ETS as part of the effort sharing decision and – a more challenging objective – agreed to increase its renewable energy consumption up to 23 per cent by 2020. The Climate and Energy package negotiations represent a case in which the French government managed to use the EU as a platform for exercising entrepreneurial leadership, in a heroic style but occasional and short-lived way.

More recently, France has been a strong supporter of a 40 per cent GHGE reduction target by 2030, adopted at the EU level by the European Council in October 2014 and enshrined in the 2015 Energy Transition Law at the domestic level. The government has been more reserved about renewable energy targets, opposing binding national commitments at the EU level – but not a binding EU target of 27 per cent by 2030 (Lindgaard 2014). French climate commitments have been driven by the will to demonstrate leadership by example in the perspective of the 2015 Paris climate conference. During the Paris summit the French government displayed strong entrepreneurial leadership, mobilizing large political resources and skills to broker an international agreement on a Post-Kyoto framework. Prepared carefully several months in advance, the conference – where the first global climate agreement was adopted – was hailed as a success and represents a tribute to French diplomacy (see below for more details).

The domestic implementation of EU and international commitments

France's leadership ambitions at the EU and international level contrast with its mixed record in terms of domestic implementation. In view of the emission reductions achieved in the 1980s and its comparatively low GHGE per capita, France was granted a relatively unambitious target as part of EU Burden Sharing Agreement (1998): a mere stabilization of its emissions at 1990 level for 2008–12. This contrasted with the large cuts accepted by Germany and the UK (see Chapters 8 and 12 respectively). In the late 1990s France's relatively strong economic growth pushed emissions up and a large overshoot was expected. This motivated the government to adopt its first Climate Programme in 2000 (Szarka 2008: 128). Enduring fears of exceeding the Kyoto target also prompted the 2004 Climate Action Programme.

Initially the government and French stakeholders had reservations about the ETS, a market-based instrument that did not seem to fit with the national policy style. Yet, implementation was easier than expected and the ETS proved to be compatible with national institutional structures and traditions (Szarka 2006: 631–2). The National Action Plan (NAP) was established by the ADEME and the register of emissions administered by the public investment bank *Caisse des Dépots et Consignations*, while the state kept the ability to set the emission cap and implement sanctions. The cap set for the second (i.e. binding) phase of the ETS (2008–12) was deemed too high by the European Commission and had to be reduced (as in most other member states). The second NAP eventually approved by the Commission set a cap slightly below 2005 emissions (–3 per cent) and required only minimal efforts from target industries. Due to the structures of its industry and energy system – and in view of its size – France has not been a major player in the ETS. In the Kyoto commitment period, it came only in sixth position in terms of emissions covered (after Spain (see Chapter 11)), the ETS representing only about 20 per cent of national emissions. As a result, adaptive pressures have been comparatively low (Szarka 2011b: 122). It is thus no surprise that the reform of the ETS (adopted as part of the EU Climate and Energy Package in 2009), which centralized its functioning at the EU level and set a European-wide emission cap, proved relatively uncontroversial with the French government.

These commitments have been facilitated by favourable dynamics in terms of GHGE reductions (see Figure 7.2). The stabilization of GHGE appeared uncertain up until the early 2000s. Since 2005, emissions have decreased sharply due to technology improvements in heavy industries and economic restructuring, a trend further accentuated by the economic crisis (Virlouvet 2015: 138–9). As a result France has overachieved its Kyoto objective, reducing its total emissions by 13 per cent in 2012. However, the evolution of emissions differs significantly across sectors. While emissions from construction, manufacturing and energy industries have followed a downward trend since the 1990s, the transport and residential sectors, which are also the largest sources of emissions, have seen

Figure 7.2 Evolution of French emissions (1990–2013).

Source: European Environment Agency/Eurostat 2016.

their emissions increase up until the mid-2000s (ibid.: 104), while in agriculture (the main emitter of NO_2 and methane) emissions have only slightly decreased since 1990. Moreover, recent analysis of the French carbon footprint reveals that consumed emissions have slightly increased over 1990–2012 due to a rise in imported carbon emissions (CGDD 2015).

Reviewing progress towards EU 2020 climate and energy targets, the European Environmental Agency finds that in sectors covered by the Effort Sharing Agreement (non-ETS), France has overachieved its interim 2013 objective by a large margin, and is well on track to meet its 14 per cent target in 2020 based on domestic reductions only (European Environmental Agency 2015). Concerning renewable energy and energy efficiency, progress is more uncertain. With a 14.3 per cent share of renewable energy in its energy mix in 2014 – mainly from large hydroelectricity – France is hardly on track towards its 23 per cent renewable energy EU target for 2020. The predominance of nuclear electricity generation and intermittent political support to renewable energies have hampered the growth of the sector. In the administration there is widespread scepticism concerning the ability of the country to meet its objective, which is widely perceived as overambitious (Bocquillon and Evrard 2016).[4] In comparison the promotion of biofuels, considered a national priority and supported by powerful agricultural interests, has been more successful. The 2004 biofuel programme triggered a rapid growth of the sector. France has achieved the 5.75 per cent target included in the 2003 EU biofuel Directive as early as 2008, and reached its 7 per cent national target in 2012. Yet, with ongoing controversy over the impact of biofuels on food crops and land use change, biofuel expansion has stalled, falling short of the 10 per cent EU target in transport.[5] Concerning the non-binding 20 per cent energy efficiency EU target, France is lagging behind its linear energy consumption reduction trajectory, a trend that reflects a long neglect of energy efficiency measures (Aykut and Dahan 2015: 557).

Climate change, from threat to industrial and diplomatic opportunity

Since the 1970s, the promotion of civil nuclear energy has been an integral part of French industrial, commercial and foreign policies (Szarka 2009: 120). The nuclear industry is considered a cutting-edge, competitive and high added-value technology which provides France with structural and cognitive leadership. As a result, the interests of the nuclear industry – notably those of the predominantly state-owned EDF and AREVA – have been consistently assimilated to the 'national interest', and the state has been closely involved in their international promotion, notably through direct contract negotiations with third countries.

During the Kyoto negotiations, the pro-nuclear elites came to perceive the rise of climate concerns as a chance to promote nuclear energy at the global level (Mülhenhöver 2002). In a difficult context for the nuclear industry, they seized the issue as an opportunity to restore the legitimacy of nuclear energy and promote its 'revival' (Szarka 2013). Nuclear technologies were framed as a solution to the problem of climate change, based on their lower GHGE intensity compared to alternative fossil fuels. They were presented as the most efficient way to curb global emissions while ensuring sufficient and cheap energy for growing economies. This new framing was resolutely anchored in a discourse on ecological modernization. The climate framing of nuclear energy has been especially directed towards developing economies and emerging markets – such as China, India and Gulf countries – emphasizing that nuclear electricity is a cheap and low-carbon technology with the potential to meet their fast-growing energy needs (see also Chapters 17 and 18). This strategy has also achieved some success in industrialized countries, Finland (2003) and the UK (2012) signing contracts with French companies partly for climate-related motivations. Yet, it has been called into question, as a result of important delays and over-costs in ongoing construction projects (notably in Finland and France), and because of the detrimental consequences of the 2011 Fukushima accident for the sector's image (Szarka 2013).

In international climate negotiations, France has pushed for the recognition of nuclear power as a low-carbon technology, to enhance its legitimacy and create new market opportunities for French technologies and expertise. At COP6 in The Hague (2000), the government pressed for the inclusion of nuclear energy projects within the Clean Development Mechanism (CDM). Although supported by the US and China, this initiative was opposed by all EU member states but Finland and eventually rejected (Mülhenhöver 2002: 175–6). At the EU level, during the negotiations on the 20–20–20 targets in 2007, President Chirac also tried to include nuclear as part of a broad low-carbon energy target, but this was too divisive and he had to back down (Euractiv 2007).

More generally, since 2007, climate change has become a prestige issue for the French government and a key area to demonstrate the country's environmental credentials and international standing. Mirroring the adoption of the Climate and Energy Package during the 2008 French Presidency (see section 4), the decision to host the COP21 in Paris in December 2015 clearly illustrates this political strategy. The decision was made in 2012, partly as a gesture towards the Green Party to forge an electoral alliance. The conference also became an opportunity for the Socialist government to 'green' its discourse after a series of environmental policy failures. The 2015 Paris climate conference reveals the importance that climate has acquired in France's diplomatic strategy of international influence. President Hollande made COP21 one of the landmarks of his term in office towards which national environmental and foreign policies have been geared. In the Environment Ministry, led by the high-profile Minister Ségolène Royal, climate issues have tended to take priority over all other environmental matters. In international environmental negotiations, the Ministry of the Environment was usually the lead negotiator. In preparation for and during COP21, it was the Ministry of Foreign Affairs – led by veteran Minister Laurent Fabius, his experienced chief negotiator Laurence Tubiana and no less than 60 members of staff – which has been at the forefront. This duo, composed of a political heavyweight and a seasoned environmental adviser, proved very efficient in leading the French team and eventually brokering an international climate agreement.

Commentators and national representatives have often hailed the summit as a diplomatic success. The French Presidency has been praised for its commitment and organization; for the dedication of its well-established diplomatic machinery; for its efforts to meet and listen to everyone, in a 'transparent' and 'inclusive' manner; and for its flexibility in brokering agreements on various drafts and on the final text (Harvey 2015; Stothard and Chassany 2015). Although the real influence of the French government may well have been overestimated in the euphoria that followed the deal, its active role in the negotiations leading to the Paris Climate Agreement reveals a national preference for (often short-lasting) entrepreneurial leadership, based on the country's negotiating skills and diplomatic resources, in a heroic style.

Conclusion: political leadership in France

France's attitude towards climate change is ambiguous. In the press and public opinion, attention to this issue has varied, depending on the national and international political contexts, as well as the economic situation. As for governmental actors, their strategies have been changing, alternating between bandwagoning at EU level, resistance to specific policy goals and instruments, and occasional bids for leadership at the EU and international levels.

Following the successful organization of the 2015 Paris climate conference, the disjunction between French humdrum domestic policies and heroic international leadership ambitions will be put to a test. The entrepreneurial role of the government during the negotiations has been widely praised. Now that the thorny process of implementing the Paris agreement begins, the focus will shift towards French domestic policies. COP21 has encouraged public authorities in their quest for exemplarity on environmental and climate issues, as shown by their support for European targets and the adoption of the Law on Energy Transition in July 2015. The analysis of the previous decades shows that France's legitimacy cannot rely only on the occasional greening of its political discourse and short-lived diplomatic efforts. It must be combined with a clarification of its domestic policy and its policy preferences regarding climate change. The effective implementation of the Law on Energy Transition will be crucial test case.

Another source of uncertainty lies in the effect of COP21 on French society as a whole. Will the media hype on climate change endure or falter after the conference? Will local and civil society actors be able to maintain their mobilization and influence? These dynamics appear more crucial to assess the long-term evolution of French climate policy than detailed accounts of the Presidency's role in international climate negotiations.

Notes

1 'Climate disruption/COP21/Major National Cause label for 2015', Communiqué issued by the Prime Minister's Office: http://fr.ambafrance-us.org/spip.php?article6696.

2 In terms of carbon intensity – both per GDP unit and per capita – the country ranks in the same category as an environmental pioneer such as Sweden.

3 The Climate plan was updated in 2006, 2009, 2011 and 2013.

4 France had already missed its indicative target of 21 per cent renewable electricity by 2010, as set in the 2001 Directive.

5 To achieve the 10 per cent EU target and curb its high emissions in the transport sector, the government has pushed for the development of electric vehicles, as reflected in the Law on Energy Transition.

References

ADEME (2014) 'Les représentations sociales de l'effet de serre et du réchauffement climatique', December 2014.

Aykut, S.C., and Dahan, A. (2015) *Gouverner le climat? 20 ans de négociations internationales*. Paris: Presses de Sciences Po.

Aykut, S.C., Comby, J.B. and Guillemot, H. (2012) 'Climate change controversies in French mass media 1990–2010', *Journalism Studies 13*(2): 157–74.

Bocquillon, P. (2016) '(De-) Constructing coherence? Strategic entrepreneurs, policy frames and the integration of climate and energy policies in the European Union', *Environmental Policy and Governance* (forthcoming).

Bocquillon, P. and Evrard, A. (2016) 'Complying with, resisting or using Europe? Explaining the uneven and diffuse Europeanization of French renewable electricity and biofuels policies', in I. Solorio Sandoval, H. Jörgens and M. Bechberger (eds) *A Guide to Renewable Energy Policy in the EU*, Cheltenham: Edward Elgar (forthcoming).

Boy, D., Brugidou, M., Halpern, C. and Lascoumes, P. (2012), *Le Grenelle de l'Environnement: Acteurs, Discours, Effets*. Paris: Armand Colin.

Comby, J.-B. (2015) *La Question Climatique: Genèse et Dépolitisation d'un Problème Public*. Paris: Raisons d'Agir.

Commissariat General au Développement Durable (2015) 'L'empreinte carbone: les émissions "cachées" de notre consommation', November.

Deroubaix, J.F., & Lévêque, F. (2006) 'The rise and fall of French Ecological Tax Reform: social acceptability versus political feasibility in the energy tax implementation process', *Energy Policy* 34(8): 940–9.

Euractiv (2007) 'EU makes bold climate and renewables commitments', 9 March.

Eurobarometer (2014) 'Climate Change', Special Eurobarometer 409, Wave EB80.2, March.

European Environmental Agency (2015) 'Trends and projection in Europe 2015: tracking progress towards Europe's climate and energy targets', EEA report n°4/2015.

Evrard, A. (2012) 'Political parties and policy change: explaining the impact of French and German Greens on energy policy', *Journal of Comparative Policy Analysis*, 14(4): 275–91.

Evrard, A. (2013) *Contre Vents et Marées: Politiques des Energies Renouvelables en Europe*. Paris: Presses de Sciences Po.

Giraud, P.-N., Collier, U., and Löfstedt, R.E. (1997) 'France: relying on past reductions and nuclear power', in U. Collier and R.E. Löfstedt (eds) *Cases in Climate Change Policy: Political Reality in the European Union*. London: Earthscan, 165–83.

Halpern, C. (2012) 'L'écologie est-elle une variable d'ajustement?' in J. de Maillard and Y. Surel (eds) *Politiques publiques sous Nicolas Sarkozy*, Paris: Presses de Sciences Po, 2012: 381–402.

Harvey, F. (2015) 'Paris climate deal: the world's greatest diplomatic success', *Guardian*, 13 December.

Lascoumes, P., Bonnaud, L., Le Bourhis, J.-P. and Martinais, E. (2014) *Le développement durable. Une nouvelle affaire d'Etat*. Paris: Presses Universitaires de France.

Lindgaard J. (2014) 'A Bruxelles, la France agit contre les énergies renouvelables', *Mediapart*, 23 January.

Mülhenhöver, E. (2002) *L'Environnment en Politique Etrangère, Raisons et Illusions*. Paris: L'Harmattan.

Poupeau, F.-M. (2014) 'Central-local relations in French energy policy-making: towards a new pattern of territorial governance', *Environmental Policy and Governance* 24(3): 155–68.

Roussel, F. (2015) 'Gouvernance environnementale: ce qui change avec la loi NOTRE', *Actu-environnement.com*, 23 July.

Schreurs, M. and Tiberghien, Y. (2007) 'Multi-level reinforcement: explaining European Union leadership in climate change Mitigation', *Global Environmental Politics* 7(4): 19–46.

Stothard, M., and Chassany, A.-S. (2015) 'COP21: Fabius lauded for successful conclusion', *Financial Times*, 13 December.

Szarka, J. (2003) 'The politics of bounded innovation: "new" environmental policy instruments in France', *Environmental Politics* 12(1): 93–114.

Szarka, J. (2006) 'From inadvertent to reluctant pioneer? Climate strategies and policy style in France', *Climate Policies* 5(6): 627–38.

Szarka, J. (2008) 'France: the search for an alternative climate policy template', in I. Bailey and H. Compston (eds) *Turning Down the Heat*. Basingstoke: Palgrave, 125–43.

Szarka, J. (2009) 'Environmental foreign policy in France: national interests, nuclear power and climate protection', in P.G. Harris (ed.) *Climate Change and Foreign Policy*. London: Routledge, 117–33.

Szarka, J. (2011) 'France's trouble bid to climate leadership', in R.KW. Wurzel and J. Connelly (eds) *The European Union as a Leader in International Climate Change Politics*. London: Routledge, 112–28.

Szarka, J. (2013) 'From exception to norm – and back again? France, the nuclear revival and the post-Fukushima landscape', *Environmental Politics* 22(4): 646–63.

Vilourvet, G. (2015) 'Vingt ans de lutte contre le réchauffement climatique en France: bilan et perspectives des politiques publiques', Rapport pour le Conseil Economique, Social et Environnemental (CESE), April 2015.

Whiteside, K.H., Boy, D. and Bourg, D. (2010) 'France's "Grenelle de l'environnement": openings and closures in ecological democracy', *Environmental Politics* 19(3): 449–67.

Yalçin, M. and Lefèvre, T. (2012) 'Local climate action plans in France: emergence, limitations and conditions for success', *Environmental Policy and Governance* 22(2): 104–15.

8 Germany

Innovation and climate leadership

Martin Jänicke

Introduction

The OECD has characterised Germany as 'a highly innovative country engaged in several initiatives to draw the maximum benefits and opportunities of globalisation to address environmental problems while boosting its environmental industry sector' (OECD 2007: 43). This seems a good characterisation especially for Germany's climate policy since 1987.

Germany has played a leadership role in climate change policy, which exhibited different types and styles of leadership (see Chapter 1) over the years. As early as 1990, it made major political contributions to the generation of a *knowledge base* for a far reaching global strategy to combat climate change, thus acting initially as an intellectual or *cognitive leader* (see Chapter 1). After 1998 the concept of ecological modernisation (*ökologische Modernisierung*) became central under a Social Democratic Party (SPD) and Green Party 'Red–Green' coalition government (1998–2005). Ecological modernisation was developed into the concept of ecological industrial policy (*ökologische Industriepolitik*) by an SPD Environmental Minister (Sigmar Gabriel) under a SPD and Christian Democratic Union (CDU)/Christian Social Union (CSU) 'Grand coalition' government (2005–9).

Germany's leadership role in international climate change politics today can be largely characterised as *leadership by example*: an interaction of demonstration effects and 'lesson-drawing' (Rose 1993). This includes the adoption of policy instruments like obligatory feed-in tariffs (*Einspeisetarife*) but partly also lessons drawn from challenges arising during the implementation of this ambitious policy (e.g. costs, NIMBY phenomenon, and the resistance of neighbouring states – such as Poland – against the export of green electricity). The overall policy outcomes demonstrate the general feasibility of an energy transition (*Energiewende*) – e.g. the over-achievement of the ambitious German Kyoto Protocol target (–21 per cent) already by 2007 (see Figure 8.1) – and a broad potential of economic co-benefits.

Germany's economic strength, advanced innovation system and political visibility were necessary *structural* conditions for its leadership role (Jänicke 2012). Germany's climate policy exerts a strong international influence as a result of its economic success and the boom in its domestic 'climate protection industry'

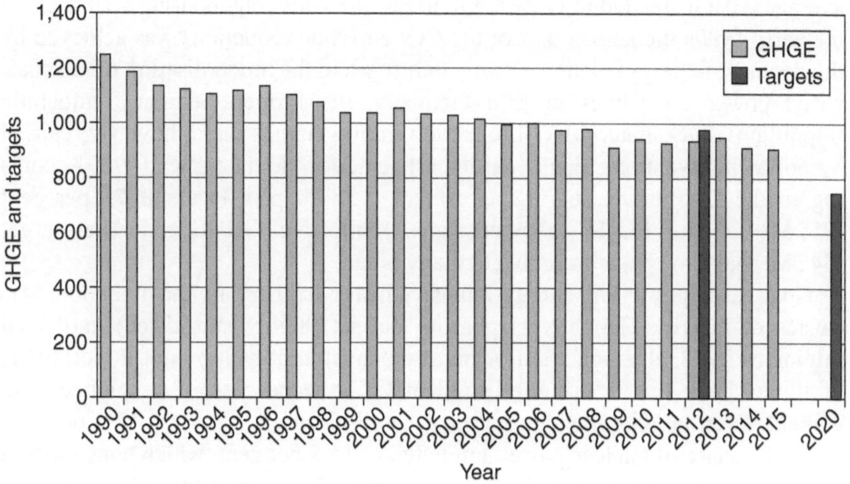

Figure 8.1 German GHGE (1990–2015) and targets (2012–20) in Mt.
Source: adapted from BMUB 2014.

(Jänicke 2011). This effect has increased with the creation of lead markets for clean-energy technologies since 2000. These initiatives helped to increase Germany's exports of climate-friendly energy technology, resulting also in competitive challenges for other countries.

The will and skill of central environmental policy actors including highly competent environmental ministers (Klaus Töpfer (CDU) 1987–94, Angela Merkel (CDU) 1994–8, Jürgen Trittin (Greens) 1998–2005, Sigmar Gabriel (SPD) 2005–9 and Barbara Hendricks (SPD) since 2013) were also important. Germany's international leadership role would not have been possible without the support of Chancellors Helmut Kohl (CDU, 1982–98), Gerhard Schröder (SPD, 1998–2005) and particularly Angela Merkel (CDU, since 2005). Here an expression of *entrepreneurial leadership* in international climate change politics could be observed from time to time.

A *transformational* policy style (see Chapter 1) can be identified for the following two time periods: First, under the Red–Green coalition (1998–2005) and the following two years (2005–7) and second, since Fukushima (2011).

The national context of climate policies

Germany is a highly industrialised country with a population of 81.5 million, or 227 people per square kilometre (Statistisches Bundesamt, 2016). GDP per capita is €34,271 (2013) (Statistisches Bundesamt 2014). Total emissions of carbon dioxide (CO_2) and other major greenhouse gas emissions (GHGEs) amounted to 912 Mt CO_2-equivalents (2014) most of which was accounted for by

CO_2 (*Umweltbundesamt* 2015). GHGE per capita in Germany are above the EU average. GHGE related to GDP are more in line with neighbouring countries. In the early 1990s the largest part of the CO_2 emission reductions was achieved by the decline of energy-intensive heavy industry and the modernisation of the coal-based power companies in East Germany, the former German Democratic Republic (GDR). Later GHGE reduction improvements were, however, caused by policy interventions; this is particularly true for the time after 1998. According to the EU burden sharing agreement (see Chapter 1) about 75 per cent (259 Mt) of the EU-15 member states' collective GHGE reduction target (342 Mt) have to be achieved by Germany alone.

Coal-based power and energy-intensive heavy industry are the main domestic causes of GHGE. Germany is phasing out its (highly subsidised) hard coal mining by 2018, although it still remains the most important producer worldwide of lignite coal: this is a highly problematic issue because of its high carbon content. Even in 2014 the share of coal-based electricity was still 43.2 per cent while the share of nuclear power amounted to 15.8 per cent, which constituted a clear reduction compared with 29.2 per cent in 2002, the year when the Red–Green coalition government adopted a law to phase out nuclear power. By 2015 the share of 'green' electricity had climbed to 30 per cent (AGEB 2015).

In the early 1970s, under Chancellor Willy Brandt (SPD), Germany developed into one of the leading European countries in environmental policy. However, during that time period Germany's environmental ambitions remained largely focused on the domestic level. In other words, Germany acted as an environmental pioneer that adopted progressive environmental measures primarily on the national level. It was only from the early 1980s onwards when Germany adopted a more active role also on the EU and international level thus developing into an environmental leader which was keen to attract followers.

Germany has a strong environmental movement (Markham 2008; Mez 2013). With five to six million members, its membership is close to that of the trade unions. Environmental reporting by the media – especially the public media – has steadily increased in its quantity, range and importance. The Green party has been represented in the German *Bundestag* (lower house) since 1983. Germany has several 'green' industry organisations such as BAUM (*Bundesdeutscher Arbeitskreis für Umweltbewusstes Management*) and Future as well as the most highly developed environmental protection industry in Europe (Ernst & Young 2006; BMUB 2014).

Phases of domestic climate change policy

German climate policy has developed under very different government coalitions including a Centre-Right coalition (CDU/CSU and Liberals, 1987–98), a Red–Green coalition (SPD and Greens, 1998–2005) and two Grand Coalitions (CDU/CSU and SPD, 2005–9 and since 2009). All five coalition governments favoured a German leadership role in EU and international climate policy. Essentially climate change constitutes a cross-party consensus issue although the

Liberals (more right) and Greens (more left) are furthest apart in climate change policy terms (see Michaelowa 2008; Oberthür and Pallemaerts 2010).

The 1987 Bundestag Enquete Commission Prevention to Protect the Earth's Atmosphere

In 1986 the influential weekly investigative magazine *Der Spiegel* published an article entitled 'Climate disaster' (*Der Spiegel* 11 August 1986). In the same year the CSU-led Bavarian state (*Land*) government launched an initiative in the *Bundesrat* (upper house) for the creation of an advisory body for climate policy (Grassl 1999: 100). This was also intended to weaken anti-nuclear arguments after the 1986 Chernobyl nuclear accident. The new issue of 'climate protection' was put on the political agenda in 1987 when the newly elected lower house (*Bundestag*) set up an Enquiry Commission called *Bundestag Enquete Commission Prevention to Protect the Earth's Atmosphere*. This commission – together with an earlier Enquete Commission on nuclear energy (Deutscher Bundestag 1980) – proved to be highly influential in the years to follow. It created the knowledge base for understanding climate change and possible policy solutions not only for the inner circle of the political elite but also for the interested wider public. Its 1990 report predicted a global temperature increase of 1.5–4.5°C in the event of a doubling of CO_2 concentrations in the atmosphere. The Climate Enquete Commission proposed a nearly utopian German CO_2 emission reduction target of 30 per cent for 2005. It also proposed minimum targets for other countries and the EU (compared to the base year 1987). The Climate Enquete Commission's proposal can be regarded as cognitive leadership (*Deutscher Bundestag*, 1990). At the first Conference of the Parties (COP) of the United Nations Framework Convention on Climate Change (UNFCCC) in Berlin in 1995, Germany proclaimed a slightly reduced CO_2 reduction target of 25 per cent by 2005 (compared to 1990). This target was based on cross-party consensus in the German *Bundestag*.

Early activities of the Kohl government (1987–98)

The government's (subsequently adopted) climate change policy was strongly influenced by the Environment Minister Klaus Töpfer. A first step constituted the federal government's 1990 *CO_2 Reduction Programme* and the *Interministerial Committee on CO_2 Reduction* which amounted to a significant break with the tradition of strong independence of Ministries in the German government. The *CO_2 Reduction Programme* contained, among others, new regulations for energy efficiency and heating. However, the most important step was the 1990 *Act on the Sale of Electricity to the Grid* which introduced obligatory feed-in tariffs for power from renewable energy. In 1991 the second *Bundestag Enquete Commission Protection of the Earth's Atmosphere* was installed.

Germany ratified the UNFCCC in 1993. The first COP, which adopted the so-called Berlin mandate (see above), was strongly influenced by the German government in which Angela Merkel, who later became German Chancellor, was the Environment Minister. Germany's 21 per cent reduction target for GHGE under the 1997 Kyoto climate change protocol was again an expression of the ambitious leadership role which the German government had adopted in international climate change politics (Schreurs 2002: 178 passim).

The Red–Green government 1998–2005

German unification, which took place in 1990, resulted in challenging economic problems. It reduced the original desire to prioritise climate change policy. The newly elected Red–Green coalition government (1998–2005), however, gave a strongly renewed impetus to climate change policy. The year 1998 became the starting point for a more ambitious climate policy which continued under Chancellor Merkel's first Grand coalition (2005–9). The main general heading for environmental and climate change policy in the Red–Green governments' coalition treaties (adopted in 1998 and 2002) was 'Ecological Modernisation' (see SPD/Bündnis 90-Die Grünen 1998). It emphasised the role of innovation and economic co-benefits in environmental and climate policy.

The 1999 *Renewable Energy Act* was an important first step by the Red–Green government to increase the obligatory feed-in tariffs to a level high enough to cause a rapid growth of 'green' power (see Figure 8.2). The Red–Green coalition government also introduced an ecological tax reform in 1999. Most of the revenue (about €18 billion) from the new energy tax was used to reduce social security contributions. A smaller part supported investment in renewable energy (*Marktanreizprogramm*). The 2000 *Climate Protection Programme* introduced new regulations and sectoral emission reduction objectives (BMU 2000). Although the 2000 *Climate Protection Programme* did not achieve the overly ambitious national 25 per cent CO_2 emission reduction target by 2005, it nevertheless provided the necessary basis for a more progressive climate policy that was announced in 2007.

Climate change was an essential factor in the re-election of the Red–Green government in 2002, following heavy flooding of the Elbe in the former East Germany a few months before the *Bundestag* elections. In 2004, a major conference on renewable energy was held in Bonn. It illustrated German entrepreneurial leadership for global diffusion of renewable energy (REN21 2008). The initiative to set up an International Renewable Energy Agency – IRENA – was a more visible result. In setting up IRENA Germany cooperated closely with Denmark and Spain in particular.

The first Merkel government (2005–9)

The first Grand coalition led by Chancellor Merkel generally followed the same path as the Red–Green coalition government in climate change policy (Jänicke 2010). The Red–Green coalition government's concept of ecological modernisation was further developed into ecological industrial policy (*ökologische*

Industriepolitik) by the Grand coalition's new Minister of Environment Sigmar Gabriel. The new formula became connected with ambitious slogans and concepts such as 'Green New Deal' and 'Third Industrial Revolution' (Gabriel 2006; BMU 2008).

The EU emissions trading scheme (ETS) became an essential policy instrument for German climate policy at this time. It was a policy instrument that had initially been opposed by Germany (Wurzel 2008) and the first German national allocation plan (NAP) (2005–7) was rather weak. It produced a significant over-allocation of emission allowances and showed a clear bias in favour of coal interests. Importantly, the coal industry and its unions had close links with the SPD. The second NAP (2008–12) was formulated by the Grand coalition government with Sigmar Gabriel as Environment Minister. It was again no pioneering proposal. The second NAP echoed the strong opposition of the main industrial organisations which had published harsh protests both against the Kyoto Protocol and Germany's leadership role in international climate policy before the 2005 federal election (DIHK 2005). However, importantly, the EU Commission now forced the German government to stick to its self-proclaimed pioneer role. The Commission refused to accept the original second NAP while demanding an additional reduction of 17 Mt CO_2. The government saw no alternative and accepted the Commission's demands. One reason was that Germany held the EU Presidency in the first half of 2007 and the G8 Presidency during 2007. Holding both these two presidencies stimulated a new political leadership ambition in international climate change politics within the Grand coalition government. Germany's international leadership role was fully re-established within the context of worrying news from the Intergovernmental Panel for Climate Change (IPCC) which published its fourth assessment report in 2007. The 2007 G8 summit in Germany (Heiligendamm) became a kind of celebration of Germany's leadership role in international climate change politics. This amounted mainly to entrepreneurial leadership which was based on the economic success of an ambitious climate policy.

The 2007 *Integrated Energy and Climate Programme* (*Integriertes Energie und Klimaprogramm* – IEKP) was the main domestic result of Germany's leadership ambitions in international climate change. A 40 per cent GHGE reduction target for 2020 was officially adopted together with a broad range of implementation measures. However, the parliamentary process again showed the strength of some veto players: the car and coal-based power industries. The 2008 global financial crisis and subsequent recession in 2009, which contributed to a 27 per cent CO_2 emission reduction in the same year, has made it easier for Germany to achieve its reduction targets. Renewable energy investment, which increased significantly under the Red–Green coalition, also contributed significantly towards CO_2 reductions.

The second and third Merkel governments (2009–13 and since 2013)

The second Merkel Government (2009–13) was a Christian Democratic and Liberal (CDU/CSU-FDP) coalition government. A significant re-orientation in climate policy was expected. And, indeed, the new Centre-Right government started by abolishing the Red–Green coalition government's phasing out nuclear energy. The new Liberal Economics Minister, Philipp Rösler, also intervened several times in Brussels to prevent what he considered to be overly ambitious EU climate targets.

However, following the 2011 Fukushima catastrophe the CDU/CSU-FDP coalition government initiated the energy transformation (*Energiewende*): a sudden decision on an (accelerated) phasing out of nuclear energy by 2021, with an immediate closure of eight nuclear power stations. This amounted to a remarkable policy U-turn which was coupled with an ambitious climate protection programme. The new Environmental Minister, Norbert Röttgen (CDU), was a clear proponent of the *Energiewende* for which he received strong support from Chancellor Merkel. Röttgen was the main driver of the climate targets for 2020, which had already been agreed in 2010: 20 per cent less primary energy consumption, 40 per cent GHGE reduction, 35 per cent share of renewable energies. The targets for 2050 were not less ambitious: 50 per cent less primary energy consumption, 80–95 per cent GHGE reduction, 80 per cent green electricity and a 'nearly climate-neutral' housing sector. However, the second Merkel Government also introduced a significant reduction of the feed-in tariffs, particularly for photovoltaic (PV) technology.

The third Merkel government (since 2013) was a coalition government of Christian Democrats (CDU/CSU) and Social Democrats (SPD). It succeeded the Centre-Right coalition government partly because the Liberals (FDP) had failed to gain representation in the national parliament (*Bundestag*). The new government tried to manage the cost increase of electricity and the back-up problems of fluctuating renewable energies (wind and PV). The feed-in tariffs for renewables were further reduced. But the growth perspective of green electricity was stabilised with the adoption of a support scheme dependent on the achievement of targets (annual increase of PV: 2.5 GW, wind onshore: 2.5 GW, wind offshore: 6.5 GW). The Grand coalition also adopted a new GHGE target for 2030 (55 per cent reduction) and new targets for green electricity (2025: 40–45 per cent; 2035: 55–60 per cent). This was criticised by green NGOs and the renewables sector as clearly below the expected potential. However, the new policy was also a reaction to the high speed of change in the energy system which had triggered unexpected high demand for adaptation (e.g. back-up and grid-connection problems and the crisis of the four big power companies). The main challenge was to find a balance between the rapid growth of decentral electricity supply (with a high share of private owners) and changes in the central supply structures (offshore wind as a substitute for nuclear power, cheap lignite coal power vs. expensive gas). As already under the Red–Green coalition (1998–2005) the Social Democrats remained strong advocates for coal and lignite coal in particular. Much of Germany's coal reserves can be found in the Ruhr area which has traditionally been the Social Democrats' heartland. In 2015 the former Environment Minister Gabriel (SPD) – now Economics Minister with responsibility for energy – made a cautious attempt to reduce CO_2 emissions of coal-based power by 22 per cent. But strong resistance within his own party together with parts of the Christian Democrats caused a modification which resulted not in a charge for old coal-fired power stations but a subsidy for their phasing out.

At the 2015 UN climate summit in Paris Germany was a strong supporter of the French Presidency in the negotiation process. The new international climate strategy with dynamic targets was the result of French cognitive leadership (see Chapter 7). However, Germany has been laying the groundwork for a commitment to decarbonisation with the G7 agreement some months before the 2015 Paris climate conference. Similarly strong was Germany's role in the preparation of the COP21. The Petersberg Climate Dialogue (May 2015) with 36 participating countries and keynote speaches of Chancellor Merkel and President Hollande discussed for instance already a '1.5 or 2°C ceiling' together with a mechanism to increase ambition over time (BMUB Press Release 111/15, 19 May 2015). There was also a leading German contribution in terms of financial commitments (€3 billion alone for African countries).

Policy instruments and policy outcomes

A broad range of policy instruments is used in German climate policy. It differs in each field of political intervention (OECD 2012). A technology-based approach and targeted support for clean energy markets have been the preferred options. Regulations and standards are often used. Three economic instruments, however, play a dominant role: the obligatory feed-in tariffs for green electricity; the ecological tax reform; and the EU ETS. The ecological tax reform has contributed to a steady reduction of CO_2 emissions. The success of the obligatory feed-in tariffs for green electricity (Figure 8.1) stimulated a rapid rate of international diffusion of this instrument (71 countries having it introduced in 2013, REN21 2014). As mentioned above, emission trading had counterproductive results in the first allocation phase of the EU ETS. Under the second NAP, however, emission trading has become a stronger instrument although it later completely failed to achieve significant CO_2 reductions. Voluntary agreements have played a minor role.

Germany's international leadership in climate policy was essentially the result of a demonstration effect: its ambitious policy turned out not only to be feasible but also to be surprisingly effective; and, above all, climate change policy has also become an economic success story. The German first-mover advantage, which is the result of Germany taking on the role of a pioneer with its energy transformation, has also caused a competitive challenge for other industrial countries.

The most important result of the German climate policy is the achievement of the Kyoto target (21 per cent GHGE reduction in 2012) already in 2007 (Figure 8.1). Other unexpected successful developments and co-benefits include:

- the rapid increase of renewable electricity (Figure 8.2)
- the booming 'climate protection industry' increasing its share of the world market since 2000
- the employment effect (Figure 8.3)
- the speeding up of the innovation process
- and the rapid increase of private investment at the local level.

The sections which follow will explain some reasons for these successes.

The rapid increase of green power

The rapidly increased share of green power reached 27.8 per cent by 2014. It avoided 148 Mt CO_2 emissions (compared to 85 Mt in 2005) (BMWi 2015). The 2010 target for green power in the Renewable Energy Act was already achieved by 2007 (Figure 8.2). The 2020 target for green power therefore has been changed from at 'least 20 per cent' to at 'least 30 per cent' and shortly later (2010) to 35 per cent, due to the unexpected rapid growth in green power.

The main instrument constituted obligatory feed-in tariffs for renewable energy, together with market incentives for investment in renewable energy. The higher price for electricity is being paid for collectively by customers (with a steady reduction over time to stimulate cost reductions).

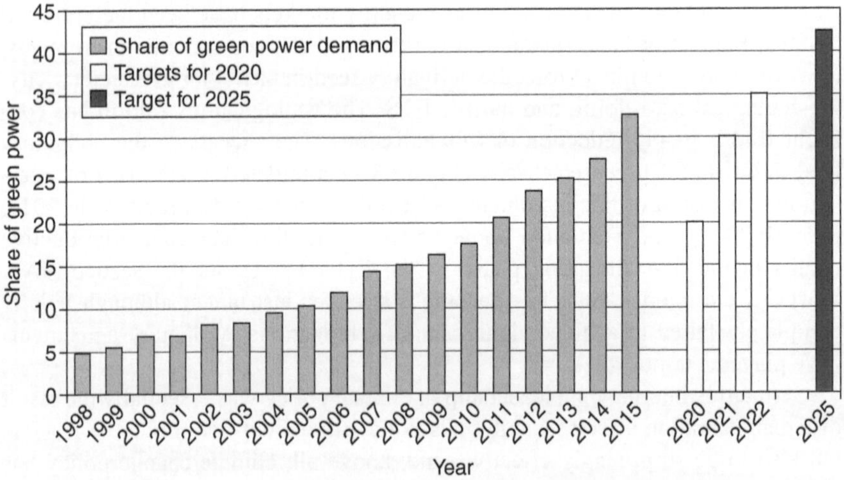

Figure 8.2 The share of green power demand (1998–2015) and targets (2020, 2025) in Germany.

Source: AGEB (2015).

Other aspects of low-carbon development showed a less impressive picture. The share of renewable primary energy was only 11.1 per cent in 2014 (AGEB 2015). Heat from renewable energy was only 9.9 per cent (BMWi 2015).

A booming 'climate protection industry'

It was only in 2008 that a study found that environmental policy had created not only a new type of 'environmental industry', but that climate protection had also very quickly become a remarkable economic factor. This kind of 'climate-protection industry' already had a 5 per cent share of German GDP in 2005 (BMU and UBA 2009).

According to a study by Roland Berger (BMUB 2014) the annual production volume and the global market share of climate-friendly technologies in Germany was as follows in 2013:

- Production, storage and distribution of clean energy: €73 billion (global market share: 17 per cent)
- Energy efficiency: €100 billion (12 per cent)
- Sustainable mobility: €53 billion (17 per cent).

The three groups of clean energy technologies (€226 billion) amount to 8.2 per cent GDP. The German 'CleanTech' sector, which also includes other environmentally friendly and resource-efficient technologies, has a volume of €344 billion and a share of 13 per cent of the German GDP (2013); its global market share is 14 per cent. With an annual growth of 6.5 per cent it is expected that the German CleanTech sector will reach €740 billion in 2025 while its GDP contribution would rise to 20 per cent (BMUB 2014).

Employment effects

The German climate policy resulted in a gross employment effect of at least 1.2 million jobs. Although the boom of renewable electricity was over after 2011 (due to a crisis of the German PV sector), about 370.000 jobs existed in 2013 (Figure 8.3). The employment effect triggered by energy efficiency measures and related technologies has to be added. Here the increase seemed to be more stable, resulting in 848.000 jobs in 2013 (DENEFF 2014).

The net employment effect is of course lower. And it cannot be ignored that in the traditional power sector some 10,000 jobs have been lost. The big four power suppliers (RWE, E.ON, VATTENFALL and EnBW) are paying a high price for their opposition to renewable energies, losing first market shares and second competitiveness at the spot market for electric power. This crisis led to a complete re-start of E.ON, EnBW and RWE (2015).

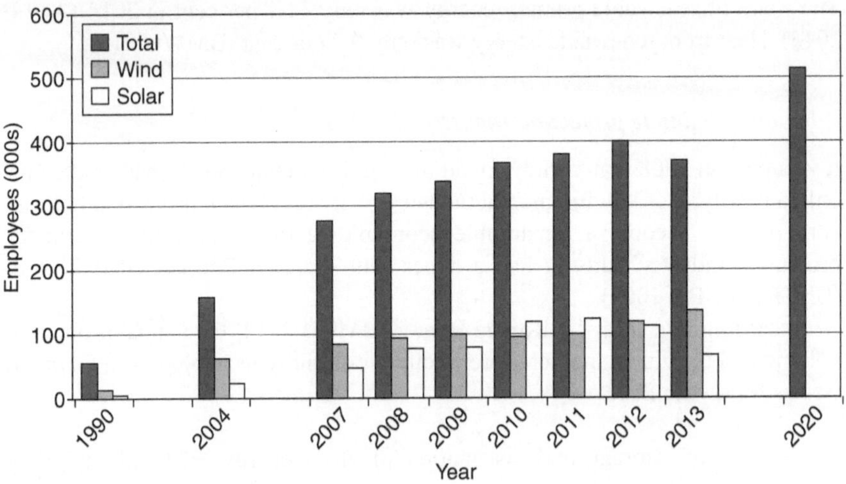

Figure 8.3 Employees in the renewable energy sector in Germany 1998–2013 and forecast 2020.

Source: BMUB, BMWi (2014).

Economic leadership: the fruits of ecological modernisation

Germany's ambitious, broadly accepted and fairly reliable climate targets have been an essential driving force for innovation. Their implementation strongly supported the development of renewable-energy markets which stimulated the speeding up of the innovation cycle. The attractive feed-in tariffs for renewable energy caused a dramatic increase in new patents for this technology. The decision to phase out nuclear power increased the pressure for innovation and demands for renewable energy.

The well-directed government support for green lead-markets has had several interesting results: The policy-driven markets for renewable energy attracted foreign investors. They also provided the conditions for successful export of climate-friendly technologies, particularly in the 2010s. Supporting the domestic markets was at the same time indispensable for paying the learning costs for better and cheaper technologies, which then could be diffused more generally, including to less developed countries.

The German climate policy has been successful as a technology-based innovation strategy; the concept of ecological modernisation (later ecological industrial policy) was a formula seeking a consensus with the modernising part of the industry after 1998. The technological pioneer role of Germany – and the EU – caused a competitive push for other industrial countries. Climate policy as a technology-based policy has become a field of regulatory and economic competition. The German (and European) economic leadership therefore may to a certain degree also result from the competitive dynamics.

Conditions of political leadership

The most important factors which help to explain German leadership in international climate change politics will now be discussed. However, a temporal condition (or *situative Bedingung*) should be mentioned beforehand. German unification was accompanied by a general modernisation drive and the rapid decline of heavy industry and lignite-dependent power production in the former East Germany. This temporal condition explains the early improvements in the 'new states' (*neue Bundesländer*) in the former East Germany. Germany's very ambitious CO_2 emission reduction targets therefore benefitted at least partly from so-called 'wall fall profits'.

In general, however, government intervention was the driving force behind the considerable GHGE reductions which Germany achieved. Its far-reaching targets may be explained by several factors. The first is a cross-party consensus based (paradoxically) on rather contradictory motives: both nuclear energy interests represented by the CSU (prior to the decision to phase out nuclear power) and the environmental concerns of the Greens can be seen as driving forces. A second factor is the German political party system, which was strongly influenced by a Green Party that entered the Bundestag in 1983, participated in two Red–Green coalition governments (1998–2005) and achieved a strong position at the state level after Fukushima. The Green Party cooperated early with the other opposition party, the Social Democrats, with ecological modernisation becoming a common project. The German Greens have not only become one of the strongest but also one of the most modern realist Green Parties in international comparison (after a fundamentalist start in the early 1980s). Coalitions even with the CDU at the state level (Hamburg, Hessen) followed in the early 2010s during which Baden-Württemberg, which is one of Germany's most highly industrialised states (*Länder*), was governed by a Green Prime Minister (*Ministerpräsident*). Largely due his popularity the Green Party became for the first time the strongest party in a state (*Land*) election in the elections in 2016.

The 1987 *Bundestag* Enquete Commission on Climate Protection created the knowledge base both for the core elite and the interested public. The Enquete Commission provided not only the necessary information about the possible long-term problems caused by climate change, but also pointed out available policy options to prevent or at least mitigate climate change. It also stipulated clear targets for the national, EU and international levels.

Institutional explanatory factors include support from Chancellors (Kohl, Schröder and Merkel) and also the role of 'strong' Environmental Ministers Töpfer, Merkel, Trittin and Gabriel, who pushed for ambitious climate change policy measures on the national, EU and international level. An important precondition for the adoption and successful implementation of ambitious climate change policy measures was the setting up of the Inter-ministerial Working Group (*Interministerielle Arbeitsgruppe* – IMA) CO_2 reduction in 1990. This constituted the first important step in the direction of integrating climate policy considerations into other policies within a domestic political system that is otherwise characterised by a high level of ministerial independence. Six different Ministries participated in the formulation of the 2007 Integrated Energy and Climate Programme.

There was also a hidden alliance for ecological modernisation (Jänicke 2006). It included not only environmental NGOs and government institutions but also parts of industry. Industrial branch organisations like the German Engineering Federation (*Verband Deutscher Maschinen- und Anlagenbau – VDMA*), the German electronics industry association (BITCOM) and parts of the construction industry became winners (or potential winners) out of the implementation of climate policy and supported a more ambitious approach. This alliance proved especially strong when the renewable energy regulation had to be defended against opponents (Jänicke 2010).

Multi-level governance

Germany has a long tradition of using both the EU and institutions at the global level for promoting domestic environmental policy innovation. Two motives are particularly salient. The international diffusion of German technology-based regulations (such as the Renewable Energy Act) can support domestic firms. A second possible motive for using the international level by showing entrepreneurial leadership is to strengthen the domestic position for more ambitious climate change policies. Of course, it is also generally attractive for governments to play a pioneer role in the global policy arena.

This approach to multi-level governance is especially important in the EU context (Knill and Liefferink 2007; Jordan *et al.* 2012; Jänicke *et al.* 2015). The EU provided not only the policy arena for 'uploading' more ambitious German targets. It also prevented domestic backlashes against ambitious climate change policy measures. A kind of 'enforced leadership' can be observed in this context. The Commission's rejection of the second German NAP (under the EU ETS) and the modification of German pro-industry positions in the EU's 2008 climate change and energy package constitute good examples. In both cases the German pioneer role was secured only by the EU.

Since 2013 it was again Germany's part to strengthen the role of the EU in climate policy. This can be explained in the context of the *Energiewende* and the change in coalition government. One important additional reason for the new ambition is, however, the influence of multi-level climate governance *within* Germany. After Fukushima the Green Party became a coalition partner in the majority of German states (*Länder*). Climate policy became a relevant policy at this level. However, the local level developed the highest speed of change at this time. The '100%-Renewable-Energy' network, which is a kind of grassroots movement of local communities and counties (mainly based on climate regulation of the national and the European level), exhibited a remarkable speed of diffusion. The number of such regions doubled between 2010 and 2014 and the population of the participating local communities increased from 7.8 to 25 million. This movement was to a high degree based on energy cooperatives and private investors.

Contradictions

The German success story is not without contradictions and anomalies. The main restrictive factors for more wide-reaching climate policy achievements were coal-based power companies and the car industry. Paradoxically it was under the climate change-conscious Red–Green coalition government that the energy sector steadily increased its CO_2 emissions, by 32 Mt between 1999–2005 (*Umwelt* 1/2008). Concessions to the power industry may have been (at least for the Social Democrats) the political price the government had to pay for tacit support for its anti-nuclear policies. The German car industry was very successful with the introduction of fuel-efficient diesel engines – creating a successful lead market for this technology. However, at the same time the engine power of the car fleet increased by nearly 40 per cent between 1995 and 2007. This made it impossible to achieve the voluntary agreement (140 g CO_2 per km by 2008) which the European Automobile Manufacturers' Association (ACEA) had put forward (SRU 2008). After 1998 Germany also remained the only country in Europe without speed limits on motorways. Here the 'automobile chancellor' (*Autokanzler*) Schröder was the main veto player. The Grand coalition (2005–9) did not introduce speed limits either, despite the fact that an oil price explosion had taken place. The Volkswagen diesel emissions scandal, which was caused by the exposure of systematic false declarations of emission performance by the American Environmental Protection Agency (EPA), was arguably only possible under conditions of partly 'captured' public institutions (Baxter 2012). It is therefore no surprise that a critical evaluation report of McKinsey (2015) expects the non-achievement of Germany's ambitious GHGE reduction target (minus 40 per cent) for 2020.

The most important contradiction of the German case is, however, the fact that the proclamation of the *Energiewende* in 2011 took place under a Conservative–Liberal coalition government, which earlier had abolished the Red–Green government's phasing-out of nuclear policy only to restart the same anti-nuclear policy one year later. The main explanation for this quasi-revolutionary change was political leadership – and in particular entrepreneurial leadership – and a clearly transformational policy style of chancellor Merkel.

Conclusion

The German leadership role in climate policy is especially visible in the over-achievement of the ambitious Kyoto Protocol second commitment period (2008–12) target for GHGE and the rapid domestic growth in renewable energy. This was essentially *leadership by example*. However, Germany has exercised all three types of leadership dimensions: It started with *cognitive* leadership with a knowledge base provided by the Climate Enquete Commission (1987–90). In the following years strong *entrepreneurial* (political) leadership could be observed, for instance at the G8 summit in Heiligendamm (2007). Organizing global coalitions (e.g. for IRENA) constitutes another example. *Structural* (economic) leadership was used for the creation of lead markets for clean energy technologies and for exerting competitive pressure on other countries after 2000. Germany has also proved that ambitious climate policy is not only feasible but also can be a clear economic advantage. The acceleration of the innovation process may have been the most remarkable effect of this policy: It enlarged not only the technical potential for climate mitigation (lower costs for and higher effectiveness of, for example, clean energy technologies) but also the political acceptance of an ambitious climate strategy. The leadership style under the Red–Green coalition and partly also in the immediate years which followed (1998–2007) was strategic and *transformational*. The same holds true for the energy transition (*Energiewende*) after the 2011 Fukushima catastrophe.

The main national conditions of climate change policy success have been: (1) the early availability of a comprehensive knowledge base for the political elite and the general public; (2) the cross-party consensus about the importance of climate protection which was strongly influenced by (3) the Green party together with the (4) fast growing new climate industry and a hidden alliance for ecological modernisation. Finally, (5) multi-level governance was an additional supporting factor. It occurred as both Europeanization of some German policy innovations and as intervention by the EU Commission, for instance when it forced Germany to live up to its self-adopted climate policy leadership ambitions. Multi-level governance at the sub-national level in Germany has become increasingly important. After Fukushima (2011) there was a Green state Prime Minister (Baden-Württemberg) and the Green Party was coalition partner in most state (*Länder*) governments. At the same time a rapid growth of '100%-Renewable-Energy Networks' took place at the local level, stabilising the political and economic basis of national climate policy.

There are, however, also challenges and contradictions in Germany's leadership role. The greatest challenge was the management of problems arising from the high speed of systemic change of power supply and from conflicting interests. The resistance of citizens to wind power installation or high-voltage grids in their neighbourhood may be to a certain degree a normal side-effect of rapid change. The resistance caused by veto players (such as the coal-based power companies and the car industry) is by far more important. It is these industries which have been largely responsible for the potential non-achievement of the 40 per cent GHGE reduction target for 2020 (McKinsey 2015). The German climate policy has partly developed its own interest base, which causes positive political feedback and may stabilise the process in the long run. However, so far it was unable to achieve the necessary structural change altogether. The phasing out of lignite coal has not been part of the *Energiewende*.

References

AGEB (2015) *Energieverbrauch in Deutschland im Jahre 2014*, Berlin: Arbeitsgemeinschaft Energiebilanzen.

Baxter, L.G. (2012) 'Understanding Regulatory Capture', in: Pagliari, S. (ed.) *The Making of Good Financial Regulation*, London; Grosvenor House, 31–9.

BMU (2000) Nationales Klimaschutzprogramm, UMWELT 11/2000.

BMU (2008) *Die Dritte Industrielle Revolution – Aufbruch in ein ökologisches Jahrhundert*, Berlin: Bundesministerium für Umwelt, Naturschutz und Reaktorsicherheit.

BMU and UBA (2009) *Umweltwirtschaftsbericht 2009*, Berlin: Bundesministerium für Umwelt, Naturschutz und Reaktorsicherheit.

BMUB (2014) *GreenTech made in Germany*. Berlin: Bundesministerium für Umwelt, Naturschutz, Bau und Reaktorsicherheit.

BMWi (2015) *Erneuerbare Energien im Jahre 2014*, Berlin: Bundesministerium für Wirtschaft und Energie.

DENEFF (2014) *Branchenmonitor Energieeffizienz 2014*, Berlin: Deutsche Unternehmensinitiative Energieeffizienz.

Deutscher Bundestag (1980) *Zukünftige Kernenergie. Bericht der Enquete-Kommission des Deutschen Bundestages*, Bonn: Deutscher Bundestag.

Deutscher Bundestag (1990) *Schutz der Erde. Eine Bestandsaufnahme mit Vorschlägen zu einer neuen Energiepolitik*, Bonn: Deutscher Bundestag.

DIHK (2005) *Für einen Strategiewechsel in der Umweltpolitik*, Berlin/Brüssel: Deutscher Industrie- und Handelskammertag.

Ernst & Young (2006) *Eco-industry, its size, employment, perspectives and barriers to growth in an enlarged EU*, Brussels: EU Commission.

Gabriel, S. (2006) *Ökologische Industriepolitik*, Berlin: Bundesministerium für Umwelt, Naturschutz und Reaktorsicherheit.

Grassl, H. (1999) *Wetterwende – Vision: Globaler Klimaschutz*, Frankfurt: Campus Verlag.

Jänicke, M. (2006) 'Trend Setters in Environmental Policy: The Character and Role of Pioneer Countries', *European Environment*, 15(2): 129–42.

Jänicke, M. (2010) 'Die Umweltpolitik der Großen Koalition', in: Egle, C., Nullmeier, F. and Zohlnhoefer, R. (eds), *Bilanz der Großen Koalition*, Wiesbaden: VS Verlag.

Jänicke, M. (2012) *Megatrend Umweltinnovation* (2nd edn), München: Oekom.

Jänicke, M. and Zieschank, R. (2008) *Structure and Function of the Environmental Industry – The Hidden Contribution to Sustainable Growth in Europe*, FFU-report 01–2008, Berlin: Environmental Policy Research Centre.

Jänicke, M., Schreurs, M. and Töpfer, K. (2015) *The Potential of Multi-Level Global Climate Governance, Policy Brief*, Potsdam: Institute for Advanced Sustainability Studies.

Jordan, A., v. Asselt, H., Berkhout, F., Huitema, D. and Rayner, T. (2012), 'Understanding the Paradoxes of Multi-Level Governing', *Global Environmental Politics*, 12(2): 43–66.

Knill, C. and Liefferink, D. (2007) *Environmental Politics in the European Union*, Manchester: Manchester University Press.

Markham, W.T. (2008) *Environmental Organisations in Modern Germany*, Oxford: Berghahn Books.

McKinsey (2005) *Energiewende-Index* (I/2015), www.mckinsey.de/energiewendeindex (accessed on 12 January 2016).

Mez, L. (2013) 'Umweltschutzverbände', in: Andersen, U. and Woyke, W. (eds), *Handwörterbuch des politischen Systems der Bundesrepublik* (7th edn) Berlin: Springer.

Michaelowa, A. (2008) 'German Climate Policy Between Global Leadership and Muddling Through', in: Compston, H. and Baily, I. (eds), *Turning Down the Heat. The Politics of Climate Policy in Affluent Democracies*, Basingstoke: Palgrave/Macmillan.

Oberthür, S. and Pallemaerts, M., with Kelly, C.R. (eds) (2010) *The New Climate Policies of the European Union*, Brussels: Vrije Universiteit Brussel.

OECD (2007) *Environmental Innovation and Global Markets*, ENV/EPOC/GSP(2007)2/REVI, OECD: Organisation for Economic Co-operation and Development.
OECD (2012) *OECD Environmental Performance Reviews: Germany 2012*, Paris: Organisation for Economic Co-operation and Development.
REN21 (2008) *Renewables 2007 – Global Status Report*, www.ren21.net (accessed on 12 January 2016).
REN21 (2014) *Renewables 2014 – Global Status Report*, www.ren21.net (accessed on 12 January 2016).
Rose, R. (1993) *Lesson-Drawing in Public Policy*, Chatham N.J: Chatham House.
Schreurs, M.A. (2002) *Environmental Politics in Japan, Germany and the United States*, Cambridge: Cambridge University Press.
SPD-BÜNDNIS 90/DIE GRÜNEN (1998) *Aufbruch und Erneuerung – Deutschlands Weg ins 21. Jahrhundert. Koalitionsvereinbarung 20. Oktober 1998.*
SRU (2008) *Umweltgutachten 2008 – Umweltschutz im Zeichen des Klimawandels*, Sachverständigenrat für Umweltfragen, Berlin: Erich Schmidt Verlag.
Statistisches Bundesamt (2014) *Statistisches Jahrbuch 2014 für die Bundesrepublik Deutschland*, Wiesbaden: Statistisches Bundesamt.
Statistisches Bundesamt (2016) *Bevölkerung*, Wiesbaden: Statistisches Bundesamt, https://www.destatis.de/EN/FactsFigures/SocietyState/Population/Population.html (accessed 25 February 2016).
Umweltbundesamt (2015) *Treibhausgasemissionen in Deutschland*, Dessau: Umweltbundesamt.
Wurzel, R.K.W. (2008) *The Politics of Emission Trading in Britain and Germany*, London: Anglo-German Foundation for the Study of Industrial Society.

9 The Netherlands

A case of fading leadership

Duncan Liefferink, Daan Boezeman and Heleen de Coninck

Introduction

This chapter analyses the relationship between the development of domestic climate policy in the Netherlands and the Dutch efforts in this field in the EU and international arena since the 1980s. Traditionally, the Netherlands has enjoyed a reputation as an environmental and climate leader, based on setting ambitious goals, experimenting with new policy concepts and actively pushing others to follow suit. In this chapter we will argue that the Dutch climate leadership has largely faded. We will discuss to what extent problems in achieving domestic climate targets have affected the Netherlands' declared ambition to act as an international leader in this particular area and identify factors that are important for understanding the current Dutch position.

We will first outline the national context of Dutch climate policy, followed by an analysis of the evolution of the policy field and the most important policy instruments and actions. After that, the question of Dutch leadership is addressed before a conclusion is put forward.

National context of climate policies

The Netherlands is a highly industrialised country with a population of 16.9 million, and a population density of 502 people per km^2 in 2015. In 2014, total emissions of carbon dioxide (CO_2) and the five other major greenhouse gases amounted to 187 Mt CO_2 equivalents. Most of this, 158 Mt, was accounted for by CO_2 (CBS, PBL and Wageningen UR 2015).

CO_2 emissions per capita in the Netherlands are among the highest in the EU, roughly 10 per cent above Belgium, Finland and Germany and twice as high as France or Sweden. Emissions related to GDP are more in line with neighbouring countries and EU averages but still significantly higher than in, for example, Denmark, Switzerland, France or Sweden (International Energy Agency, 2014). It was not until 2005 that total greenhouse gas emissions (GHGE) reached a level below that of 1990, and this was due to the reduction of non-CO_2 greenhouse gases rather than to the slight decoupling of CO_2 emissions and economic growth that could be observed in the same period (CBS, PBL and Wageningen UR 2015).

At first sight this is surprising, considering the large natural gas reserves in the northern part of the country and the Dutch parts of the Wadden Sea and North Sea. Natural gas, a low-carbon fossil fuel, accounts for almost 50 per cent of domestic energy use. This natural advantage is partly outdone, however, by the presence of relatively energy-intensive industries in the Netherlands, such as a number of big refineries and chemical plants and massive greenhouse horticulture. A second factor is the virtual absence of nuclear and renewable energy. Only a few per cent of Dutch electricity demand is covered by a single 450 MW nuclear plant. Renewables play a modest role with a contribution to the total Dutch energy supply of 2.7 per cent in 2006, growing to 5.6 per cent in 2014 (CBS, 2015).

Historical development of Dutch climate change policy

In the late 1980s, when the issue of climate change first entered the political agenda, environmental politics ranked high in the Netherlands. Aiming at a stabilisation of CO_2 emissions in 2000, the 1989 National Environmental Policy Plan (NEPP) (VROM 1989) set the first nation-wide CO_2 target in the world (cf. Rowlands 1995: 77). One year later, ambitions were even further raised to a reduction target of 3 to 5 per cent by 2000 (VROM 1990), with an additional long term goal of minus 60 per cent over the next 100 years. To fulfil this mission, the government opted for the introduction of an energy/CO_2 tax.

Its national schemes placed the Netherlands in the league of the most ambitious EU-countries at the time, and its international intentions surely did not lag behind. The Netherlands actively contributed to setting the agenda and building up political pressure in the run-up to the UNFCCC (United Nations Framework Convention on Climate Change), signed in Rio de Janeiro in 1992. It saw no fault in going for a 'unilateral' EU-wide tax on CO_2, rejecting a US equivalent as a precondition for its introduction (N.N. 1992).

In the early 1990s, climate change policy, like other areas of environmental policy making, became part of an effort to move responsibilities from government to private actors. Anticipating the turn away from traditional top-down regulation which spread over the EU in later years, the Dutch government was looking for alternative measures, based either on the market mechanism or on negotiation and consensus (VROM 1993).

A prime example of this shift was the introduction of 'covenants', i.e. negotiated, not strictly legally binding agreements between the government and various industry sectors, aimed at improving the latter's energy efficiency (see below). In the same vein, the Netherlands was among the first to investigate emissions trading (VROM 1993: 183), to explore (what is now called) carbon dioxide capture and storage (CCS) and to consider fulfilling part of its reduction obligations by stimulating and financing projects abroad (VROM 1993: 74ff., 133, 220).

Despite considerable support from the Netherlands and a number of other member states, the EU-wide CO_2 tax never materialised. In order to avoid 'going it alone' – with its potential negative impact on economic competitiveness – as long as possible, the Dutch made a final attempt in 1995/1996 to revitalise the issue at the EU level by convening a meeting of eight like-minded countries in The Hague, exploring the possibility of co-ordinating national CO_2 taxes outside the formal EU framework. This 'club', however, never took off either. Eventually, a purely national tax scheme was established, but only for small consumers and at a merely symbolic level.

This course of events reflects the overall development of climate change policy in the Netherlands in the mid-1990s. A re-orientation towards the economy and employment led to the general energy reduction goal for 2000 being lowered from 20 to 17 per cent and subsidies for energy reduction and renewable energy being cut back. The Netherlands also readjusted its vision on its role within the EU. At the international level, the Netherlands put increasing emphasis on 'the national interest' and 'the competitive position of Dutch industry' (VROM 1998). Making an active contribution to international efforts, and the intention to take, if needed, unilateral action for demonstration effects, gave way to making multi-lateral commitment a precondition for national activities. This marked a shift away from an active and constructive leadership ambition to a more conditional type of leadership.

The tensions that this caused on the ground were reflected in the role played by the Netherlands in the context of the Kyoto conference. Holding the EU Council Presidency in the first half of 1997, the Netherlands ensured an EU-wide agreement on a common negotiating position for Kyoto by putting forward the so-called 'triptych' approach (Phylipsen *et al.* 1998). This approach provided a method for sharing the 'burden' of emission reductions among the member states in a 'scientific' or at least criteria-based manner. It formed the basis for the EU's opening offer of reducing GHGE by 15 per cent in 2008–12 (relative to 1990) for the EU as a whole, provided that other industrialised countries would commit to comparable reductions. For the Netherlands, a target of minus 10 per cent was envisaged. This strategy was to no avail, however: the delegation returned home from Kyoto with a target of only minus 8 per cent for the EU, in the face of the low commitments of other parties including in particular the US (see Chapter 16). What subsequently happened in Brussels was guided less by the triptych approach than by the basic principles of political horse-trading. Given the decreasing domestic enthusiasm for taking substantive reduction measures, the Netherlands joined the ranks of those countries that wanted to keep their national obligations as low as possible (Van den Biggelaar and Wams 1998) and eventually left the arena with a national target of minus 6 per cent.

Nevertheless, the Netherlands was still struggling with how this target could possibly be reached. The covenants with industry and measures in the building sector had enhanced energy efficiency, but not reduced absolute emissions. For transport, hardly any policy measures were in place. Renewable energy in the power sector showed slow growth rates.

The Netherlands consequently first turned towards other means to achieve its targets, notably by reducing emissions abroad. The Netherlands' initial intention to reduce more than half of its emission reductions through Joint Implementation (JI) projects in Eastern European countries and the Clean Development

Mechanism (CDM) in developing countries, however, was received with scepticism by other EU member states (N.N. 1999).

The Sixth UNFCCC Conference of Parties (COP-6) in The Hague in November 2000 and the active role of the Dutch president of COP-6bis in Bonn several months later at first seemed to pave the way for a reinvigoration of climate change policy. However, the 2001 Fourth NEPP (VROM 2001) proved differently. As before, it considered unilateral action off-limits. Policies for industry kept relying heavily on negotiated agreements. The Fourth NEPP first introduced the idea of an energy 'transition' (*Energietransitie*), i.e. a radical shift towards a more sustainable energy system. From the beginning, however, this trajectory was criticised for its strong technological bias, its continued (and therefore contradictory) dependence on the fossil fuel industry, and the prevalence of the short-term efficiency goals of the energy liberalisation agenda over long term sustainability objectives (Kern and Howlett 2009).

The 2000s developed into a decade of political turmoil in Dutch politics, fed by the rise of several populist parties de-emphasising environmental themes. Cabinet periods were short and policy more volatile, adding to the lack of coherence and stability of policies. In 2002 the right-wing coalition government substituted the office of a Minister of the Environment by that of a State Secretary, or junior minister. The first victim of the renewed focus on reducing government intervention was road pricing, which had been discussed and investigated for a long time, but was not introduced after all. In its 2004 report, the Netherlands' Environmental Assessment Agency concluded that, of all budget cuts in the field of the environment, most were undertaken in the domain of climate change policy (Milieu- en Natuurplanbureau 2004: 31).

The surge in climate attention following Al Gore's *Inconvenient Truth* and the Stern Report temporarily turned the tide. In 2007, the new Christian Democratic/Social-Democratic government reinstalled a Minister for Environment. A new climate programme aimed at an ambitious 30 per cent reduction in GHGE by 2020 (base year: 1990), a 20 per cent share of renewable energy by the same year, and an annual energy saving rate of 2 per cent (VROM 2007). These objectives were to be bolstered both by the provision of financial resources and by new policy measures, such as – once again – road pricing and an additional fiscal greening package including an intensification of the emissions-based vehicle tax, an aviation tax and a waste package tax (the latter two were later abolished). For the first time since the early 1990s, the Netherlands decided on a programme with long-term goals even before the European Commission had presented its proposals on future effort sharing. The Netherlands supported the 2008 EU Climate and Energy Package, yet pushed for consideration of competitiveness and the issue of carbon leakage in the EU 2020 package. It was sceptical of the Renewable Energy Directive, but advocated its sustainability criteria for biofuels (Gulbrandsen and Skjaerseth 2016).

Equally notable was the acknowledgement, at least on paper, that an active role within the EU can only be played if substantial steps are taken at home (VROM 2007: 53). Action at the international level was also agreed: the Minister for International Cooperation could spend €500 million on sustainable energy access in developing countries, over and above the 0.8 per cent Official Development Assistance that the Netherlands was already providing. In the absence of clear goals on energy in the Millennium Development Goals, the Netherlands was the first country explicitly committing itself to access to sustainable energy as a prerequisite for sustainable development of the world's poor.

However, the new episode of ambition was short-lived. Following the controversy over hacked emails from the University of East Anglia's Climate Research Unit in 2009 ('Climategate'), both the Dutch populist and conservative parties became increasingly climate sceptic (see also Chapter 12). The centre-right right minority cabinet taking office in 2010 lowered its ambitions to the EU goals of 20 per cent reduction in GHGE compared to 1990 and 14 per cent share of renewable energy by 2020 (Ministerie van Infrastructuur en Milieu 2011). The EU goal of 20 per cent energy saving, part of the EU's 2020 climate and energy package, was formally retained, but hardly elaborated. The subsequent Liberal/Social Democratic cabinet, in power from 2012, announced that a fully sustainable energy system should be in place by 2050 and raised the renewable energy ambition from 14 to 16 per cent by 2020, but reconfirmed its commitment to an EU level-playing field. Rather than acting as a leader in the context of the EU, it explicitly subscribed to the policy principle of 'no domestic mark-ups on European legislation' (Hoogervorst and Dietz 2015: 25).

In the 2000s, the frustration grew in the environmental movement and in industrial circles about the capricious energy policies of the Netherlands. In 2011, it resulted in an initiative by the Social-Economic Council (SER), an influential discussion body between societal organisations and the private sector, to forge a broadly supported *Energy Agreement for Sustainable Growth* (EA). The negotiation process was concluded in 2013 and the EA was considered to constitute the balance of a wide range of actors and interests, from heavy industry associations to Greenpeace. It set out the general direction of Dutch energy policy until 2020, including a renewable energy target of 16 per cent in 2023, i.e. postponing the 16 per cent target set in the 2012 government agreement for 2020 but retaining the EU target of 14 per cent for 2020. The long list of measures included closure of five older coal-fired power plants and a plan to speed up the construction of onshore and offshore wind parks. However, the EA simultaneously also abolished a recently introduced coal tax and aimed at protecting heavy industry against potentially increasing CO_2 prices in view of international competitiveness. Referring to the broad support by industry and societal organisations, the EA was hailed as a great achievement and hardly criticised. However, independent evaluations by the Energie Centrum Nederland/Netherlands Environmental Assessment Agency (Schoots and Hammingh 2015) and the Court of Audit (Algemene Rekenkamer 2015) concluded that the measures included in the agreement would not be sufficient to realise its goals. The strength of the EA – the broad support and attention for implementation – was also its greatest weakness: it ended up with the lowest common denominator and became, although a useful and arguably necessary implementation step, void of leadership.

The EA reflects a long-term struggle between two ministries with conflicting goals: the Ministry of Environment, responsible for climate change matters, and the Ministry of Economic Affairs, responsible for energy and industry (Köper, 2012; Duyvendak, 2011). According to a study by Notenboom *et al.* (2012), the drivers behind Dutch national plans for decarbonisation pathways are, in decreasing order: affordability, industrial opportunities, greenhouse gas emissions and security of supply. The Ministry of Economic Affairs is very much aligned with the interests of the gas sector in the Netherlands, in all its incarnations including fossil fuel companies such as Shell and NAM, energy-intensive industry, and other sectors depending on oil and natural gas, such as transportation and horticulture. A sizeable chunk of Dutch employment and the government's budget relies on fossil energy. The influence of the fossil fuel sector can arguably be recognised in the EA.

Shortly after the EA, the cabinet presented the *Climate Agenda*. It reflected the rapidly increasing importance of climate adaptation. In this field, the government saw a potential for leadership and exporting Dutch water technology (Ministerie van Infrastructuur en Milieu 2013: 28–29), an important argument for pushing the EU's Climate Adaptation Strategy. In the context of EU negotiations towards its 2030 Climate and Energy Package, the Netherlands backed the more ambitious EU-wide GHGE target, but preferred to combine it with EU ETS only. Specifying targets for renewable energy and energy saving at the member state level was seen as reducing the cost efficiency of the ETS and as mingling with matters perceived as 'national business'. Together with the UK, among others, the Dutch actually played an important role in removing targets for renewables and energy saving from the 2030 Climate and Energy Package (cf. Gulbrandsen and Skjaerseth, 2016).

In 2012, unhappy with the lack of both the domestic and international ambition of Dutch climate policy, the activist organization Urgenda initiated a lawsuit against the Dutch State. On 24 June 2015, the District Court of The Hague decided that the government had to increase its efforts to ensure a GHGE reduction of at least 25 per cent by 2020, relative to 1990. This was necessary, according to the Court, to limit the global increase of temperature to the UNFCCC-agreed maximum of 2°C. The Court based its judgement on IPCC-based evidence combined with the state's general duty of care for its citizens. The verdict sparked a lively debate, not only in relation to climate policy itself, but also in view of its more principal implications for the *trias politica*. In September 2015 the government announced an appeal.

In the run-up to the Paris COP in December 2015, the idea of a Climate Act, which had been discussed in the Dutch Parliament before but foundered on party political log-rolling, was revived. In addition, a parliamentary majority proposed to close down *all* coal-fired electricity plants in the Netherlands between 2020 and 2030.

Policy instruments and actions

In the 1980s, and in line with the country's neo-corporatist tradition, Dutch environmental policy shifted its focus from direct 'command-and-control' regulation to negotiation and consensus between the state and polluting sectors (Liefferink 1997; Liefferink and Mol 1998). In climate policy, this led to voluntary or negotiated agreements (also known as covenants) and strengthening of market-based instruments. Nevertheless, the domestic use of different instruments has been volatile and inconsistent (cf. Kern and Howlett 2009).

Long-term agreements

From 1991 onwards, the Ministry of Economic Affairs negotiated so-called Long-Term Agreements (LTAs) – sometimes also referred to as voluntary agreements – for improving energy efficiency with all major industrial branches. From 2005, the larger installations were included in the EU ETS, while the remaining companies continued agreeing on further LTAs.

Industry so far met the 45 per cent efficiency improvements included in the LTAs between 1998 and 2020 (SenterNovem 2008) at very limited and often negative costs. In 2000, however, the Central Planning Bureau argued that LTAs accounted for only 30 per cent of the energy efficiency improvements (Centraal Planbureau 2000: 13, 90). Moreover, improvements in energy efficiency were outdone by the growth in production volume and a slight structural shift from low to heavy energy consuming industries (Enevoldsen 2005: 170–174).

Comparing the smooth LTA process with the difficult negotiation of the Energy Agreement suggests that non-costly coordination problems are considerably easier to agree on through a consensual process than collective action entailing considerable effort, investment and, for some, loss of market share.

Energy taxation

Compared to other EU member states, the Netherlands raises a relative high percentage of its tax income, 10 per cent, with green taxes (Vollebergh 2014: 5). Within the EU, the Netherlands was able to flaunt the unilateral introduction of an energy tax after fruitless efforts to raise a similar tax at the EU level. The Regulatory Energy Tax (*Regulerende Energie Belasting*, REB) was imposed in January 1996 and applies to the consumption of gas and electricity. In view of competitiveness considerations, forcefully brought to bear by Dutch business, the tax is primarily directed to households. Large users (i.e. agriculture and especially industries) are largely exempted. In addition, subsidy schemes exist for compensating internationally competing large-scale electricity users. While the actual effect of the tax on behaviour, even for households, has been called into question (cf. Joosen *et al.* 2004: xiii), overall energy taxes are now quickly rising due to an additional levy, charged since 2013, to finance a growing subsidy scheme for sustainable energy (see below).

Flexible mechanisms: emissions trading, JI, and CDM

From very early on, the Netherlands showed much affinity with carbon emissions trading. In early 2000, when the European Commission proposed launching the EU ETS, the Dutch government had already installed a special committee investigating the introduction of an ETS on a national basis. Nevertheless, actual plans had hardly crystallised and could not sufficiently be pushed in Brussels to effectively act as an example for the EU scheme (Veenman and Liefferink 2005).

The first phase of the EU ETS, starting in 2005, included a few hundred of the Netherlands' large emitters. Nowadays, the ETS has taken over the role of the LTAs as the prime policy instrument stimulating CO_2 reductions in the domestic industry and energy sectors. Following the economic crisis, however, the carbon price plunged to only a few Euros per tonne of CO_2-equivalent. Proposals to remove allowances from the market led to protests from industry, which regarded the low allowance prices as an advantage in difficult economic times.

The Netherlands engaged in Activities Implemented Jointly (the predecessor of JI) already before the 1997 Kyoto Protocol was agreed. After Kyoto, JI and CDM were seen as important instruments to achieve the national reduction targets.

Carbon dioxide capture and storage (CCS)

Before most other Member States, and probably because of the low renewable energy potential in the Netherlands, Dutch governments have shown a special interest in carbon dioxide capture and storage (CCS). With the intention of becoming a leader in CCS development, the Dutch government invested in the demonstration of CCS and lobbied for it internationally. However, the mood turned rapidly sour on CCS after 2009. CCS inevitably involves additional cost, it thus always requires some kind of incentive. The failure to reach an ambitious agreement at the 2009 Copenhagen COP hurt the prospects of the technology and reduced the appetite of industry all over the world to invest significantly in CCS.

In the Netherlands, moreover, the key demonstration project at Barendrecht, where CO_2 from a refinery was going to be stored in a depleted gas field onshore under a highly populated neighbourhood, provoked vehement protests from citizens due to a naïve and clumsy engagement process (Brunsting *et al.* 2011). CCS thus became a political minefield which resulted in a de facto moratorium on onshore CCS in the Netherlands. At the time of writing, one CCS demonstration project – entailing offshore storage of CO_2 from a coal-fired power plant out of view of citizenry – was still alive and awaiting a go/no-go decision. Despite €150 million of subsidy from the Dutch government and €180 million from the European Commission, this project was not economically viable for Eon, the owner of the power plant, because of the low CO_2 prices in the ETS.

Renewable energy

Compared to other member states, renewable energy has not featured prominently on the Dutch climate policy agenda. With a 5.6 per cent share in primary energy supply in 2014, renewable energy in the Netherlands remains far below the EU28 average of 15 per cent as well as its national 2020 target of 14 per cent. Traditionally, this is related to the limited potential for renewable energy, in particular hydropower, but inconsistent policies have not made things better (Milieu- en Natuurplanbureau, 2007: 70). The main achievement of the environmental organisations in the EA was an agreed tenfold increase of wind power offshore, i.e. from around 300 MW in 2013 to 4,450 MW in 2023.

Reaching this target will require very considerable investments. So far, however, efforts by the authorities to foster the investment climate in the renewable energy sector have been rather fickle. Subsidy schemes have been introduced, abrogated and re-introduced within only a few years' time (see also Kern and Howlett 2009). In 2008, a new scheme, the *Stimuleringsregeling Duurzame Energieproductie* (SDE), was introduced to ensure investors more planning reliability. From 2011, the scheme excluded private persons while focusing more on efficiency. In 2016, it had a budget of approximately €8 billion (up from 3.5 billion in 2014). For small users (such as households) energy cooperatives and SMEs, various tax exemptions and 'net metering' regulations remained in place to stimulate renewable energy generation, but discussions about the flexibility and the scope of the regulations are ongoing.

Multi-level governance and Dutch leadership

For most of the history of climate change politics, the Netherlands has striven to acquire a leadership role in global and EU climate change policy. An international leadership role tends to be strengthened by a successful, or at least a forceful and consistent policy at the domestic level. As we have seen, however, domestic climate policy in the Netherlands can hardly be described in these terms. This section will explore the sometimes ambiguous links between the Dutch ambitions and efforts in climate policy at the domestic, the EU and the international level.

Already before climate change appeared on the political agendas worldwide, the Netherlands had built up a reputation as one of the environmentally progressive countries in the EU (Liefferink 1997). Climate change politics initially presented itself as a field where the Netherlands, as a geographically small state, might be able to play in the premier league with big states – by compensating its lack of structural leadership with fervent entrepreneurial and cognitive leadership.

Cases in point were the early nation-wide CO_2 target in the First NEPP, the active role of the Dutch in the run-up to the UNFCCC, the efforts to establish an EU-wide CO_2 tax, the introduction of the 'triptych approach' in the context of EU negotiations about sharing the Kyoto 'burden', and the pioneering role in developing and operationalising the mechanisms of JI and CDM. Dutch international efforts were based mainly on diplomatic skills, networking and coalition building – in short: entrepreneurial leadership – with the Netherlands often trying to find the common ground between 'extreme' positions. Complementarily, however, the Netherlands tried to persuade by expertise and good arguments. In this respect, the Netherlands' reliance on cognitive leadership needs to be emphasised. Scientific experts were both engaged by the Netherlands itself and seconded to international institutions such as UNFCCC and IPCC. Dutch diplomats and researchers played key roles in these institutions. For instance, two of four Executive Secretaries of the UNFCCC were Dutch – Joke Waller-Hunter (2002–2005) and Yvo de Boer (2006–2010).

The adage of 'active environmental diplomacy' based on consistent and credible domestic policies (e.g. VROM 1993: 53; 2007: 57) proved hard to fulfil in climate policy. Dutch per capita GHGE continue to be high in comparison to most other EU Member States and hardly more than stabilised since 1990. The share of renewables in the total Dutch energy production, although increasing in recent years thanks to the EA, remains very modest. In the face of this, domestic targets for emission reduction and renewables were repeatedly tempered or shifted to a later point in time. The continuous revision of targets and measures also led to considerable inconsistency in the long-term development of Dutch climate policy.

The fact that many Dutch efforts at the international level failed (e.g. the efforts to introduce a common EU-wide CO_2 tax and the proposal for an EU burden sharing agreement on the basis of the 'triptych' approach) can obviously not be blamed entirely on the lack of credible domestic policies to back up these efforts. Failing diplomatic efforts are part and parcel of long and highly complex international negotiation processes.

Much more interesting is what happened at home *after* these somewhat disappointing outcomes had been reached at the EU/international level. The Dutch government, quite irrespective of its party political composition, more than once used these disappointing outcomes as an excuse to water down domestic commitments. First, although a national energy/CO_2 tax was established after it had become clear that an EU-wide tax was definitively doomed to fail, this unilateral tax was hardly more than symbolic. Thus, despite having lost the battle in Brussels, the Netherlands could boast that it was a pioneer in this field. Second, one can hardly avoid the impression that the Dutch government was actually quite happy with the failure of the 15 per cent GHGE reduction scenario in Kyoto. Why else did it undertake such efforts to reduce to a minimum its share in the final EU burden? Finally, the initial Dutch intention of realising a considerable part of its Kyoto target through JI and CDM was seen by many other member states as a cheap escape from the obligations to which the Netherlands had committed itself and, more importantly, to which it had tried to push other countries in the first place. Several Nordic countries also announced the use of JI and CDM, but in a manner additional to their domestic emission reductions.

Already in the 1990s, these examples in combination with the ambivalent performance of Dutch domestic climate policies started to convey the impression of symbolic or 'cost-free' leadership (Huber 1997; Liefferink and Birkel 2011). By the mid-2010s, the question ought to be raised if the Netherlands still qualifies as a climate leader, even if only as a 'cost-free' leader.

As discussed above, the Netherlands subscribed to the latest EU GHGE targets although it tried to avert EU-wide – and even more so domestic – targets for renewable energy and energy saving for 2030. This appears to be mainly inspired by an ideological conviction of economic efficiency which also prominently features in the 2016 Energy Report by the Ministry of Economic Affairs.

Instead of pushing for tougher EU targets for 2030, the Netherlands preferred to rely solely on the domestic EA, signed in 2013, and to focus on its implementation. Its most conspicuous measures included the closure of five old coal-fired electricity plants – perhaps to be followed by the remaining ones over the next 15 years thanks to a sudden and unexpected parliamentary effort (see above) – and a tenfold increase of offshore wind power until 2023. However, these measures could only be agreed domestically by abolishing a coal tax and granting full compensation for emissions trading costs for energy-intensive industries which face international competition. The overall target for renewable energy was watered down to the EU minimum at a relatively modest 16 per cent by 2023. And although the EA, which was initially welcomed by all participants as a further great achievement of the Dutch consensus culture, ought to guarantee implementation of the existing targets, authoritative assessments (Schoots and Hammingh 2015; Algemene Rekenkamer 2015) question whether the measures, both in place and proposed together with the EA, are actually sufficient to reach the goals. These assessments criticise the lack of a long-term climate and energy vision for realising an energy transition. The Energy Report, which is supposed to outline a longer-term vision of the Dutch government on the energy transition and which is led by the Ministry of Economic Affairs (responsible for energy, industry and innovation), reiterates the CO_2-only target-setting, the measures in the EA, and adds an 'Energy dialogue' with societal partners that is supposed to work out feasible measures. The only aspect of climate policy in which the Netherlands may perhaps still be considered a pioneer is that of energy and development. At the Paris climate conference (COP21) the Netherlands announced a €50 million extension of its energy and development programme, although this is now part of the Official Development Assistance and not additional to it anymore.

As analysed above, the reasons behind the stagnation of Dutch climate policy can be found primarily in the dominance of the traditional fossil fuel sector, a firm belief in market efficiency which is derived from classic economic (policy) theory, and the change of the political tide in the 2000s. The dominance of fossil fuels is rooted in the country's strong reliance on its large gas reserves and its considerable heavy industry sector, much of it related to the Rotterdam harbour as a major oil and coal hub. It is mirrored politically in the persistently strong position of the Ministry of Economic Affairs vis-a-vis the Ministry of Environment and economically in the weakly developed sustainable energy sector, especially compared to countries like Germany and Denmark (see Chapters 8 and 6 by Jänicke and Andersen and Nielsen respectively). The rise of climate-sceptic populist parties and, partly related to that, a series of short-lived cabinets both legitimised continued reliance on fossil fuels and further contributed to the inconsistency of Dutch climate policies and the increasingly weak basis for playing an international leadership role.

Conclusion

This chapter has examined to what extent, how and why the Netherlands has acted as a leader in climate policy. The Netherlands initially pioneered in this area and formulated ambitious GHGE goals beyond EU targets. However, this has, except for a short period following 2007, faded over time. It has been argued that, generally speaking, domestic performance did not live up to domestic, EU and international ambitions and expectations.

Three general observations can be made in this regard. First, it appears that high ambitions and an active role at the international level may, at least for the foreseeable future, be quite unrelated to policy performance at home. In the case of Dutch climate policy, the international leadership role was based primarily on expertise combined with diplomatic efforts (i.e. cognitive and entrepreneurial leadership) respectively. In the absence of good domestic examples, 'leading by example' was not sought by the Netherlands, except perhaps in the early 1990s, when domestic circumstances of a strong fossil sector, limited renewable energy potentials and the rise of populism were not as prevalent as in the 2000s, and the hope of achieving domestic targets was still alive. Second, in the longer term, the modest performance of domestic climate policy undermined Dutch credibility as a climate leader and raised the suspicion of the Netherlands cultivating merely a symbolic or 'cost-free' leadership.

Finally, the chapter shows that where domestic policies fail, EU and international policies take over. With emissions trading largely replacing negotiated agreements with domestic industry, with the government retaining general targets for renewable energy and energy saving more or less against its will, and with a judge reminding the government of its international obligations, climate change probably ranks among the most Europeanised areas in Dutch policy.

Acknowledgement

This chapter partly builds upon earlier work by Kathrin Birkel. We would like to thank her for her generous permission to make use of it. Ton van Dril's (ECN) comments are gratefully acknowledged.

References

Algemene Rekenkamer (2015) *Stimulering duurzame energieproductie (SDE+) – haalbaarheid en betaalbaarheid van beleidsdoelen*, The Hague: Netherlands Court of Audit.

Brunsting, S., de Best-Waldhober, M., Feenstra, C.F.J. and Mikunda, T. (2011) 'Stakeholder participation practices and onshore CCS: lessons from the Dutch CCS case Barendrecht', *Energy Procedia*, 4: 6376–6383.

CBS, PBL, Wageningen UR (2015). *Emissies broeikasgassen, 1990–2014 (indicator 0165, versie 27, 29 september 2015)*, The Hague: CBS/The Hague and Bilthoven: Planbureau voor de Leefomgeving/Wageningen: WUR. www.compendiumvoordeleefomgeving.nl.

CBS (2015) *StatLine – Renewable Energy*, http://statline.cbs.nl/Statweb/publication/?DM=SLEN&PA=83109eng&D1=0-3&D2=0,2-5,20&D3=0&D4=23-24&LA=EN&VW=T (accessed 02 September 2015).

Centraal Planbureau (2000), *Naar een efficiënter milieubeleid. Een maatschappelijk-economische analyse van vier hardnekkige milieuproblemen*, The Hague: CPB.

Duyvendak, W. (2011) *Het groene optimisme*, Amsterdam: Bert Bakker.

Enevoldsen, M. (2005) *The theory of environmental agreements and taxes. CO2 policy performance in comparative perspective*, Cheltenham: Edward Elgar.

Gulbrandsen, L.H. and Skjaerseth, J.B. (2016). 'The Netherlands' in J.B. Skjærseth, P.O. Eikeland, L.H. Gulbrandsen and Jevnaker, T., *Linking EU climate and energy policies: decision-making, implementation and reform*, Cheltenham: Edward Elgar.

Hoogervorst, N. and Dietz. F. (2015) *Ambities in het Nederlandse milieubeleid toen en nu*, The Hague: WRR working paper nr 3.

Huber, M. (1997) 'Leadership in EU climate policy: innovative policy making in policy networks', in D. Liefferink and M.S. Andersen (eds) *The innovation of EU environmental policy*, Oslo/Copenhagen: Scandinavian University Press, 133–155.

International Energy Agency (2014) *Key World Energy Statistics 2014*, Paris: IEA.

Joosen, S., Harmelink, M. and Blok, K. (2004) *Evaluatie van het Klimaatbeleid in de gebouwde omgeving 1995–2000*, Utrecht: Ecofys.

Kern, F. and Howlett, M. (2009) 'Implementing transition management as policy reforms: a case study of the Dutch energy sector', *Policy Sciences*, 42 (4): 391–408.

Köper, N. (2012) *Verslaafd aan energie*, Amsterdam: Business Contact.

Liefferink, D. (1997) 'The Netherlands: a net exporter of environmental policy concepts', in M.S. Andersen and D. Liefferink (eds), *European environmental policy: the pioneers*, Manchester: Manchester University Press, 210–250.

Liefferink, D. and Birkel, K. (2011), 'The Netherlands: a case of "cost-free leadership" ', in R. Wurzel and J. Connelly (eds), *The European Union as a leader in climate change politics*, London: Routledge, 147–162.

Liefferink, D. and Mol, A. (1998) 'Voluntary agreements as a form of deregulation? The Dutch experience', in U. Collier (ed.), *Deregulation in the European Union: environmental perspectives*, London: Routledge, 181–197.

Milieu- en Natuurplanbureau (2004) *Milieubalans 2004*, Bilthoven: Milieu- en Natuurplanbureau.

Milieu- en Natuurplanbureau (2007) *Milieubalans 2007*, Bilthoven: Uitgeverij RIVM.

Ministerie van Infrastructuur en Milieu (2011) *Kabinetsaanpak Klimaatbeleid op weg naar 2020*, The Hague: Ministerie van I&M.

Ministerie van Infrastructuur en Milieu (2013) *Klimaatagenda*, The Hague: Ministerie van I&M.

Ministry of Economic Affairs (2016) *Energy report 2016*, The Hague: Ministerie van Economische Zaken.

N.N. (1992) 'Kabinet: ook energieheffing zonder de VS', *NRC-Handelsblad*, 18 May, p. 3.

N.N. (1999) 'Nederland dwarsboomt klimaatbeleid', Trouw, *18 May*, p. 16.

Notenboom, J., Boot, P., Koelemeijer, R. and Ros, J. (2012) *Climate and Energy Roadmaps towards 2050 in north-western Europe*, The Hague: PBL.

Phylipsen, G.J.M., Bode, J.W., Blok, K., Merkus, H. and Metz, B. (1998) 'A triptych sectoral approach to burden differentiation; GHG emissions in the European bubble'. *Energy Policy* 26 (12): 929–943.

Rowlands, I. (1995) *The politics of global atmospheric change*, Manchester: Manchester University Press.

Schoots, K. and Hammingh, P. (2015) *Nationale Energieverkenning 2015*, Petten: Energieonderzoek Centrum Nederland.

SenterNovem (2008) *MJA3: intensivering, verbreding en verlenging afspraken*, The Hague: SenterNovem, Factsheet no. 2MJAF0803.

Van den Biggelaar, A. and Wams, T. (1998) 'Streng milieubeleid is geen weggegooid geld', *Algemeen Dagblad*, 16 June, p. 11.

Veenman, S. and Liefferink, D. (2005) 'Different countries, different strategies. "Green" member states influencing EU climate policy', in F. Wijen, K. Zoeteman and J. Pieters (eds) *A handbook of globalization and environmental policy. National government interventions in a global arena*, Cheltenham: Edward Elgar, 519–544.

Vollebergh, H (2014) *Green tax reform: Energy tax challenges for the Netherlands*, The Hague: PBL.

VROM (Ministerie van Volkshuisvesting, Ruimtelijke Ordening en Milieubeheer) (1989) *Nationaal Milieubeleidsplan (NMP)*, The Hague: SDU.

VROM (Ministerie van Volkshuisvesting, Ruimtelijke Ordening en Milieubeheer) (1990) *Nationaal Milieubeleidsplan Plus (NMP Plus)*, The Hague: SDU.

VROM (Ministerie van Volkshuisvesting, Ruimtelijke Ordening en Milieubeheer) (1993) *Tweede Nationaal Milieubeleidsplan*, The Hague: SDU.

VROM (Ministerie van Volkshuisvesting, Ruimtelijke Ordening en Milieubeheer) (1998) *Nationaal Milieubeleidsplan 3*, The Hague: VROM.

VROM (Ministerie van Volkshuisvesting, Ruimtelijke Ordening en Milieubeheer) (2001) *Een wereld en een wil: Nationaal Milieubeleidsplan 4*, The Hague: VROM.

VROM (Ministerie van Volkshuisvesting, Ruimtelijke Ordening en Milieubeheer) (2007) *Nieuwe Energie voor het klimaat: werkprogramma schoon en zuinig*, Den Haag: VROM.

10 Poland's clash over energy and climate policy

Green economy or grey *status quo*?

Karolina Jankowska

Introduction

With regard to its attitude to (and implementation of) climate change policy measures aimed at reducing greenhouse gas emissions (GHGE) at the domestic, EU and international level, Poland has mostly played the role of either policy taker, policy shaper or veto player. This chapter deals with the complex relationship between Poland, the EU and the international community in the field of climate and energy policy. It argues that the most relevant factors for explaining the Polish role in EU and international climate change politics are the importance of the energy sector for Poland's emissions profile and the dominance of coal in Poland's energy mix to such an extent that it is an outlier both in Europe and globally (World Bank 2015: 33). Hard coal and lignite constitute 57 per cent of gross inland energy consumption (World Bank 2015: 33) and almost 90 per cent of electricity consumption. The combustion of fossil fuels causes over 75 per cent of GHGE (Szewrański 2012: 10). Overall Poland's share in GHGE is low and amounts to only 1 per cent (World Bank 2015: 30). However, the share in the total EU GHGE amounts to 8 per cent (Szewrański 2012: 10). Moreover, Poland's economy remains among the least carbon-efficient in the EU and its energy intensity is almost twice as high as the EU (World Bank 2015: 30; Szewrański 2012: 1). Another explanatory factor is the centralized energy sector structure with the Polish government owning the majority in the biggest energy and fossil fuels companies such as the Polish Energy Group (*Polska Grupa Energetyczna* – PGE) and Lotos Group.

National attitudes to climate change

In Poland, the late 1980s and early 1990s were a time of 'ecologic enthusiasm' linked to a broader enthusiasm about socio-economic transformation as well as to acute environmental threats that had resulted from the forced industrialization of the country during the communist period (Karaczun 2012: 88–89). This enthusiasm declined relatively quickly when economic reforms introduced free market principles, resulting in high unemployment and an increase in the gap between the rich and poor. As a consequence, public environmental awareness was very low by the end of the 1990s and remained low until at least the second half of the 2000s (Bokwa 2000: 121). However, a remarkable change in attitudes of Poles could be observed by the early 2010s.

According to a 2014 survey carried out for the Polish Ministry of Environment (TNS Polska 2014: 59) 86 per cent of Poles perceived climate change as a very important problem. Seventy-four per cent of Poles pleaded that Poland should reduce GHGE (TNS Polska 2014: 62). Only 9 per cent of respondents were against measures or policies to reduce GHGE (ibid.: 62). In 2013 only 5 per cent of respondents opposed these measures (ibid.: 64). Still, a change in attitudes can be observed, because in 2008 almost half of the respondents (48 per cent) stated that the intensive development of Poland's economy justifies the increase of Poland's CO_2 emission limits (Jankowska 2011: 163–164). It means that public awareness has risen compared to 2008 and Poles started to accept the necessity of making sacrifices for the environment.

In comparison to 2008 (see Jankowska 2011: 164), the attitudes of political parties and important stakeholders to climate change did not change. Among political parties represented in the lower house of the Polish Parliament (*Sejm*), climate change is still largely unimportant and hardly treated as a serious political matter, unless climate change policy appears in the discussion as a threat to the Polish economy. There are some exceptions regarding the development of renewable energy at the local level (see below).

Interestingly, awareness about climate change seems to be higher among the representatives of Polish municipalities. According to a study carried out by the University of Warsaw and the Norwegian Institute for Urban and Regional Research in 2014, more than 52 per cent of respondents noticed the severity of the intensity of extreme weather events in recent years while 50 per cent considered these events to have been caused by climate change (Swianiewicz 2014: 2). What is more, 69 per cent of municipal officers and politicians claim that they have recognized the human impact on climate change (ibid.).

Climate change as a threat or opportunity?

Until at least the late 1990s climate change had been perceived in Poland merely as an environmental threat. Since then climate change has started to be perceived as a threat to the economy. The policy developments at the EU and domestic levels since the adoption of the EU's energy and climate package in 2008 have even shown that Poland has started to perceive climate change policy as one of the main threats to its economic development (Karaczun 2012: 88; Bukowski and Śniegocki 2014; Adamczewski 2015), although there have also been signs of change (see below). The reason for perceiving climate change as a threat is not that it may cause severe damage to the environment or economy: the reason is that Poland as a member state of the EU is obliged to implement its climate change policy, which it perceives as generally economically disadvantageous, because it causes excessively high costs for the energy and other industries as well as for private households. The Institute for Sustainable Development and the Warsaw Institute for Economic Studies produced in 2012 a report entitled *Low-emission Poland 2050* (Adamczewski 2015) which showed that in the long-term perspective Poland could benefit from a strong climate policy (Adamczewski 2015). It has not had, however, any significant impact within the policy process, because it has not been in line with Poland's main goal in energy policy, namely to protect the domestic, coal-based industry.

During the negotiations of the EU energy and climate package in 2008, Polish authorities often declared that Poland was capable of meeting its climate change policy obligations (Jankowska 2011: 169). They claimed, however, that the EU measures leading to the implementation of these obligations should have been modified in a way that would not raise the cost of electricity on the Polish market (ibid.: 170). A few years later Poland was no more willing to accept the EU climate policy goals, which have become more ambitious. During the Environment Council the Polish Environment Minister, Andrzej Kraszewski, vetoed on his own the adoption of the conclusions of the Communication of the European Commission *Low Carbon Roadmap for Moving to a Low Carbon Economy* (Malko 2011: 288). In March 2012 the new Polish Environment Minister, Marcin Korolec, once again vetoed on his own the *Roadmap* in the Environment Council. He also stated that Poland would in future veto any binding targets for renewable energy for 2030 (Ancygier and Jankowska forthcoming). The reason was rather slow development of renewable energy in Poland based mainly on biomass co-firing. In June 2012, the Polish Minister for Economic Affairs, Waldemar Pawlak, vetoed on his own the European Commission's Communication *Energy Roadmap 2050* during the Energy Council. The Polish government criticized the above-mentioned Commission Communications for not having been preceded by an analysis of different measures or options for emission reductions as well as for their detrimental economic and social impacts on EU member states, private households and different industry sectors (Malko 2011: 286).

Importantly, parallel to the increasing opposition of the Polish government at the European level, green or low-carbon economy concepts and measures have been becoming increasingly popular, especially at the local level. For example, according to a Greenpeace survey conducted in 2013 the support for different sources of renewable energies amounted to 89 per cent (Greenpeace 2013). The Polish People's Party (*Polskie Stronnictwo Ludowe* – PSL), which formed part of the coalition government from 2007 to 2015, emphasized in its programme for the parliamentary election in 2011 the need to speed up the development of renewable energy at the local level (PSL 2011). In 2014 one of the Polish People's Party's Members of Parliament, Artur Bramora, proposed the so-called 'prosumer amendment' to the draft Polish Renewable Energy Law in order to facilitate the development of the local renewable energy sources. It consisted of feed-in tariff scheme for the smallest renewable energy sources up to 10 kW of installed capacity.[1] This amendment was widely supported by environmental NGOs and different citizens' groups as a measure to create a local green economy and democratize the energy system. It was adopted in 2015 by the majority in parliament against votes of the PSL's government coalition partner, the Civic Platform (*Platforma Obywatelska* – PO). It demonstrated the existing discrepancy about how climate change policy is perceived between the national government and society as well as between different political parties even within the same government coalition in Poland.

Phases of domestic climate change policy

Based on Karaczun's (2012: 88) classification, two main phases of Polish climate change policy can be distinguished: First, climate change policy was regarded merely as an environmental issue; second, climate change policy was seen merely as an economic issue and the link between climate change and energy policies has been established.

The first phase began in the late 1980s/early 1990s on the wave of 'ecologic enthusiasm' (see section 2). In 1991 the *Sejm* adopted the first *National Environmental Policy* (*Polityka Ekologiczna Państwa*) in the post-Communism period that included a GHGE reduction target in accordance with international agreements (Jankowska, 2011: 165; Karaczun, 2012: 88). It legitimized the Polish government to participate in the negotiation of the UN Framework Convention on Climate Change (UNFCC) signed a year later and ratified by Poland in 1994. At the time Poland had a very strong political will to contribute to and participate in policy changing decisions with long-term objectives and effects. What is more, the first *National Environmental Policy* was the first strategic environmental action programme in Central and Eastern Europe to treat environmental protection in an integrated manner (Bokwa 2007: 123). Thus, by adopting, implementing and referring to it at the international level Poland exhibited cognitive leadership.

In the mid-1990s social and economic problems led to a decreased interest in environmental issues and increased attention was instead put on economic growth (Karaczun 2012: 89). During that time the dominant understanding in the energy policy discourse in Poland was that economic growth needs to be accompanied by growing energy demand as well as growing GHGE (ibid.). This understanding, which is still dominant in Poland, led to the adoption of a very low Kyoto Protocol target by Poland which agreed to a 6 per cent reduction by 2012 in comparison to 1988 which was the last year before Poland's economy and GHGE rapidly declined due to the economic transformation process. By 1997, however, Poland had already reduced GHGE by about 25 per cent. The main reason for this was the collapse of many energy intensive industries after 1989 rather than climate change policy measures.

The second half of the 1990s was a period of further environmental and climate policy developments – although merely at formulation and not implementation stage. During that time the EU and international climate change policies played a very important role in the political debate, mainly due to the Polish will to prepare for EU accession. In 2000 an amended second *National Environmental Policy* was adopted that highlighted the need for climate protection and the aim to prepare a national strategy to reduce GHGE (Karaczun, 2012: 89). With the strategy *Climate Policy of Poland* (*Polityka klimatyczna Polski*), which was adopted in late 2003, Poland fulfilled the requirement of its participation in the UNFCCC (ibid.: 89–90; Jankowska, 2011: 165). The most important element of this strategy was the adoption of a more ambitious GHGE reduction goal than the Kyoto Protocol's target for Poland, namely to reach a 40 per cent reduction by 2020 compared to 1988. The reason was that already by 2003 Poland had reduced its GHGE by more than 30 per cent (Jankowska, 2011: 165). However, the aims of the strategy did not translate into action.

Since Poland joined the EU in 2004, its attitude to climate policy has changed and the second phase of domestic climate change policy began. Climate change policy has been perceived not merely as an environmental issue but primarily as an element of the economic development (Karaczun 2012: 90). Therefore climate policy has become more strongly linked to energy policy and all climate policy measures (such as GHGE reduction targets and the development of renewable energy) have been included in the national energy policy programmes. There were two main reasons for this. First, the high costs of EU environmental law transposition. Second, after Polish accession the EU started to formulate more ambitious climate policies, requiring more economic efforts from the member states (ibid.). These facts have led Poland to pay more attention to the economic impacts of climate policy measures.

In the Polish energy policy strategy *Energy Policy of Poland until 2030* (*Polityka energtyczna Polski do 2030 r.*), which was adopted by the government in 2009, the GHGE reduction target for Poland was formulated in a new way to achieve 15 per cent reduction by 2020 compared to 1999. This document introduced also a new 'climate policy measure', namely the building of a nuclear power plant, in addition to the modernization of conventional plants, energy efficiency, energy saving and the development of renewable energy. The main reason has been the perception of nuclear power as one of the cheapest climate policy measures, which may also contribute to energy source diversification and increase the energy independence (mainly from Russian gas). Also in the draft energy policy strategy until 2050 the Ministry for Economic Affairs expressed its will to continue the nuclear energy programme. The share of nuclear, renewables and gas should each achieve 15 per cent by 2050 in the preferred scenario. Importantly, both of these documents emphasize the dominant role of coal until 2030 or 2050; they also assume the exploitation of new domestic hard coal and lignite deposits. However, this would be economically viable mainly in the case of availability of the high-efficiency low-emissions coal combustion or so-called 'clean coal' technologies.

On 4 August 2015, the Polish government adopted the draft *National Programme for the Development of Low-Carbon Economy* (*Narodowy Program Rozwoju Gospodarki Niskoemisyjnej*), which showed positive economic impact linked to climate action, especially by improving energy efficiency in residential and non-residential buildings, in industry and waste management as well as the proliferation of fuel-efficient vehicles. This document seems to be the first sign of a changing discourse in Poland towards perceiving climate change also as an economic opportunity, not only as a threat.

Institutional responses, policy instruments and programmes

Poland does not follow any overall strategy for climate protection. Several energy and climate policy documents were adopted, the most important of which is the quite ambitious *Climate Policy of Poland* (see below). Nevertheless, none of the strategies has really been implemented or has had a significant impact on policy formulation and implementation in other fields (such as economic or technological policy). The *Climate Policy of Poland* has often been described as a 'dead document' (ibid.: 39). Until now Poland has not managed to present its EU partners with any alternative climate policy proposals that would take into account its specific economic and social conditions (as the Polish government often requires) and, at the same time, contribute to reaching the overall EU climate policy goals (Bukowski and Śniegocki 2014).

The main driver for the formulation and adoption of climate change policy measures in Poland has been the EU both during the accession process and afterwards. In that sense Poland, which is sometimes a policy taker, has accepted EU structural, entrepreneurial and cognitive leadership in this area. Policy taking, however, does not mean necessarily proper policy implementation. For example, consider renewable energy policy. In the preparation period for EU accession Poland formulated a renewable energy sources target for electricity production, which was included in the EU Enlargement Treaty and subsequently also incorporated into Directive 2001/77/EC (Jankowska 2011: 167). In order to reach this legally binding target Poland adopted a mixed system of support for renewable energy sources – a quota mechanism combined with the tradable green certificates. Nevertheless, the way this support scheme was implemented in Poland has not enabled a strong increase of renewable electricity production (e.g. Jankowska 2012). The same concerns the implementation of Directive 2009/28/EC, which allowed only moderate alterations in the energy system that would not reduce the economic and political advantages of its main actors – the state-owned conventional energy companies (Ancygier and Jankowska forthcoming). For example, the *National Renewable Energy Action Plan* (*Krajowy Plan Działań w zakresie energii ze zródeł odnawialnych*) adopted in December 2010 includes mainly measures (such as development of wind energy, biomass co-firing and the building of one additional large hydro power plant with 100 MW capacity), which may involve the conventional energy industry in the development of renewable energy (ibid.). In July 2013 the government amended the existing energy law introducing a support mechanism for certain small scale renewable technologies up to 40 kW of installed capacity based on a feed-in tariff scheme. However, the tariffs amounted to only 80 per cent of the average electricity price from the last year (ibid.). A long expected separate renewable energy law adopted by the government in April 2014 still included this disadvantageous support scheme for the small scale installations. For all the other renewables it introduced the so-called competitive bidding process (or auctioning) in the form that seemed to be beneficial only for the big energy companies due to the planned capacity and price caps (ibid.). A positive improvement to this law was imposed by parliament which supported the so-called 'prosumer amendment' (see section 3).

The reason Poland has adopted renewable energy policies at a rather unambitious level that mainly serve the conventional, coal-based energy industry is, first, Poland's strategy to keep the dominant position of state-owned energy company PGE on the market (Ancygier and Caspar 2014). Another reason is the very strong political influence and political closeness of the conventional energy industry, especially of the state-owned companies, in contrast to rather week and not very well organized RES industry (cf. Jankowska 2012).

Generally it can be stated that Poland, when it formulates policies to tackle climate change, is in favour of 'new' environmental policy instruments (NEPIs). The above-presented policy in the field of renewable energy is a good example for this thesis. Another example constitutes the white certificates scheme introduced in Poland in 2011 by the *Energy Efficiency Law* (*Ustawa o efektywności energetycznej*). Out of NEPIs, Poland prefers by far market-based instruments. For the first time market-based instrument were recommended in the *Climate Policy of Poland* as incurring the lowest costs for achieving the policy goals (Jankowska 2011: 166). Importantly, feed-in tariffs for renewables are not perceived in Poland as a market-based instrument. In contrast, tradable green certificates are perceived as a typically market-based instrument and have therefore always been preferred to feed-in tariffs. On the other hand emissions trading, which was also recommended in the *Climate Policy of Poland*, is a typical market-based instrument, but it has been criticized by some Polish officials as artificial, bureaucratic and redistributive (cf. Jankowska 2011: 168).

Multi-level and polycentric climate governance

As mentioned above, the 'ecologic enthusiasm' of the early 1990s led Poland to actively participate in the UNFCCC negotiations and to support an ambitious international climate change agreement. During that period Poland offered cognitive leadership. Since the mid-1990s and especially after the EU accession, Poland has changed its performance and started to act as a veto player or policy taker at best. It has not offered any type of leadership in EU and/or international energy and climate change politics, apart from some considerable entrepreneurship developed more recently (Jankowska 2011: 174). But this entrepreneurship has been provided 'only in a narrow sense of using diplomatic, negotiating and bargaining skills (and opportunities) to facilitate agreements which do not primarily promote the overall aspiration of the EU climate change policy' (ibid.: 175–176) or the international climate policy. Agreements have been preferred that take sufficiently into account the economic and social conditions in Poland or other countries.

More precisely, one can distinguish three different types of Polish positions at the EU and international level with regard to climate and energy policy:

1. *Policy taker*: Poland implemented policies adopted at the EU or international level. For example, the directives on energy efficiency (Directive 2006/32/EC and Directive 2012/27/EU) and the directives on renewable energy (Directive 2001/77/EC and Directive 2009/28/EC).
2. *Policy shaper*: Poland tried to reshape the EU or international policy by making amendments in order to make it possible for Poland and other countries to attain the ambitious EU or international targets. In so doing, however, it watered down EU and/or international climate change policy measures/agreements. For example, the EU 2020 energy and climate package adopted in 2008, the EU 2030 energy and climate package adopted in 2014 and the negotiation at the EU level before and during the 21st Conference of the Parties (COP21) in Paris in 2015.
3. *Veto player*: Poland blocked the EU or international policy/negotiations. For example, the EU roadmaps on energy and climate in 2011 and 2012, the EU GHGE reduction commitment (based on the roadmaps already vetoed by Poland) during the COP18 in Doha in 2012 and the negotiation during the COP19 in Warsaw in 2013.

The first time when Poland transformed from EU climate policy taker to EU climate policy shaper was during the formulation and adoption of the EU 2020 energy and climate package (Jankowska 2011: 175). During this process Poland took an active, agreement-oriented and entrepreneurial role in proposing amendments to the European Commission's proposal of the package (ibid.: 170). Many Polish politicians underlined it as an answer to the allegations that Poland was against the package: 'That is incorrect. What we want is to amend the package to make it acceptable for all' – said Jerzy Buzek, the former Polish Prime Minister and President of the European Parliament between 2009 and 2012 (*Polish Market* 2008: 29 as cited in Jankowska 2011: 170). To support its position on the EU level Poland initiated an interest coalition, which consisted of the Visegrad four nations[2] and the Baltic States (Lithuania, Latvia and Estonia) (ibid.: 171). Its main amendment to the Commission's proposal was several years' derogation from entering the auctioning system for countries in which more than 50 per cent of electric and thermal energy is produced from coal (*Polish Market* 2008: 28 as cited in ibid.: 170). By taking into account the Polish amendment the European Commission watered down its initial draft.

By applying for the organization of the COP19 in Warsaw in 2013 Polish officials openly admitted that this should have enabled Poland to gain more influence in global climate change politics. The aim was, however, not to stand up for a more ambitious international climate policy agreement, but to prevent targets and measures that would force coal-dependent countries such as Poland to transform their economics and energy sectors away from fossil fuels. This well explains why the Polish government facilitated access of the biggest fossil fuel lobby groups and companies to formal negotiations and top officials. For example, it invited business to join the official negotiations during the pre-COP meeting of Environmental Ministers on 2–4 October 2013 in Warsaw (Corporate Europe Observatory and Jankowska 2013). In parallel to COP19 ran the International Coal and Climate Summit by the World Coal Association with the support of the Polish Ministry for Economic Affairs (ibid.). The Association together with the Polish Ministry issued a joint *Warsaw Communiqué*, which proposed the so-called 'clean coal' technologies, to fight climate change (ibid.). It raised questions of whether COP19 was 'being taken seriously as an international meeting of paramount importance, or a facilitated lobbying opportunity for those who have a direct commercial interest in burning more fossil fuels' (ibid.). All in all, the COP19 in Warsaw, as expected, did not achieve any major breakthroughs in the negotiations nor did it adopt any substantial decisions.

After the period of vetoing the EU climate change policy plans since 2011, Poland changed its strategy towards reshaping the EU policy proposals. During the energy and climate summit in Brussels on 23–24 October 2014, Poland, for the first time since 2012, has not simply blocked the EU agreement – the 2030 energy and climate package. The Polish government's new strategy was first and foremost to fight for conditions as good as possible for the Polish economy. A similar stance was taken also by other countries of the Visegrad Group as well as Bulgaria and Romania. As in 2008 (see Jankowska 2011), for the 2020 EU energy and climate package most of the Central and Eastern European member states spoke with a common voice with Poland being their informal leader. They rejected binding renewable energy and energy efficiency targets and made their approval to the binding 40 per cent GHGE reduction target dependent on whether the economic and energy systems' situation of each member state would be adequately taken into account in designed policies (Visegrad Group Countries 2014). In order to obtain at least agreement on binding GHGE reduction targets significant compensation measures were adopted for the Central and Eastern European countries. Countries whose income per capita is less than 60 per cent of the EU average were granted additional funds from the solidarity mechanism for modernization of energy infrastructure as well as free emission allowances beyond 2020 until 2030. Moreover, no binding targets for renewable energy and energy efficiency for 2030 were included in the package. According to the Polish government the agreement was thus a huge success. The EU energy and climate policies provided clear evidence for the ability of the Central and Eastern European member states to reshape or weaken common policies.

This strategy of policy reshaping or weakening was continued during the negotiations of the conclusions of the European Union Council, which dealt with the COP21 in Paris that were adopted on 18 September 2015. The Polish government realized that it would be isolated if it resisted EU climate policy (Crisp 2015). It therefore did not demand any substantial changes to the agreement of October 2014. One of its major requests, accepted by the other member states, was the introduction of 'climate neutrality' into the language of the climate agreement as a step away from the commonly used terms 'decarbonisation' or 'carbon neutrality' (Adamczewski 2015). Adamczewski (2015) interprets Poland's intention 'in a way that allows for further use of coal, as long as its emissions are offset (for example by planting forests)'.

The change in government which resulted from the election on 25 October 2015 did not result in any substantial changes to the Polish stance. The new majority government of the Law and Justice Party (*Prawo i Sprawiedliwość*, PiS – since 16 November 2015) has seemed to continue the strategy of not blocking but fighting for the best possible conditions by weakening the initial policy proposals. This was shown in the negotiation during the COP21. Afterwards Poland planned to convince the EU to renegotiate and water down its energy and climate policy. But it therefore also decided to 'tread delicately' during the COP21 (Cienski and Kureth 2015). Poland accepted the final 2015 Paris Agreement even if the goal of 'greenhouse gas neutrality', which had been stated in the agreement draft, was finally replaced by the 'aim to reach global peaking of greenhouse gas emissions as soon as possible' (Reuters 2015). The reason for this is that in the opinion of the Polish COP21 delegates the 2015 Paris Agreement does not rule out the use of coal by the end of the century and in that sense it is the best Poland 'could have hoped for' (IAR, *Gazeta Wyborcza*, 2015). The final 2015 Paris Agreement is less ambitious than what could possibly have been adopted without Polish attempts to assert its interests.

The domestic implementation of EU and international commitments

Poland has always implemented EU and international climate change commitments and policies in a way that has not required too much far-reaching changes to the *status quo* in its energy sector or high costs (although the Polish government has never explained what costs would be too high apart from those leading to increased energy prices). In the *Climate Policy of Poland* it adopted even much higher GHGE reduction target than that required by the Kyoto Protocol, but it has had in reality only a declarative character. Nevertheless, according to the European Environment Agency, by 2012 Poland reduced its GHGE by 30 per cent compared to 1988 (PAP 2014) and thus exceeded its Kyoto target without making any additional effort. It means also that since at least 2003, Poland had not further reduced its GHGE. Poland is also not planning to reduce them as much as agreed at the EU level, which can be clearly seen from different scenarios included in the draft of the energy policy strategy until 2050.

Interestingly, in the period 1990–2009 GHGE from coal combustion fell indeed quite strongly, i.e. by 32.1 per cent (Szawrański 2012: 10). At the same time, however, GHGE from oil combustion rose by 84.9 per cent, which can be attributed to the dramatic growth of the transport sector (ibid.). In fact, no strategy exists for a climate-friendly transport sector in Poland and each Polish government has given preferential treatment to private over public transport.

Concerning other policy fields, only with regard to first-generation biofuels promotion Poland has implemented policies which will enable it to achieve a much higher share than the EU level on average (Ancygier and Jankowska forthcoming). But the main reason has not been climate protection but job creation in the farming sector. Surprisingly, this argument does not play any relevant role in the case of other renewable energies. The much smaller potential of the second-generation biofuels for jobs creation has been the reason why Poland has opposed their stronger promotion at the EU level and domestically (ibid.).

Concerning the implementation of the 2001 and 2009 EU renewable energy directives and the target of Directive 2001/77/EC to increase the share of electricity produced from renewable energy sources to 7.5 per cent by 2010 Poland failed to achieve it. Although Poland formally transposed the directive properly, the adopted domestic policy measures were not sufficient (Ancygier and Jankowska forthcoming). With regard to the implementation of the second renewable energy directive it is also quite doubtful that Poland will achieve its target to increase the share of renewable energy in gross final energy consumption of 15 per cent by 2020. Although the share of renewable energy in the electricity sold to the end consumers had increased to over 12 per cent in 2012, almost half of it was produced by co-firing of biomass in old, inefficient coal-fired power plants, many of which have to be switched off due to EU air pollution obligations (ibid.). Therefore Poland has not supported the proposal to adopt binding EU renewable energy targets for 2030.

The implementation of the directives on energy efficiency has not been as controversial as policy implementation in other fields. Since 2011 Poland has a separate law on energy efficiency that introduced the above mentioned white certificates scheme. There are also several state financial programmes for modernization of buildings and a special Modernization Fund. Energy efficiency and saving have always played quite an important role in Polish climate and energy policy, as these measures are generally perceived as less costly than others. Nevertheless, the overall ambition and progress in this field is also rather low and slow.

Conclusions and outlook: Polish leadership

Poland's performance in climate and energy politics has changed over time. On the wave of 'ecologic enthusiasm' in the early 1990s, Poland tried actively to participate in the UNFCCC negotiations and to support an ambitious international climate change agreement. Therefore it offered, for a short period of time, cognitive leadership. Later and especially since its EU accession in 2004, Poland has started to act as policy taker, policy shaper or veto player. While doing so it has not offered any type of leadership in EU and/or international energy and climate politics. However, in several cases Poland provided considerable entrepreneurial leadership by using diplomatic, negotiating and bargaining skills (and by exploiting opportunities) to facilitate agreements, but only at a level of ambition which was considerably lower than that of the climate leader states within the EU. It supported only those agreements, which sufficiently took into account the economic and social conditions in different countries. In this regard, after its EU accession Poland mainly tried to reshape relatively ambitious EU climate change proposals by watering them down, or even vetoing them. Importantly, Poland has been 'in favour of the EU becoming an important actor in international climate change politics, although not necessarily a leader' (Jankowska 2011: 173). This means that Poland has been, at best, willing to accept a humdrum leadership style from the EU and strongly opposed any attempts by the EU to adopt a heroic leadership style. Nevertheless, the EU has been able to offer structural and entrepreneurial (and to a lesser degree cognitive) type of leadership for Polish domestic energy and climate policy. Being a policy taker has required Poland to accept this EU leadership.

Poland's position in the EU and international climate change negotiations has been mainly to protect the *status quo* in its coal-based energy sector or to enable only slight improvements, which are not too costly (although it is not actually clear, what not too costly means for the Polish government). Therefore Poland has supported internationally and domestically the development of 'clean coal' technologies and plans to build a nuclear power plant as they are, in the view of the Polish government, less expensive technologies than, for example, renewables. Poland has also tried to implement EU and/or international climate change commitments and policies in a way that has not caused far-reaching changes to the *status quo* and costs for its economy and society. This is also the reason why the link between climate change and energy policies has become more important for Poland especially after EU accession as it was around that time when the EU started to propose more ambitious climate and energy policies leading potentially to higher costs (especially for Poland) at least in the short term.

The main obstacles to Polish climate leadership are the dominance of the energy sector over Poland's emissions profile and the dominance of coal in its energy mix as well as its centralized, coal-based and state-owned energy industry structure. The relatively low standard of living of the majority of the Polish population, the increasing gap between the rich and poor, and the high level of structural unemployment are important factors which help to explain why it is so difficult to introduce changes to the *status quo* in Poland that would require making sacrifices for the environment. Nevertheless, an understanding has slowly been emerging in Poland that climate protection is not only a sacrifice but may also be turned into an advantage by developing a green economy, which creates jobs that are more attractive and (economically and environmentally) sustainable than those in coal mining. The Polish population seems to be ready and eager to reap such benefits. Now it is time for Polish politicians to recognize it and to formulate appropriate domestic policies as well as policy proposals for the EU level, which would allow Poland to make a stronger contribution to the EU climate policy goals.

Notes

1 The feed-in tariff scheme should have been valid from January 2016, but the new PiS government decided to postpone its introduction until spring 2016.
2 The Visegrad countries include Poland, Hungary, Slovakia and the Czech Republic.

References

Adamczewski, T. (2015) 'Poland's approach to the Paris COP', www.boell.de/en/2015/11/24/background-polands-approach-paris-cop (accessed 13 December 2015).

Ancygier, A. and Jankowska, K. (forthcoming) 'Poland at the energy policy crossroads – an incongruent Europeanization?', in Sandoval, I.S., Jörgens, H. and Bechberger, M. (eds), *A Guide to Renewable Energy Policy in the EU*, Cheltenham: Edward Elgar.

Bokwa, A. (2007) 'Climatic issues in Polish foreign policy', in P.G. Harris (ed.) *Europe and Global Climate Change. Politics, Foreign Policy and Regional Cooperation*, Cheltenham: Edward Elgar, 113–138.

Bukowski, M. and Śniegocki A. (2014) 'Stanowisko polskiego biznesu wobec unijnej polityki energetyczno-klimatycznej do 2030 r.', http://konfederacjalewiatan.pl/legislacja/wydawnictwa/_files/publikacje/2014/Przelamujac_impas.pdf (accessed 26 June 2015).

Cienski, J. and Kureth, A. (2015) 'Poland takes a tough line ahead of COP21', www.politico.eu/article/poland-tough-line-cop21-paris-climate-summit/ (accessed 14 December 2015).

Crisp, J. (2015) 'Ministers Unite on mandate for Paris climate talks', www.euractiv.com/sections/energy/eu-ministers-unite-mandate-paris-climate-talks-317778 (accessed 14 December 2015).

Corporate Europe Observatory and Jankowska, K. (2013) 'Trouble always comes in threes: Big polluters, the Polish government and the UN', http://corporateeurope.org/blog/trouble-always-comes-threes-big-polluters-polish-government-and-un (accessed 4 December 2013).

Greenpeace (2013) 'Badanie opinii publicznej: Zdecydowana większość Polaków woli energię odnawialną od energii z węgla i atomu', www.greenpeace.org/poland/PageFiles/564046/Energia_badanie_opinii_publicznej_briefing.pdf (accessed 01 July 2015).

IAR, *Gazeta Wyborcza* (2015) 'Poland successful at COP21, environment minister says', http://thenews.pl/1/10/Artykul/232693,Poland-successful-at-COP21-environment-minister-says (accessed 14 December 2015).

Jankowska, K. (2011) 'Poland's climate change policy struggle: Greening the East?', in Wurzel, R. and Connelly, J. (eds), *The European Union as a Leader in International Climate Change Politics*, London: Routledge, 163–178.

Jankowska, K. (2012) *Die Kräfte des Wandels. Die Wandlung Polens von einer auf Kohle basierenden zu einer an erneuerbaren Energien orientierten Gesellschaft*, Dissertation FU Berlin (Dissertations online), www.diss.fu-berlin.de/diss/receive/FUDISS_thesis_000000037268 (accessed 26 June 2015).

Karaczun, Z., Kassenberg, A. and Sobolewski, M. (2009) 'Polityka klimatycza Polski – wyzwanie XXI wieku', Warszawa: Instytut na rzecz Ekorozwoju, Polski Klub Ekologiczny.

Karaczun, Z. (2012) 'Polska polityka klimatyczna. Próba analizy', *Studia BAS*, 29(1): 85–108.

Malko, J. (2011) 'Klimatyczne aspekty polityki energetycznej', in *Polityka energetyczna*, www.min-pan.krakow.pl/Wydawnictwa/PE142/19-malko.pdf (accessed 1 July 2015), 273–290.

Ministry of the Environment (2013) 'Polish National Strategy for Adaptation to Climate Change (NAS 2020) with the perspective by 2030', https://klimada.mos.gov.pl/wp-content/uploads/2014/12/ENG_SPA2020_final.pdf (accessed 1 July 2014).

PAP (2014) 'Korolec: możemy poprzeć wzrost efektywności energetycznej UE', www.pap.pl/palio/html.run?_Instance=cms_www.pap.pl&_PageID=1&s=infopakiet&dz=gospodarka&idNewsComp=&filename=&idnews=174566&data=&status=biezace&_CheckSum=-1056264664 (accessed 3 July 2015).

PSL (2011) 'Człowiek jest najważaniejszy. Program Wyborczy', http://psl.pl/upload/pdf/Program_Wyborczy_PSL.pdf (accessed 1 July 2015).

Reuters (2015) 'How the world found common ground in landmark climate accord', www.trust.org/item/20151212140846-nt8d2/?source=gep (accessed 14 December 2015).

Swianiewicz, P., Gendźwiłł, A., Lackowska, M., Szajewska, N. and Szmigiel-Rawska, K. (2014) 'Perception of climate change by politicians and municipal officials. Press release, http://polcitclim.uw.edu.pl/wp-content/uploads/sites/25/2014/04/press_release_polcitclim_27-10-2014.pdf (accessed 28 June 2015).

Szewrański, S. (2012) 'Resource Efficiency Gains and Green Growth Perspectives in Poland', Friedrich Ebert Stiftung: http://library.fes.de/pdf-files/id-moe/09381.pdf (accessed 3 July 2015).

Tansey, R., Holland, N., Balanyá, B. and Jankowska, K. (2013) 'The COP19 Guide to Corporate Lobbying. Climate crooks and the Polish government's partners in crime', Corporate Europe Observatory and Transnational Institute.

TNS Polska (2014) 'Badanie świadomości i zachowań ekologicznych mieszkańców Polski, Badanie trackingowe – pomiar: październik 2014, Raport TNS Polska dla Ministerstwa Środowiska', www.mos.gov.pl/g2/big/2014_12/fe749deb7e1414bf1c4afbc6548300f9.pdf (accessed 28 June 2015).

Visegrad Group Countries (2014) 'Joint Statement of the 21st Meeting of the Ministers of Environment of the Visegrad Group Countries, the Republic of Bulgaria and Romania', www.mos.gov.pl/g2/big/2014_09/d6cdc9370500325c6f02a77f46f9d1c5.pdf (accessed 14 December 2015).

World Bank (2015) 'Poland's greenhouse gas emissions', http://siteresources.worldbank.org/ECAEXT/Resources/258598-1256842123621/6525333-1298409457335/chapterA.pdf (accessed 28 June 2015).

11 Spanish climate change policy in a changing landscape

Israel Solorio[1]

Introduction

Spain provides a good case for assessing the EU's climate policy's strengths and weaknesses. EU membership has produced considerable pressures at the national level for adapting climate and other related policies to the ambitious European goals (see Chapter 1). This relationship initially followed a mostly top-down dynamic, where Spain was a more passive taker of European goals and policies (Costa 2011: 182–183), but then turned into a two-way interaction in which reducing the level of ambition of EU policies and targets has been a constant task for Spanish negotiators over the last years. In this regard, Europeanization is key – although not the only variable for understanding the contradictions of climate policy in Spain, which is reflected in the growing social preferences for climate change abatement that have not been fully materialized in the policy-making (Hanemann *et al.* 2010: 2).

The fact that Spain is a decentralized state with powerful sub-national authorities has accentuated the complexities around climate policy. Therefore, Spanish climate policy is characterized by multi-level governance that inevitably needs cooperation and coordination between different levels of government in order to reach its goals. Last but not least, it is important to take into consideration that Spain has faced during the last years a situation of economic crisis that affected the national performance in this field. On top of that, austerity policies led to an outcry of Spanish society for change that has completely shaken up the political landscape. The sum of these factors helps to explain how Spain became one of the EU member states with the poorest results in the 2016 Climate Performance Index (Germanwatch 2016), but at the same time has a society engaged against climate change.

This chapter deals with Spanish climate policy in a changing political landscape. It argues that contrasts in Spain's climate change policy come precisely from this multi-causal background. The EU empowered domestic actors to promote change, but they have come up against structural resistance to it. The case that best exemplifies this is renewable energy, but it is not the only one. Within the framework of multi-level climate governance in Spain, the role of the Autonomous Communities is also crucial for understanding the implementation of climate policies. To complete the picture, the economic and political turmoil that Spain experienced since 2008 has given rise to green shoots of transformational leadership in Spanish society and new political forces that are increasingly challenging the status quo in climate affairs and beyond.

National attitudes to climate change

One of the most comprehensive studies on Spanish society and climate change depicted, in 2013, the existence of a country divided into four: (1) the *disaffected Spain*, characterized by its indifference towards climate change (59.4 per cent of the population sample); (2) the *concerned Spain*, with 29.6 per cent of the respondents worried about climate change; (3) the *committed Spain*, with only 9.3 per cent of the population who are directly taking action to combat global warming; and, (4) the *sceptic Spain*, representing the 1.8 per cent of the respondents who were incredulous of climate change risks (Meira 2013). However, national attitudes to climate change have been evolving over the last years, demonstrating a more engaged society and a tendency towards considering that climate policies and economic welfare are not necessarily mutually exclusive.

Days before the start of the 2015 Paris climate change conference (COP21), Madrid witnessed its biggest demonstration for the climate in history, with the participation of around 20,000 people, and similar demonstrations in all the main cities of the country. And the streets are not the only indicator of the growing salience of climate change in Spanish society. The 2015 Barometer of the Elcano Royal Institute reflected that Spaniards perceived climate change as the second most important threat to Spain – just after international terrorism – and considered that climate change should be a top priority for Spanish foreign policy (Escribano 2015). When asked about whether they have taken personal action to fight climate change over the last six months, 66 per cent of the Spanish respondents replied yes (Eurobarometer 2015: 29). Spaniards occupied the fourth place in the list of EU citizens actively engaged against global warming.

In brief, Spanish society is concerned with climate change but without losing sight of its socio-economic needs. When questioned about the cost of environmental measures, almost 50 per cent of Spanish respondents considered that they should be adopted even when implicating high costs; in contrast 20 per cent believed that those measures should not involve any additional cost for citizens (CIS 2015: 8). All these indicators point to the fact that a majority of Spaniards are committed to tackling climate change in spite of the economic difficulties. In fact, the 2015 Eurobarometer indicated that 88 per cent of Spaniards consider that climate change policies can actually boost the economy and generate jobs at the same time (Eurobarometer 2015: 43). These figures represent a major shift from the former perception that preoccupation with climate change was second to the preoccupation with the consequences of climate policies (Costa 2011: 180).

The predominant political approach to climate change has been strongly informed by a fear of the loss of competitiveness, as expressed mainly – although not exclusively – by the Conservative Popular Party (*Partido Popular* – PP) and the Centre-Left Spanish Socialist Workers Party (*Partido Socialista Obrero Español* – PSOE). In 2015, Greenpeace denounced the alliance between politicians and business which it claimed was sacrificing the climate in favour of corporate interests (Greenpeace 2015a). But the change in the Spanish political arena is also affecting attitudes towards climate change. While the 2015 Paris Climate Conference played a role in the re-emergence of climate change among the political discourses, another factor to be taken into consideration is the rise of the new Left party PODEMOS (Spanish for 'We Can'). Having channelled the social discontent expressed by the Spanish *indignados* in 2011, this new party appeared with a more radical discourse nourished by social movements. In alliance with the regional Catalonian Green Party for the 2015 elections, PODEMOS called in its programme for a renewable-based energy transition by 2050 and a new climate change law including measures such as a new green tax. In a Greenpeace evaluation of the environmental proposals, PODEMOS turned out to be the most committed political party with the environment (Greenpeace 2015b).

Perhaps the best representation of this shift in the discourse among politicians is Mariano Rajoy's changing attitude. When he was in opposition, Rajoy presented himself as a climate change denier in 2007 (*El País* 2007). As Spanish Prime Minister, he announced during the 2015 Paris Climate Conference a new climate change law in case he would remain in office after the 2015 general elections. Interestingly, the PP was not the only party raising the flag of climate change during the 2015 election campaign. The PSOE also proposed the adoption of a new law for Spain. At least on paper, energy transition and climate change have been considered as a fundamental part of the government agenda by all the main political parties. The salience of climate change in the political discourse points to a possible reduction of the breach between social preferences for climate change abatement and policy-making.

Phases of domestic climate change policy

The history of Spanish climate policy is intertwined with the development of EU action in this field. Given that multi-level governance plays a fundamental role in Spain, it is almost impossible to distinguish between domestic, European and foreign climate policies (Costa 2011: 181). Instead, the most appropriate way to understand the phases of Spanish climate policy is to follow its interrelationship with the EU.

Phase 1: the defence of Spanish specificities within the EU (up until 1997)

Until 1997, Spain's approach to climate change in international and (especially) European arenas gave priority to the defence of the demand that Spain be given special treatment with regard to the implementation of climate policies and to the definition of greenhouse gas emission (GHGE) targets (Costa 2006: 226). Spain's position was based on the understanding that national interests should be defended within the EU, while in international fora the role of Spain was reduced to one of monitoring negotiations, defending European positions and, at best, offering technical input (Costa 2011: 181–182). For the negotiations of the United Nations Framework Convention on Climate Change (UNFCCC) in 1992, the negotiating position of Spain was aligned with the EU's official position. It defended the necessity of a strongly regulated international regime capable of guaranteeing compliance with targets and timetables. However, inside the EU, Spain was reluctant to accept the objective of stabilizing emissions during this period, and as such looked for guarantees that the less developed countries would be allowed to implement such commitments with more flexibly (Costa 2006: 227). Spain also insisted on ensuring a favourable share of emissions within the 15 per cent overall reduction proposed for the EU in the negotiations that leading up to the Kyoto Protocol, as it did for the EU's final Kyoto target of 8 per cent. Indeed, on the road to Kyoto the most delicate issue for Spain was to ensure that there would be special treatment within the EU (Costa 2011: 182). This period was characterized by the lack of institutional capacities on climate change in Spain, which translated into a strong top-down relationship between the EU and Spain, only nuanced by Spanish efforts to defend the national specificities within the European arrangements.

Phase 2: passive 'Kyotoism' and active pro-EU stance (1997–2000)

Once the debate on the EU's burden sharing agreement (see Chapters 1 and 5) was over, Spain felt comfortable with the commitments derived from the Kyoto Protocol. The agreement reached by the Environment Council of June 1998 allowed Spain to increase its emissions by 15 per cent compared to the base year (1990) until the compliance period of 2008–2012. None of the major political parties thought the 15 per cent increase to be overly restrictive. Spain was therefore able to add its voice to the European consensus without any problems. This Spanish position lasted until the 2000 The Hague climate conference (COP6), where Spain took the side of the most environmentally ambitious countries. Thus, during the COP6 the Environment Minister, Jaume Matas (PP), pointed out that the entry into force of the Kyoto Protocol could not come about at 'any price' and that the mechanisms of flexibility should not become 'formulae to evade commitments' (Costa 2006: 229).

Climate change negotiations were mainly perceived by (and often presented to) policy-makers as an opportunity for the EU to act as an international leader (see also Chapters 1 and 2). The economic and more delicate dimensions of climate negotiations were under-emphasized or simply ignored. As a result, the ambition displayed by Spain in the international and European arenas was not corresponding with the actual development of internal policies to limit GHGE or by the building of institutional capacities able to bring about such policies (Costa 2011: 182). During these years, Spain displayed a symbolic cost-free leadership.

The only remarkable exception was related to renewable energy that, by then, already included the adoption of feed-in tariffs for its promotion. All in all, this period was characterized by a top-down relationship with the EU, where Spanish officials considered that the best option was to hide behind the shield of the positions adopted by the EU.

Phase 3: a two-way adjustment and a new pro-EU stance (since 2001)

The third phase of Spanish climate policy can be characterized by three main features: (1) the construction of institutional capacities; (2) the initiation of the debate on policies to reduce emissions; and, (3) a reaction against such policies (Costa 2006: 230–231). Despite some nuances, these characteristics still define the relationship of Spain with the EU policies.

With the ratification of the Kyoto Protocol by the EU and its member states in 2002, the adoption of concrete measures to fight climate change accelerated at the EU level (Morata and Solorio 2012). In light of the adoption of the EU's internal climate policies, given the global trend towards a more flexible international regime, and considering the evolution of its emissions (which by 2000 had already more than doubled the 15 per cent increase allowed for 2008–2012), Spain developed a more pragmatic approach to international and EU climate policies. It adopted a position of strong support for market mechanisms and favoured flexible instruments including the Clean Development Mechanism (CDM). Finally, climate change was perceived as an important internal issue with relevant economic and social implications. For Spain this was a period for a 'reality check' where institutional capacities had to be developed in order to be able to participate in the definition of EU and international climate change policies. Moreover, the mere existence of EU climate policies obliged the more reluctant actors (mainly Spanish industries, the Economics Ministry and, from 2002 onwards, the Conservative PP and some minor parties) to address the issue.

From 2004 onwards, Spain started to show a trend of two-way adjustment to EU policies. First, Spain had to comply with its EU and international commitments. Second, while not entirely opposing EU policies, Spain tried to reduce the ambition of future commitments. Over this period, Spain became more realistic in its estimation of its capacities to reduce GHGE, but at the same time made an effort to comply with them. Drawing on the relatively new institutional capacities on climate change, Spain has also attempted to be more influential in the European arena. The country has adopted a more realistic approach, where adapting EU commitments to the national conditions is part of the deal (Costa 2011: 183). In other words, Spain has developed an entrepreneurial leadership in order to curb down the national commitments with the EU.

Undoubtedly, the economic crisis was a game changer for Spanish climate policy. Given its impact on the economy, Spain achieved a reduction of GHGE between 2008 and 2013. But this reduction was largely circumstantial. In 2014 Spain presented the largest absolute growth in GHGE occurred in the recent history (+3.5 Mt CO_2-eq.), produced by an increase of solid fossil fuel consumption (EEA 2015a: 10). These results were partly caused by austerity policies which led to a retrenchment of the up to then successful renewable energy policy, provoking a decline in the growing rate of renewables within the electricity mix and thus triggering an increase of fossil fuels consumption. Boosted by international low prices, coal consumption has dramatically increased in this context. Climate policy was a big loser of the economic crisis. Yet, the economic crisis even accentuated the relevance of the two-way adjustment relationship with the EU and its climate change policies. This can be best illustrated by the Spanish position on the EU's 2030 climate goals. Spain was in line with those member states which demanded a binding 40 per cent reduction of GHGE, but was a fierce opponent to binding targets for renewable energy and energy efficiency (*Euractiv* 2014). Still, EU climate goals and policies continue to being a driving force at the national level.

Institutional responses, policy instruments and programmes

Compared to many other EU member states, Spain can be considered a latecomer on climate action. It was not until 2001, when Spain started to respond actively to developments at the EU and institutional level. Prior to 2001, Spain lacked the necessary institutional capacities to even be an active part of the EU climate policy-making processes or in the international negotiations.

The first administrative unit entirely dedicated to climate policy was established only in January 2001. The creation of the Spanish Office on Climate Change (*Oficina Española de Cambio Climático* – OECC), a collegiate body of the Environment Ministry, was intended to meet the challenges of the Spanish EU Presidency of the Council during the first half of 2002 with an agenda full of climate-related policies (Costa 2006: 230). The growing volume and relevance of the tasks carried out by the OECC made necessary to upgrade its rank to a General Directorate in 2006. By 2008, a specific Secretariat General on Climate Change was established, including within its structure the General Directorate on Climate Change, but budget reductions in the context of the economic crisis marked its short existence in the Spanish institutional architecture.

In 2007 the Spanish government finally approved the *Climate Change and Clean Energy Strategy*. This strategy was adopted after several drafting processes in the previous years that at least helped to raise interest among stakeholders (Costa 2011: 185). Before its adoption, Spain had only had in place pieces of strategic planning on specific climate-related issues such as the *Plan on Renewable Energy 2005–2010*, the *2005 Action Plan on Energy Efficiency and Savings*, the *National Allocation Plan (NAP) 2005–2007* and the *Plan for the Adaptation on Climate Change* adopted in 2006. The *Climate Change and Clean Energy Strategy* was a long-awaited document by environmental NGOs (ENGOs) that had been demanding the need for the adoption of an overall strategy document which included all governmental climate change actions. However, the *Climate Change and Clean Energy Strategy*, which was supposed to be the governmental roadmap until 2020, fell below environmentalist expectations. Importantly, the economic crisis greatly constrained the potential effect of this strategy. However, specific climate-related measures have been updated. This is especially true for renewable energy and energy efficiency action plans and for NAPs, which have been constantly brought up-to-date in order to comply with EU targets and policies.

Obviously, most of the Spanish legislation on climate change transposes EU directives or fulfils obligations under them (Costa 2011: 185). A distinction can be made between those Spanish climate policy measures which are the result of EU impulses and those that were adopted before any EU legislation. The latter measures thus responded to specific Spanish reasons. Policies and instruments on emission trading are part of the first group. The Spanish law 1/2005, which promotes the reduction of CO_2 emissions in the industrial sector and of electricity generation, was adopted in order to transpose the EU emission trading directive (2003/87/CE). The same applies to the Building Technical Code which transposes the directive on energy efficiency on buildings (2002/91/CE).

While emission trading was a new policy instrument for Spain, the Building Technical Code had not been updated since 1977 and it was only after EU legislation that Spain incorporated energy efficiency into building norms (Aranda *et al.* 2007: 12). Both policy measures respond to an Europeanization logic and have been updated accordingly. The Spanish modifications on emission trading, which are the result of the transposition of the revised EU directive (2009/29/EC) were incorporated into the national legislation by the Law 13/2010. As a result, the aviation sector became part of the emission trading system in Spain. Under this new regime, auctions acquired a central role in the allocation of emissions allowances and the income from them should be allocated to climate policies. Spain included within the Law 18/2014 on urgent measures for the growth, competitiveness and efficiency, the necessary instruments in order to comply with the revised EU policy on energy efficiency (Directive 2012/27/EU). In line with this EU directive, in 2014 Spain adopted a *National Fund for Energy Efficiency* and also a system of obligations.

On the other hand, Spain is a pioneer in Europe when it comes to renewable energy promotion (Solorio 2011: 110). However, the starting point of this policy in 1982 had nothing to do with climate change, but was instead motivated by energy security concerns. It was only at a later point in time when renewable energy policy became an important part of Spanish climate policy (Solorio 2013: 132–133). In contrast to the experience with emission trading and energy efficiency, Spain already had in place legislation for the implementation of the first EU Directive on renewable energy promotion (2001/77/EC). During the 2000s, Spain displayed cognitive leadership based on the recognition of the national feed-in tariffs as one of the most effective systems to promote renewable energy (Commission 2005: 7). In 2004, the Royal Decree 436/2004 modified several elements of the Spanish feed-in tariffs. As Jacobs explained, the '2004 amendment established the dual remuneration system, comprising the fixed tariff payment option and the market sales option' (Jacobs 2012: 79). But the economic crisis was a game-changer for renewable energy in Spain. In spite of the mandatory targets for 2020 in Directive 2009/28/EC, the Spanish government has been constantly altering the Spanish feed-in tariffs. Examples include the Royal Decrees 1565/2010 and 1614/2010, which established cutbacks to wind and solar thermoelectric energy, and the Law 14/2010 which applied cutbacks to both existing and new photovoltaic installations. The latest stab at the once successful Spanish feed-in tariffs was the Royal Decree 413/2014. The new austerity-driven economic regime, which also affected already functioning plants, changed the conditions under which investments were originally made and introduced a new cutback to the support of renewable energy.

In 2011 Spain created the Carbon Fund for a Sustainable Economy. This instrument was conceived in order to encourage the participation of Spanish companies in carbon markets. In addition, this fund could also be used for acquiring international credits from projects developed under the flexibility mechanisms of the Kyoto Protocol. Spain has been an active promoter of CDM within the EU and beyond. Given the extensive participation of Spanish companies in CDM projects abroad, it has become a key instrument in Spanish climate change strategy. The implementation of this instrument has resulted in different reactions. On the one hand, ENGOs have blamed the Spanish government for financing with public funds the development of compensation projects (which form part of polluting industries) in the South instead of tackling domestic GHGE. On the other, this instrument has shown signs of success in some regions of the South. Particularly in Latin America, governments have been receptive to CDM projects. This form of structural leadership in Latin America based on a financial stream from Spain to Latin America has been reinforced with an entrepreneurial leadership from Spain in the region. The Spanish role was decisive in the adoption of the Ibero-American Plan on Adaptation to Climate Change (Costa 2011: 189). In December 2014 the Ibero-American Network of Climate Change Offices (RIOCC) reached its 10th anniversary. The Spanish Minister for the Environment, García Tejerina, took advantage of the occasion to 'reaffirm Spain's commitment on promoting and supporting cooperation and climate actions in Latin America and the Caribbean' (García Tejerina 2014). But the Spanish leadership in Latin America cannot be understood without taking into account both Latin American governments' perception that it offers a window of opportunity to promote development in the region and the interest of Spanish companies in expanding their businesses in Latin America and other Southern regions.

The domestic implementation of EU and international commitments

The Spanish experience of dealing with EU and international obligations has been a story full of contrasts. Starting with the Kyoto Protocol, Spain initially experienced some problems to meet its EU and international commitments. From 2001 to mid-2004, Spain failed to comply with important UNFCCC deadlines including the delivery of National Communications and the submission of data on emissions (Costa 2011: 189).

In meeting its substantive commitments, Spain had an even harder time. In its report on progress towards the EU's 2008–2012 Kyoto targets, the EEA pointed out that GHGE remained above targets by more than 9 per cent in Spain and that the flexibility mechanisms were crucial for Spain to be able to reach its Kyoto targets (EEA 2014: 6–16). In the end, Spain was able to comply with its commitments within the first commitment period of the Kyoto Protocol, but only thanks to an agreement with Poland from which Spain bought almost 100 million tons of CO_2 allowance for around €40 million under the EU ETS. The agreement was strongly criticized by ENGOs for being a 'false solution' to the challenge of climate change (*Ecologistas en Acción* 2012). Importantly, it was the drop of carbon prices in the EU ETS by 2012 that made the agreement possible.

With the ratification of the so-called Doha Agreement, which established a second commitment period (2013–2020) for the EU under the Kyoto Protocol, Spain has continued to struggle to meet its international commitments. The EEA positioned Spain as the sixth largest emitter in the EU-28, accounting for 7 per cent of total EU-28 GHGE in 2013 and having one of the highest absolute increases (EEA 2015a). As Spain has been able to meet the substantive international commitments only thanks to the EU's flexibility arrangements it could be argued that Spain's alleged climate leadership position actually hides a laggard profile which overall more accurately describes Spain's climate change policies.

The difficulties of meeting its Kyoto targets are a symptom of the Spanish performance on EU climate-related policies. When criticizing the Spanish way of complying with the Kyoto Protocol, ENGOs argued in favour of a different energy policy model. In 2015 the EEA reported that the evolution in Spanish GHGE was 'largely due to emission increases from road transport, electricity and heat production, and households and services' (EEA, 2015a: xii). In 2015 the EU Commission placed Spain within the group of member states needing to 'assess whether their policies and tools are sufficient and effective in meeting their renewable energy objectives' (Commission 2015: 5). In spite of the fact that Spain still is one of the top three EU producers of wind power, it seems that the feed-in tariffs cut down is taking its toll in the national performance on renewable energy. In this context, wind producers have argued that 'Spain is in danger of missing its 2020 renewable energy target because of its inadequate subsidy regime' (*ENDS Europe* 2016). The dilemma for Spain is whether to use the EU's climate change policies to boost a renewable-based energy transition in line with a green economy or to continue with business-as-usual just to meet the EU and international climate change commitments.

On energy efficiency, the EEA's 2015 report on the progress towards EU's climate and energy targets placed Spain on track to achieve its energy efficiency targets. Moreover, the EEA considered that maintaining the pace of reductions in primary energy consumption should allow Spain to reach the 2020 target. According to the EEA, the challenge is to stabilize energy consumption as the Spanish economy recovers from the crisis (EEA 2015b: 54). However, this seems a very difficult, if not impossible, task as the largest part of Spain's GHGE reduction and primary energy consumption in previous years was the result of the economic recession rather than due to climate change policy measures. In this context, the Spanish government has complained that the 2012 directive 'places a disproportionate burden on countries which, like Spain, have already made significant efforts in this area in previous years and are going through a phase of economic recovery' (MINETUR 2014: 3). Moreover, questions about Spain's performance on renewable energy and energy efficiency have increased since the 2015 Royal Decree legislated for regulating self-consumption of energy. Criticism has come from both ENGOs and the building industry because of the extra taxes for photovoltaic installations (the so-called 'sun tax'). This means that self-consumption system owners pay more for maintaining the electricity grid than other users, even when they are less reliant on it.

Multi-level and polycentric climate governance

In order to explain the interplay between the different actors on the domestic, EU and international levels, it is important to recall that in Spain the EU has functioned as a filter for international pressures. National authorities primarily negotiate at the EU level which share Spain will be responsible for out of the EU's collective commitments. It is also mainly due to the EU authorities that the national governments have to report and are accountable for their climate change related actions.

Overall Spanish climate policy has been strongly Europeanized (Costa 2006). The EU has altered the distribution of resources among domestic groups by strengthening pro-climate positions and it has fostered the emergence of new understandings of what constitutes proper behaviour (Costa 2011: 187). During the 1990s the EU moulded national discourses on climate change and determined which ones were legitimate or illegitimate, providing cognitive leadership to Spain. Processes of institutional adaptation in Spain can also be explained by EU initiated processes. In brief, the EU is a powerful explanatory factor for understanding Spanish climate policy. The EU facilitated the emergence of new approaches to climate policies and empowered domestic actors to promote change, even when they have met with structural resistance. In the framework of EU climate targets, an important part of the debates on Spanish policy has been shifted to the European level and national debates are frequently limited to how best to comply with EU goals. The EU has also become the arena to which ENGOs and 'green' industries have turned to protect the ground gained in Spanish domestic climate policy. This can be well illustrated by the renewable industry's constant lobbying of Brussels and its frequent complaints to either the European Commission or the Court of Justice of the European Union with the aim of achieving modifications to the Spanish electricity market.

Inherent to the Spanish semi-federal political system is the challenge that the regional authorities of the Autonomous Communities are also competent authorities for implementing EU climate policies. As a result, the national government has 'only limited means for directly intervening in the way in which regions apply and enforce European law' (Börzel 2000: 21), giving to the Autonomous Communities an important role to play in the formulation and implementation of Spanish climate policy. The most important Autonomous Communities (including Andalusia, Catalonia, the Basque Country and Valencia) already have plans or strategies to mitigate climate change. As Börzel points out, the national administration has attempted to promote the 'effective coordination of competencies and resources between the levels of government' (Börzel 2000: 28). The National Climate Council, which includes representatives from the national government, Autonomous Communities, municipalities and provinces, is the best representation of this effort. Still, cooperation and coordination on climate issues have proved to be not an easy task for Spanish authorities. For example, the dispersion of competencies between national, regional and local authorities facilitated the uncontrolled 'boom' of renewables during the past decade (Solorio 2011), leading to the retrenchment of that policy since the 2010s. The reason behind this is that Autonomous Communities have competencies to authorize the building of renewable energy plants, provoking a competitive environment among regional authorities in a race to be recipient of investments.[2] In light of this, the national government implemented in 2014 a differentiated register system for the renewable energy plants in order to improve the control over its expansion within the electricity system.

The changing political environment in Spain has facilitated the emergence of new local governments formed by a coalition of political and social actors not connected to traditional ruling parties as PP and PSOE. The so-called 'new municipalism' won in 2015 local government elections in Spain's principal cities such as Madrid, Barcelona and Valencia, among others. Given its progressive agenda, these governments have been receptive to social demands related to climate change and energy policy. At the same time, the composition of these new governments, close to social organizations and movements, has facilitated its engagement with polycentric governance. For example, in November 2015 the Barcelona local government presented the city's Commitment to the Climate, elaborated together with 800 citizen associations as a roadmap for climate action for the period 2015–2017 with the goal of reducing 40 per cent the GHGE by 2030. This experience can be considered as local and transformational leadership with a huge potential of transformative effects if replicated in all of Spain's main cities.

Finally, together with the growing social awareness on climate change (see above), there has been an increase in climate initiatives taken by non-state actors which could be said to amount to emerging polycentric governance. The *Alliance for the Climate*, which was the organizer of the largest demonstration against climate change in Spain's history, constitutes a good example. It is made up of 400 organizations representing the environmental movement, development organizations, trade union and consumers. Part of this alliance is the *Platform for a New Energy Model* which is a citizen initiative that appeared in 2012 that demands the transition to a more sustainable and socially fair energy model. It has become one of the main voices against the government's energy policy.

Another expression of this polycentric governance is the emergence of renewable energy cooperatives in Spain. The economic crisis, rise in electricity prices and the resultant high levels of energy poverty in Spain has led to a remarkable expansion of this new model of participation in the energy market. In 2015 the energy cooperative *Som Energia* – one of the largest in Spain – signed more than 30,000 contracts to provide electricity to households and had more than 20,000 shareholders. While these actors are still marginal in terms of their participation in the energy market, they have the potential to contribute cognitive and transformational leadership to transform the energy model in Spain.

Conclusion: political leadership in Spain and climate change as an opportunity for change

At the governmental level, climate change has been largely perceived as a threat to Spain's economic development. This has been expressed both by political discourses and policy measures. Renewable energy is the best example for illustrating the lack of vision of Spanish government to boost a low-carbon economy. Once an EU role model, Spain is now in the spotlight for the dramatic changes to the scheme for renewable energy promotion. So much so that the EU Commission has put in question the Spanish capacity to reach the 2020 renewable energy goals. The challenge for Spain is to convert the transactional leadership which is exhibited in its initial attempts to implement the EU's renewable energy policy into a transformational leadership oriented towards long-term goals. In the meantime, Spain has lost its cognitive leadership in the field as reflected in the cutbacks to the previously successful national feed-in tariffs and the problems this country is facing to meet the 2020 renewable energy goals. The evolution in domestic GHGE demonstrates that Spain has not yet embarked on structural change in the economy which aims to achieve a low-carbon model. Instead Spain has made heavy use of the flexibility mechanisms to meet its EU and international climate change commitments. On climate change, Spain has largely been a laggard which sheltered behind the EU's ambitious position. The way in which Spain tries to influence the dynamic of two-way adjustment works as follows: Catching up with EU goals while simultaneously attempting to curb the level of ambition at which they are aimed. Until now, Spain has been quite successful in this strategy. Quite surprisingly, Spain has nevertheless been able to promote (structural and entrepreneurial) leadership on climate change-related issues in Latin America.

However, the conventional approach to climate change traditionally adopted by Spanish authorities is being tested by the changing political landscape. Not only is the emergence of new political formations that are confronting the existing consensus within Spanish politics, but also the promoters of the 'new municipalism' are demanding the reduction of the gap between social preferences and policy-making. In this changing political landscape, it is no coincidence that substantive long-term changes in climate and energy policies are demanded by the promoters of political change in Spain.

Notes

1 The author is grateful to Oriol Costa for his comments on earlier drafts of this chapter. In particular the section on phases of domestic policy is based on his previous work on Spanish climate change policy (Costa 2011).

2 'Cambio ladrillo por huerto solar' *El País*, 7 December 2008. Online http://elpais.com/diario/2008/12/07/sociedad/1228604401_850215.html (accessed 2 March 2016).

References

Ajuntament de Barcelona (2015) 'Compromis de Barcelona pel Clima', Online: http://premsa.bcn.cat/wp-content/uploads/2015/11/Compromis_Bcn_Clima.pdf (accessed 5 January 2016).

Aranda, A., Zabalza, I., Llera, E., Martínez, A., Scarpellini, S. and Barrio, F. (2007) *El ahorro energético en el nuevo código técnico de la edificación*, Madrid: FC Editorial.

Börzel, T. (2000) 'From Competitive Regionalism to Cooperative Federalism: The Europeanization of the Spanish State of the Autonomies', *The Journal of Federalism*, Spring 2000, 30(2): 17–42.

CIS (2015) 'Barómetro de resultados de diciembre 2015', Online: http://datos.cis.es/pdf/Es3121mar_A.pdf (accessed 5 January 2016).

COM (2015) *Renewable energy progress report*, {SWD(2015) 117 final}. Online: http://eur-lex.europa.eu/resource.html?uri=cellar:4f8722ce-1347-11e5-8817-01aa75ed71a1.0001.02/DOC_1&format=PDF (accessed 5 January 2016).

Commission (2005) *The support of electricity from renewable energy sources*, SEC(2005) 1571 627 final. Online: http://eur-lex.europa.eu/legal-content/EN/TXT/PDF/?uri=CELEX:52005DC0627&from=EN (accessed 16 March 2016).

Costa, O. (2011) 'Spanish EU and International Climate Change Policies' in R.K.W. Wurzel and J. Connelly (eds), *The European Union as a Leader in International Climate Change Politics*, London: Routledge, 179–194.

Costa, O. (2006) 'Spain as an Actor in European and International Climate Policy: From a Passive to an Active Laggard?', *South European Society and Politics*, 11(2): 223–240.

Ecologistas en Acción (2012) 'El engañoso cumplimiento español del protocolo de Kioto'. Online: www.ecologistasenaccion.org/article24609.html (accessed 16 March 2016).

EEA (2015a) *Annual European Union greenhouse gas inventory 1990–2013 and inventory report 2015*, Technical report No 19/2015, Copenhagen: European Environment Agency.

EEA (2015b) *Trends and projections in Europe 2015 – Tracking progress towards Europe's climate and energy targets*, Report No 4/2015, Copenhagen: European Environment Agency.

EEA (2014) *Progress towards 2008–2012 Kyoto targets in Europe*, Technical report No 18/2014, Copenhagen: European Environment Agency.

ElPaís (2007) 'Rajoy cuestiona el cambio climático', Online: http://sociedad.elpais.com/sociedad/2007/10/22/actualidad/1193004007_850215.html (accessed 5 January 2016).

Escribano G. (2015) 'Cambio climático y seguridad energética en la opinión de los españoles'. Online: www.blog.rielcano.org/cambio-climatico-y-seguridad-energetica-en-la-opinion-de-los-espanoles (accessed 5 January 2016).

ENDS Europe (2016) 'Wind industry slams new Spanish system'. Online: www.endseurope.com/article/45005/wind-industry-slams-new-spanish-system. (accessed 17 March 2016).

Euractiv (2014) 'Member states' positions on 2030 climate and energy targets revealed', Online: www.euractiv.com/section/energy/news/member-states-positions-on-2030-climate-and-energy-targets-revealed/ (accessed 17 March 2016).

Eurobarometer (2015) Report Climate Change, Online: http://ec.europa.eu/clima/citizens/support/docs/report_2015_en.pdf (accessed 5 January 2016).

García Tejerina, I. (2014) 'Compromiso de España para seguir promoviendo acciones de cooperación en América Latina y Caribe', Online: www.magrama.gob.es/es/prensa/noticias/garc%C3%ADa-tejerina-reafirma-el-compromiso-de-espa%C3%B1a-para-seguir-promoviendo-acciones-de-cooperaci%C3%B3n-y-cambio-clim%C3%A1tico-en-am%C3%A9rica-latina-y-caribe/tcm7-356953-16 (accessed 5 January 2016).

Germanwatch (2016) 'The Climate Change Performance Index 2016', Online: https://germanwatch.org/en/download/13626.pdf (accessed 5 January 2016).

Greenpeace (2015a) '*El Monstruo de la Energía*'. Online: www.greenpeace.org/espana/Global/espana/2015/Report/energia/MonstruoEnergia12claves.pdf (accessed 5 January 2016).

Greenpeace (2015b) '*Compromisos Ambientales*'. Online: www. Greenpeace.org/espana/Global/espana/2015/imgs/radiografia/Evaluaci%C3%B3n%20programas%20electorales.doc.pdf (accessed 5 January 2016).

Hanemann, M., Labandeira, X. and Loureiro, L. (2010) 'Climate Change, Energy and Social Preferences on Policies: Exploratory Evidence for Spain', Economics for Energy, WP 03/2010, Online: http//:labandeira.eu/publicacions/ClimateChange.EnergyandSocialPreferencesonPolicies.pdf (accessed 5 January 2016).

Jacobs D. (2012) *Renewable Energy Policy Convergence in the EU. The evolution of Feed-In Tariffs in Germany Spain and France*, Farnham: Ashgate.

MINETUR (2014) *2014–2020 National Energy Efficiency Action Plan*, Online.: https://ec.europa.eu/energy/sites/ener/files/documents/2014_neeap_en_spain.pdf (accessed 5 January 2016).

Meira, P. (2013) La Respuesta de la sociedad española ante el cambio climático, Fundación Mapfre, Online: www.fundacionmapfre.org/documentacion/publico/i18n/catalogo_imagenes/grupo.cmd?path=1074055 (accessed 5 January 2016).

Morata F. and Solorio I. (eds) (2012) *European Energy Policy: An Environmental Approach*, Cheltenham: Edward Elgar.

Solorio, I. (2013) *'La Política Medioambiental Comunitaria y la Europeización de las Políticas Energéticas Nacionales de los Estados Miembros. La Política Europea de Renovables y su Impacto en España y el Reino Unido'*, PhD in International Relations, Universitat Autonoma de Barcelona UAB. Online: www.tdx.cat/bitstream/handle/10803/117262/ifss1de1.pdf?sequence=1 (accessed 5 January 2016).

Solorio, I. (2011) 'La Europeización de la Política Energética en España: ¿qué sendero para las renovables?', *Revista Española de Ciencia Política*, 25(4): 105–123.

12 The United Kingdom

A record of leadership under threat

Tim Rayner and Andrew Jordan

Introduction

The UK's engagement with the issue of climate change has been rather paradoxical. On the one hand it has been widely regarded as a global leader, both in terms of its international diplomatic effort and its domestic performance on greenhouse gas emission (GHGE) reduction (IEEP 2006). The UK reached its effort-sharing target under the EU's Kyoto Protocol emission reduction commitment ahead of schedule, and went on to over-deliver – one of only a handful of EU-15 member states to do so (EEA 2013). To help achieve this, it deployed several innovative policy instruments, such as emissions trading. Internationally, the UK has shown entrepreneurial leadership, principally during the Rio and Kyoto climate summits, and when chairing the G8 and acting in its Presidency of the EU. The celebrated Stern Review of the economics of climate change (Stern 2006) and decades of scientific research offer evidence of cognitive leadership. With the adoption of the 2008 Climate Change Act, the UK was widely lauded as the first country to enshrine in law carbon emission reduction targets – of up to 80 per cent by 2050 – with a view to removing policy from the vagaries of the issue attention cycle. The same Act also heralded wide-reaching five-yearly climate change risk assessments, to inform a rolling national programme of adaptation to climate impacts, confirming the UK as a leader in that domain as well.

And yet the emission reductions achieved during the 1990s were to a large extent a fortuitous by-product of unrelated energy policy reforms (Kerr 2007). In the 2000s, CO_2 emissions started to rise again, and a unilateral 1997 commitment to reduce them by 20 per cent by 2010 was quietly dropped. Moreover, it is questionable whether any politician in the UK has fully grasped the enormity of the challenge of radically decarbonising the national energy system in just over a generation, widely regarded as necessary to avoid the worst effects of climate change. Finally, the UK remains in many respects 'critically unprepared' to adapt to predicted climate impacts (EAC 2015).

In this chapter we outline the history of the UK's response to climate change, tracing the waxing and waning of leadership in the face of vested interests, economic and financial pressures, concerns over energy security and, latterly, a growing degree of 'luke-warmism' over the seriousness of the problem (Ridley 2015). The overall story is one of repeated attempts by policy entrepreneurs in governments of both right and left to 'join-up' government to deliver deeper emission cuts, but a rather piecemeal and often incoherent introduction of new policy instruments. In other words, the ability of governments to lead a low-carbon transition has been rather more fitful and halting than the UK's international reputation might suggest.

National attitudes to climate change

Public awareness of climate change is comparatively high in the UK. Polling suggests that by 2014, only 13 per cent of the population believed that climate change is not caused by human activity, down from 2011 when 21 per cent said it was mainly or entirely the result of natural processes (Capstick *et al.* 2014). Most people also reported willingness to reduce their own emissions, and supported government and businesses doing likewise. Fifty-three per cent were willing to make significant lifestyle changes; only 7 per cent were opposed to the UK signing international emission reduction agreements (ibid.).

There appears to be a 'disconnect', however, with electoral behaviour. At the 2005 election, for example, only 2 per cent of voters considered climate change the most important issue (Whiteley *et al.* 2005). After 2006, climate change did rise dramatically up the party-political agenda, for reasons which are disputed. While Khatri (2008: 575) suggests that the main driver was Prime Minister Blair, who 'undoubtedly [stole] a march' on the other parties by elevating the issue during the UK's Presidencies of the European Council and G8 and commissioning the Stern Review, others such as Carter and Jacobs (2014) argue that a wave of inter-party competition, in which each attempted to 'out-green' the other, was triggered by David Cameron's call (as part of a broader attempt to detoxify the Conservative party's image) for the country to 'Vote Blue, Go Green'.

Governments of right and left have been spurred into action by pressure from a range of stakeholders. Significantly, these have not only included the 'usual suspects' in the environmental movement, but increasingly business groups advocating the benefits of moving towards a low-carbon (or green) economy. Friends of the Earth's 'Big Ask' campaign, encouraged by New Labour government ministers personally committed to the agenda, brought together an unprecedented coalition of environment and development charities, trade unions, faith, community and women's groups, in support of the Climate Change Act (Carter and Jacobs 2014).

Phases of domestic climate change policy

1988–92

In this early phase, Prime Minister Margaret Thatcher's high-profile speeches to the Royal Society and UN General Assembly offered leadership, albeit rather symbolic. The most concrete outcomes were the establishment of the Hadley Centre for Climate Prediction and Research and an environment White Paper which advocated stabilisation of carbon emissions and the adoption of market-based instruments (HM Government 1990). This new-found concern, however, did not translate into support for a strong EU response: UK opposition was instrumental in the failure of the Commission's carbon-energy tax proposal. At the same time, the then-Environment Secretary Michael Howard is credited with negotiating a deal to ensure United States participation in the UN Framework Convention on Climate Change (UNFCCC) of 1992 (Haigh 1996).

In terms of emission reductions delivered, this period is more significant for developments in 'non-climate' policy. The privatisations of the oil, gas, coal and electricity sectors (many pushed through against concerted political opposition) and the ensuing 'dash for gas' had the completely unintended effect of lowering the UK's emissions throughout the 1990s (Kerr 2007).

1992–7

The Major government (1990–7) ratified the UNFCCC in 1993 and then set about returning GHG emissions to their 1990 levels by 2000 by means of the UK's first Climate Change Programme (1994). This relied heavily on voluntary measures, in what was termed a 'partnership approach', although the fuel duty 'escalator' (introduced in 1993 at a rate of 3 per cent per year above inflation) and the imposition of value added tax (VAT) on domestic fuel in 1994 were more interventionist. However, these fiscal instruments were arguably motivated primarily by revenue raising considerations.

1997–2006

The 1997 election of the New Labour government raised hopes among environmentalists. A manifesto commitment to a 20 per cent reduction of CO_2 emissions by 2010 was adopted, while the Treasury pledged to implement ecological tax reform, shifting the burden of taxation from 'goods' such as employment to 'bads' including waste and pollution. At Kyoto, Blair's deputy, John Prescott, helped to drive through an agreement based on the EU's favoured 'targets and timetables' approach. At home, he promised a more integrated and sustainable approach to national transport planning, eschewing large-scale road building. After a landmark Royal Commission report on the subject (RCEP 2000), the pace and urgency of climate policy making began to increase. The Climate Change Programme of 2000 included a Climate Change Levy on industry, linked to a system of domestic emissions trading: a high-profile manifestation of ecological tax reform. However, the so-called fuel-tax rebellion – a concerted and economically damaging protest against the high price of transport fuels – led to the suspension of the automatic fuel duty escalator in 2000. By 2007, the proportion of all taxation made up by green taxes was markedly less than it had been in 1997 (EAC 2007).

2006–10

Although the 2006 revamp of the Climate Change Programme was widely condemned as a missed opportunity – a 'model of incrementalism' (Carter and Jacobs 2014: 125) – a period of increased government ambition had begun. Climate change began to be framed as an economic opportunity, as Chancellor Brown's decision to commission the Stern Review testifies. In his final years in power, Blair became more committed to international climate diplomacy, partly to win over critics of his hugely unpopular role in Iraq. In 2008, Blair's successor Gordon Brown established the Department of Energy and Climate Change (DECC) and adopted the landmark Climate Change Bill, in part to respond to Tory attempts to claim the issue as their own (Carter and Jacobs 2014). A broad-based Low Carbon Transition Plan (HM Government 2009) showed far greater ambition than the 2006 programme. At EU level, the UK was broadly supportive of the Commission's 2008 climate and energy package. Globally, the UK worked actively for a deal at the December 2009 Copenhagen summit, and was among those pressing to raise the EU's unilateral emission reduction commitment from 20 to 30 per cent (Skovgaard 2014).

2010–15

This period covers the term of the Conservative–Liberal Democrat government in which climate commitments formed a key element of the coalition agreement and which, according to the newly-elected Prime Minister Cameron, was to be 'the greenest government ever'. ENGOs assumed that having the 'Lib Dems' in the coalition (running DECC) would keep the Conservatives 'honest' to their apparent conversion to climate protection. Tensions were soon evident, however, in Conservative Chancellor Osborne's pronouncement that '[w]e're not going to save the planet by putting our country out of business' (*Daily Telegraph* 2011) and then in Cameron's apparent instruction to ministers to 'get rid of all the green crap' (Mason 2013). Fierce Cabinet-level debate surrounded the adoption of the Climate Change Act's Fourth Carbon Budget, which required a 50 per cent reduction in emissions by 2025, against a 1990 benchmark. The budget was only finally adopted in July 2014 when Cameron over-ruled Osborne (*Carbon Brief* 2014).

Despite these clear fault-lines, the lead up to the 2015 election saw the three main Party leaders reaching a joint agreement (brokered by the Green Alliance NGO) that recognised the importance of domestic carbon budgets, a legally-binding global climate deal to limit temperature rise to below 2°C, and the need to accelerate the transition to a competitive, energy-efficient, low-carbon economy, phasing out unabated coal for power generation.

2015

In 2015, the Conservatives were freed from the constraints of coalition with the Lib Dems, to govern as a majority party. A series of decisions collectively referred to as an energy policy 'reset' – including the removal of financial support for renewables and energy efficiency measures while offering tax breaks for oil and gas extraction, and the axing of a planned £1 billion competition to commercialise carbon capture and storage (CCS) – provoked public warnings from the Committee on Climate Change about the achievability of carbon budgets (CCC 2016). They also threatened the UK's international diplomatic credibility in the run-up to the 2015 Paris climate conference (COP21) (Witte 2015). A decision to revive large-scale road building had already been taken in 2014.

Institutional responses, policy programmes and instruments

The overall story of the last two and a half decades has been one of repeated attempts by policy entrepreneurs in governments of all political colours to improve 'joined-up' government, often compromised by the unwillingness of the most powerful departments and politicians to cooperate (Lockwood 2013). Policy innovation has clearly been evident in the mix of policies deployed, but has mostly been of a rather piecemeal and incremental kind, leaving a degree of incoherence. More enduring has been the emergence of a comparatively sophisticated set of institutions for appraising and evaluating the UK's performance.

Institutional responses

The UK has been willing (and able) to readjust the 'machinery of government' in pursuit of the goal of environmental policy integration comparatively frequently; the incorporation in 1997 of the powerful transport department into a Department of Environment, Transport and the Regions 'super-ministry' providing a prominent case. However, ambitious pledges were subsequently diluted and the department was eventually dismembered in 2001. Thereafter, the responsibility for climate issues was scattered across a number of departments.

In 2006, an influential new Office of Climate Change (OCC) was introduced as a shared resource across the main departments with climate change-related responsibilities, working closely with the Treasury, Cabinet Office and No. 10. The subsequent creation of DECC was widely praised by both ENGOs and business groups, who had long called for greater coordination between the energy and climate change briefs. Arguably, the primary benefit of this was that climate considerations got a stronger airing in Cabinet, from both DECC and the longer-standing Department of Environment, Food and Rural Affairs (DEFRA).

Most significant – and potentially radical – of all, the Climate Change Act created new mechanisms for long-term governance, designed to shield climate policy from the vagaries of the issue attention cycle and to some extent from party-political conflict. It put into statute national targets to reduce greenhouse emissions through domestic and international action by at least 60 per cent by 2050. It also required binding limits known as carbon budgets on aggregate GHGE over five-year periods. To advise on these, and on the pathway to the 2050 target, a new, independent Committee on Climate Change was created.

The Committee subsequently recommended that the 2050 GHGE reduction target be increased to 80 per cent (Lockwood 2013). A new system of annual Government reporting to Parliament in response to progress reports by the Committee was designed to improve transparency and accountability.

National climate change policy programmes

The first national climate change programme of 1994 was reviewed and updated by Labour administrations in 2000 and 2006. The 2006 programme, just predating rise of greater party competition over climate policy, was criticised for taking insufficient steps to deliver the UK's domestic 2010 target, neglecting potentially effective measures that were deemed politically unacceptable (e.g. lowering speed limits) or requiring a longer timeframe, and neglecting new fiscal measures (NAO 2007; EAC 2007). The more ambitious 2009 Low-Carbon Transition Plan, by contrast, aimed to transform the UK's energy system, including a seven-fold increase in renewable supply to 15 per cent (spurred on by targets set at EU level). In establishing a 'low-carbon industrial strategy', it introduced a brand new field of activity, seeking to increase technology, output, and employment gains from meeting the new targets. Though a commitment to effectively decarbonise the electricity sector by 2030 failed to make it into the final text, the Energy Act of December 2013 altered much of the landscape, establishing a legislative framework for delivering secure, affordable and low-carbon energy and including a wide range of measures, under a programme of Electricity Market Reform (see below).

Policy instruments

The UK's preference for market-oriented policy instruments that can reduce emissions at least cost is well known, though sometimes exaggerated. The UK emissions trading scheme (ETS) introduced in 2002, however, constitutes one highly innovative attempt to devise a cost-effective, market-based system for reducing industrial emissions, that found favour with many sections of industry (Wurzel 2008). Arguably, by developing the instrument, the UK was providing a degree of cognitive leadership to the rest of Europe.

The introduction of the Climate Change Levy (CCL) in 1999 represented the most significant eco-taxation measure to date. The tax was levied on all non-household uses of coal, gas, electricity, and non-transport liquid petroleum gas. However, significant concessions were deemed necessary, including an 80 per cent discount for energy-intensive users in return for the adoption of voluntary targets for improving energy efficiency through Climate Change Agreements (CCAs). In keeping with the recommendations of a report on economic instruments from the industrialist Lord Marshall (1998), eco-taxes (i.e. the CCL), voluntary agreements (i.e. the CCAs) and emissions trading became interlinked to form a mixed approach, but one which was criticised for being too responsive to industry pressure (Helm 2007a).

In the case of renewable energy, instruments have been designed to encourage the increased supply of electricity from renewable sources in the context of a liberalised market. From 2002, the Renewables Obligation required suppliers to purchase and supply a certain amount of electricity generated from renewable sources. This was a form of renewable portfolio standard, but the market-based element came from the tradeable nature of the Renewable Obligation Certificates (ROCs) issued by the regulator. The instrument's lack of flexibility and versatility, however, caused the UK to miss its target of 10 per cent of electricity from renewable sources by 2010 (Szarka 2007). In more recent years, and faced with the need to meet EU targets, governments have shown a greater willingness to adopt policy approaches – namely feed-in tariffs – that have proved successful in member states with less liberalised energy sectors.

The Electricity Market Reform proposals first set out in 2010, and incorporated in the 2013 Energy Act, consisted of four key mechanisms, intended to address the need for both cost-effective decarbonisation and new generation capacity. First, 'contracts for difference' are the government's new system for supporting low-carbon (including nuclear) energy. These long-term contracts between Government and a low-carbon generator give a guaranteed tariff or price for electricity over a defined period of time. A top-up payment between a market-based reference price and a pre-defined 'strike price' reduces price volatility. Together with a system of feed-in tariffs for smaller generators, these contracts will replace the Renewables Obligation by 2017. As a second mechanism, the Capacity Market provides a regular payment to reliable forms of capacity (both demand and supply), in return for the capacity being available when the system is tight. Third, an emission performance standard prevents the construction of new unabated coal power stations. Finally, a carbon price floor augments the (inadequate) carbon price incentive generated by the EU emissions trading scheme (EU ETS) by extending the CCL to the fossil-fuel inputs to electricity production. For all its rhetoric in favour of market mechanisms, the UK has been alone in implementing such an interventionist measure.

Compared to industry and energy, measures covering the domestic and transport sectors have taken longer. One of the very first acts of the New Labour government was to cut the VAT levied on domestic fuel from 8 to 5 per cent. Although this positively affected fuel poverty,[1] it *increased* carbon emissions. It was not until 2002 that the main policies for the household sector – the Energy Efficiency Commitment and reform of the Building Regulations – were introduced. Arguably the Government has mistakenly regarded energy efficiency as a no-cost option, whereas an effective approach requires significant up-front expenditure.

In the transport sector, despite increasing rhetoric throughout the 1990s regarding the need to shift journeys from private cars to public transport, governments have been unwilling to address the long-term decline in the real costs of motoring since the 1970s. With the demise of the road fuel duty escalator (see above), road pricing has offered the only potential alternative. On this, with the notable exception of London, politicians have shied away from the political risk, preferring a series of incremental supply-side initiatives instead.

Climate change: a threat or an opportunity?

During the 1990s and 2000s, 'ecologically modern' storylines were invoked as a form of cognitive leadership to win support for more ambitious climate policies. Several policy documents (and speeches) could be cited, including the Treasury's Statement of Intent on Environmental Taxation (HM Treasury 1997) and the Low Carbon Transition Plan of 2009. The UK ETS, introduced in 2002, was explicitly intended to deliver 'first mover advantages' to UK companies by giving them experience prior to the anticipated launch of the EU's scheme. Prime Minister Blair (2004) declared that 'the very act of solving [climate change] can unleash a new and benign commercial force ... providing jobs, technology spin-offs and new business opportunities'.

The clearest and most authoritative expression of the ecologically modern framing of climate change – and act of cognitive leadership – came from the Treasury-commissioned Stern Review:

> Tackling climate change is the pro-growth strategy for the longer term, and it can be done in a way that does not cap the aspirations for growth of rich or poor countries. The earlier effective action is taken, the less costly it will be.
>
> (Stern 2006: ii)

Although subsequently criticised for cherry picking the most favourable evidence (Helm 2007b), it had the politically intended effect of demonstrating that Labour was taking climate policy seriously (Jordan and Lorenzoni 2007) and galvanising support for stronger action at global level.

Interestingly, at times it has been industry, or at least sections of it, that have been the most emphatic exponents of moving to a low-carbon economy. In June 2006, the Corporate Leaders Group on Climate Change sent a public letter to the Prime Minister calling for government to provide a clearer and more ambitious framework – even if that involved much more challenging targets and tougher regulations. Even the CBI shifted its stance, noting that 'alongside the risks, the shift to a low-carbon economy offers the UK a unique opportunity to develop innovative environmental technologies of the future and prosper in new, multi-billion-dollar world markets' (CBI 2007). In 2015, sections of industry expressed serious concern about the effects on investor confidence of moves to dismantle several key policies (Petry *et al.* 2015). But such moves by government have highlighted how tenuous was the hold achieved by ecologically modern storylines among senior policy makers in the wake of the international financial crisis.

Multi-level governance and leadership

In general, the relationship between UK national, EU and international policy should be seen against the backdrop of two underlying currents. The first is the

UK's instinctive preference to exercise leadership at the international level (where its seat on the UN Security Council provides much political weight) and in scientific fora. By contrast, it has been less comfortable exercising leadership in the EU (where voting rules are less favourable and public opposition to deeper European integration is profound). That said, from the mid-1990s, Labour politicians did became much more willing to lead the EU towards deeper harmonisation in areas like energy where national sovereignty has normally been jealously guarded, with Blair vying with Merkel to be climate champion on the European Council (Boasson and Wettestad 2013). The second current has been the steady flow of new environmental policies from the EU, which have transformed national policy and politics (Bache and Jordan 2006). However, in climate policy, Europeanization has been a less significant driver, not least because EU policy was slower to evolve, only really taking off in the period after *c.*2000 (Jordan *et al.* 2010). In the remainder of this section we therefore mainly discuss the UK's leadership at the global level, then its record of leadership in the EU.

Diplomatic leadership

As noted above, the UK played a much more active role in achieving agreement on the UNFCCC than the European Commission (Haigh 1996), with then-Environment Secretary Michael Howard brokering a compromise with Washington that became enshrined in Article 4 (2) of the Convention; undeniably a case of skilful diplomacy – or entrepreneurial leadership – by a country that at the time did not hold the EU Presidency. A similar point can be made about John Prescott's role in negotiations at Kyoto, which contributed to (a brief) acceptance of targets and timetables by the US. Subsequently, Prime Minister Blair has been credited with securing an unprecedented acknowledgment of the need to act on climate change from the George W. Bush administration.

The UK has also achieved some notable international diplomatic successes acting in its capacity as the EU's Presidency. During the second half of 2005, faced with the real prospect of stalemate at the Montreal conference of the parties, the combined efforts of the EU and the Canadian hosts facilitated a decision to initiate talks on the future development of the UNFCCC, ensuring that the momentum from the entry into force of the Kyoto Protocol was maintained (IEEP 2006). Failure to keep the process moving would have dented the prospects for the Clean Development Mechanism (CDM) in particular.

Since 2006, the UK has also sought to 'shift climate change into the realms of high politics' (Khatri 2007: 591), linking it to related challenges such as poverty alleviation and international security. In April 2007, for example, the UK raised climate change for the first time at the UN Security Council, despite stiff opposition from the US, Russia and China (see also Chapters 16 and 17). Since 2010, however, deep financial cuts have severely affected the Foreign Office's ability to conduct climate diplomacy (Darby 2014).

EU leadership

As already noted, since the mid-1990s the UK has consistently been at the more ambitious end of the scale regarding the EU's overall level of mitigation ambition. And at times, it has furthermore shown itself willing to take on a relatively large share of total effort, under the EU's internal burden sharing agreements. In 1997 it offered to take on a 10 per cent reduction, but agreed to increase this to 12.5 per cent after the Kyoto accord – the only state prepared to shoulder a greater burden (Jordan *et al*. 2010).

Having gradually adopted a more pro-European stance in the late 1990s (Jordan 2002), the UK was better placed to set the climate policy agenda in the EU through the development of domestic policy instruments that could be 'uploaded' to the rest of Europe. The secondment of officials with expertise on the UK ETS to the Commission to assist drafting of the EU ETS Directive can be regarded as a case of both cognitive and entrepreneurial leadership. These efforts were not entirely successful, however, and UK policy makers quickly learned the lesson that in a Union of 27 states, potential first mover advantages can be hard to translate into concrete policy outputs. Nevertheless, the UK played a positive role during the preparations for the EU ETS's third trading phase beyond 2012, agreeing on the need for greater harmonisation, increased transparency and clearer allocation rules (Wurzel 2008). Although the Climate Change Act was not something that could be readily scaled-up to the EU level, in 2009 an unprecedented partnership between government and ENGOs toured other member state capitals to promote it (Murray 2009).

In areas where EU policy had yet to substantially develop, the UK has also shown leadership. In December 2005, for example, the Environment Council under the UK Presidency gave the Commission a clear mandate to include aviation in the EU ETS (IEEP 2006). Meanwhile, the UK government's early start on developing its own adaptation policy meant that it was broadly supportive of Commission efforts to mainstream consideration of the implications of climate impacts into relevant policy sectors, such as water and agriculture.

While these examples tell a positive story about the UK's attitude to leadership in the EU, far less flattering cases are not hard to find. The UK's reaction to key aspects of the Commission's 2008 climate and energy package was far less enthusiastic. Most notably, it pressed for a weakening of the revised ETS Directive by allowing greater use of external credits (*ENDS Report* 405: 46–47), and sought to weaken proposed renewable energy policies. In the development of post-2020 targets, the UK successfully opposed the adoption of binding national renewable energy targets for 2030 (Harvey 2013). Similarly, the non-binding nature of the proposed 27 per cent energy efficiency target for 2030 owes much to UK reluctance. Thus overall, the pattern of leadership has been rather more variable at the EU level than it has at the global.

The domestic implementation of EU and international commitments

The UK has been in the somewhat fortunate position of being able comfortably to meet its burden-sharing target of –12.5 per cent as its share of the EU's Kyoto commitment, and ultimately to over-deliver without the use of carbon sinks or flexible mechanisms (EEA 2013). Emissions in the traded sector have declined by 15 per cent from 2005 to 2012, thanks to reduced demand for electricity and increased generation from wind power. Furthermore, emissions in industrial sectors, which had grown until 2008, had also dropped by 7 per cent below 2005 levels by 2012 (EEA 2013). Emissions covered by the EU's Effort Sharing Decision (ESD) (i.e. in the sectors not covered by the ETS) are projected to be below the 2020 target with the measures currently in place (EEA 2014).

In signing up to 2007 EU 2020 targets, profound changes were set in train. Carter and Jacobs (2014: 136) highlight how these targets, which required a 34 per cent reduction in GHG emissions and a 15 per cent share for renewable energy from the UK, meant that governments 'were effectively compelled to adopt a much more interventionist energy policy. The result was a major overhaul that included new and increased subsidies, new industrial incentives, and a new planning regime'.

The costs of mitigation have turned out to be far higher than anticipated. A core element of the climate change programme was the domestic renewables target of 10 per cent by 2010 (inspired by EU legislation). Not only was this impossible to achieve, but the cost per tonne of carbon saved was extremely high, at between £280 and £510 (NAO/DTI 2005). An equally problematic issue is the emission profile of the transport sector. Given the background of rapidly rising traffic growth, concessions on fuel duty and reliance on an ineffective EU-level voluntary agreement with car manufacturers to deliver more efficient vehicles, it is hardly surprising that UK transport sector carbon emissions declined by only 2.3 per cent since 1990 (*ENDS Report* 483).

As far as transposition is concerned, the UK has generally implemented its EU obligations. One prominent exception has been the UK's implementation of the EU ETS. In November 2004, the Government attempted to revise the National Allocation Plan (NAP) submitted in January, claiming it had simply been a draft. Whether the British government's revised NAP – which increased the total number of allowances by 20 million – was primarily driven by industry lobbying, or due to the publication of updated CO_2 emission projections, remains open to dispute. The UK subsequently took the Commission to the European Court of Justice (ECJ), only to abandon proceedings to prevent uncertainty undermining investor confidence and the fledgling EU ETS, which the government strongly supported in principle (Wurzel 2008).

The targets for 2020 provide a sterner test of the UK's commitment to emission reduction, especially as they relate to renewable energy – despite the country's excellent potential for wind and wave power. In order to reach its target, the UK will need an average annual growth rate for renewable energy consumption 17.6 per cent in the run-up to 2020. In absolute terms, this is equivalent to 4.1 times its cumulative effort so far (EEA 2014). In late 2015, the Climate and Energy Secretary Amber Rudd was forced to acknowledge that the UK was not on course to meet its 2020 target (Kaminski 2015), but despite this, proceeded as planned with the removal of subsidies.

Conclusion

In this concluding section, we reflect on the extent to which the UK has taken on a sustained leadership role in climate politics, both in terms of emission reduction domestically, and influencing political processes at EU level and beyond. Where it has been able, we suggest how this proved possible. We also reflect on the prospects for a more 'heroic' style of leadership, following the adoption of long-term targets in the Climate Change Act.

While the UK has not exercised the kind of structural leadership exhibited by Germany (a much larger, higher emitting economy), it has been more able to demonstrate entrepreneurial and cognitive leadership, particularly on the international stage. This has to be seen against an underlying trend away from defensive and introspective attitudes towards EU environmental policy since the UK's accession in 1973 (Jordan 2002).

In terms of leadership by example, the UK's record of emission reductions achieved appears fairly impressive. Without its contribution (and that of Germany) the EU as a whole would have faced a much tougher task to meet its 8 per cent reduction target. However, as we have noted, this has to a large extent been what Liefferink and Birkel (2011) refer to as 'cost-free' leadership. According to critics, emission reductions achieved serendipitously – primarily by unrelated reforms to the energy sector – have allowed UK politicians to posture on the climate issue, rather than develop a coherent strategy more in line with IPCC recommendations (Kerr 2007). On the occasions when real effort has been required, as when projections revealed a gap between the UK's domestic emission reduction target and what was likely to be achieved, leadership has been shown to be lacking. Specifically, the revamp of the Climate Change Programme in 2006 was widely perceived to have been a missed opportunity (IEEP 2006). The government's approach in critical areas such as eco-tax reform, renewable energy deployment and road pricing has fallen well short of environmentalists' expectations. All too often, the government has failed to make a coherent case to the public, arguably making future leadership in this area even more politically challenging.

The Climate Change Act has provided a major demonstration of renewed leadership over the long term. In terms of the stringency and legal force of the 80 per cent target and the comprehensiveness of its associated evaluation machinery, the Act undeniably puts the UK in a world-leading position. It is tempting to classify the Climate Change Act as a case of 'heroic' leadership, in that it 'sets explicit long-term objectives to be pursued by maximum coordination of public policies and by an ambitious assertion of political will' (Hayward

2008: 7). However, it was only possible once key stakeholders had persistently lobbied the government to lead and, in effect, expressed *their willingness to be led*. But it remains questionable whether any politician in the UK has fully grasped the enormity of the challenge of radically decarbonising the energy system in just over a generation, and the extent to which governments will continue to heed the advice of the Committee on Climate Change remains unclear (Lockwood 2013). After the Act's adoption, policies that undermine long-term decarbonisation efforts did not take long to surface, including the 2009 decision to allow a third runway at Heathrow airport and the apparent energy policy 'reset' of 2015. A complex and frequently changing policy landscape has resulted in confusion about available incentives and planning requirements, significantly delaying necessary investment decisions in renewable energy and energy efficiency.

That said, the UK has demonstrated greater commitment, and offered a far greater degree of cognitive and entrepreneurial leadership, at international level. In terms of entrepreneurial leadership, Howard's intervention at Rio, Prescott's role at Kyoto and Blair's industrious elevation of the issue during the UK's chairmanship of the G8 in 2005 were especially notable (IEEP 2006).

In terms of cognitive leadership, the UK's sponsorship of major scientific events such as the symposium on avoiding dangerous climate change at Exeter in 2005, and its ongoing efforts to reframe climate change as an issue that extends well beyond the narrow environmental 'ghetto', deserve similar recognition. The Stern Review was deliberately and very consciously targeted at a global audience, and represents an attempt by the UK to act as a global leader, by helping to break the logjam on commitments after the expiry of the Kyoto Protocol.

Exactly how far the UK's record will be weakened by resurgent Tory scepticism over the need to 'burden' industry and taxpayers with climate policy and remain a full member of the EU, remains unclear at the time of this writing. The apparent pre-election joint commitment to the Climate Change Act is reassuring, but will be useless without a consistent and coherent policy mix. Agreeing the fourth carbon budget was a struggle due to Treasury opposition, despite the Lib Dems running DECC. In the absence of strong and sustained governmental leadership, new coalitions and sources of societal pressure will most likely be needed if the Act's 80 per cent (or potentially higher) reduction target is to be delivered.

Afterword

The momentous result of the UK's 'In-out' referendum on EU membership, held in June 2016, throws the future of climate policy into question at national, EU and indeed global scales (Rayner and Moore 2016). Much will depend on the type of new relationship that is now forged: one in which the UK remains within the European Economic Area (EEA), like Norway (which implements a great deal of EU legislation in return for access to the single market), or one where even that degree of involvement is rejected in favour of some kind of 'free-trade' option. It will also depend on the complexion of the UK government that oversees the Brexit process. Should the UK eventually opt to dispense with EU climate and energy-related targets (rejecting what might be called the Norwegian option), some suggest that the 2008 Climate Change Act would guarantee the continuation of ambitious decarbonisation efforts. But without the reinforcing effect of EU targets, critics of the Act could be emboldened in their efforts to undermine the UK's own, world-leading system of carbon reduction target setting. Moreover, even if it adopts EEA status, without the UK advocating high ambition from its seat on the Council, EU targets would be less likely to keep pace with the UK's, and pressure to weaken the latter would only multiply. In a world where the success of the Paris Agreement is said to depend on countries voluntarily 'ratcheting up' ambition in periodic review cycles in the knowledge that others are doing likewise, foot-dragging on the part of both the UK and the EU would slow progress at precisely the time it is most needed.

Note

1 A household is said to be in fuel poverty when its members cannot afford to keep adequately warm at reasonable cost, given their income.

References

Bache, I. and Jordan, A.J. (eds) (2006) *The Europeanization of British Politics*, Basingstoke: Palgrave.

Blair, T. (2004) 'International action needed on global warming', speech at the Banqueting House, 14 September 2004.

Boasson, E and J. Wettestad (2013) *EU Climate Policy*, Farnham: Ashgate.

Capstick, S.B., Demski, C., Sposato, R., Pidgeon, N., Spence, A. and Corner, A. (2015) 'Public perceptions of climate change in Britain following the winter 2013/2014 flooding'. Understanding Risk Research Group Working Paper 15–01, Cardiff: Cardiff University.

Carter, N. and Jacobs, M. (2014) 'Explaining radical policy change: the case of climate change and energy policy under the British Labour Government 2006–10'. *Public Administration* 92(1): 125–141.

Carbon Brief (2014). 'Government decides not to amend UK's fourth carbon budget'. 22 July, www.carbonbrief.org/government-decides-not-to-amend-uks-fourth-carbon-budget.

CBI (2007) *Climate Change: Everyone's Business*, London: Confederation of British Industry.

CCC (2016). 'Implications of the Paris Agreement for the fifth carbon budget'. https://www.theccc.org.uk/publication/implications-of-the-paris-agreement-for-the-fifth-carbon-budget/.

Daily Telegraph (2011). 'Conservative Party Conference 2011: George Osborne speech in full', www.telegraph.co.uk/news/politics/georgeosborne/8804027/Conservative-Party-Conference-2011-George-Osborne-speech-in-full.html.

Darby, M. (2014) 'UK slashes climate diplomacy budget'. *Climate Home*. 6 August, www.climatechangenews.com/2014/07/31/uk-slashes-climate-diplomacy-budget/.

DETR (2000) *Climate Change – the UK Programme*, Norwich: HMSO.

EAC (2007) *Pre–Budget 2006 and the Stern Review, Fourth Report of Session 2006–7*, HC 227, London: Stationary Office.

EAC (2015) *Climate Change Adaptation. Tenth Report of Session 2014–15*, HC 453, London: The Stationery Office.

EEA (2013) *Trends and Projections in Europe 2013*, Copenhagen: European Environment Agency.

EEA (2014) *Country Profile – United Kingdom*, Copenhagen: European Environment Agency.

ENDS Report, various issues, London: ENDS (Environmental Data Services).

Haigh, N. (1996) 'Climate change policies and politics in the European Community', in O'Riordan, T. and Jaeger, J. (eds), *Politics of Climate Change*, London: Routledge, 155–183.

Harvey, F. (2013) 'Britain resists EU bid to set new target on renewable energy', *Guardian*, May 25th, www.theguardian.com/environment/2013/may/25/uk-blocks-eu-target-renewable-energy.

Hayward, J. (ed.) (2008) *Leaderless Europe*, Oxford: Oxford University Press.

Helm, D. (2007a) 'Climate change policy: lessons from the UK', *Economists' Voice*, www.dieterhelm.co.uk/publications/Climate_change_policy_lessons_from_UK.pdf.

Helm, D. (2007b) 'Climate change: sustainable growth, markets, and institutions', www.dieterhelm.co.uk/publications/HDR_climate_change_apr_07.pdf.

HM Government (1990) *This Common Inheritance*, Cm 1200, London: HMSO.

HM Government (2009) *The UK Low Carbon Transition Plan*, Norwich: The Stationary Office.

HM Treasury (1997) 'Statement of intent on environmental taxation', London: HM Treasury.

IEEP (2006) *Climate Change Action. The UK: Leader or Laggard?* London: Institute for European Environmental Policy.

Jordan, A.J. (2002) *The Europeanization of British Environmental Policy*, Basingstoke: Palgrave.

Jordan, A.J., Huitema, D., van Asselt, H., Rayner, T. and Berkhout, F. (eds) (2010) *Climate Change Policy in the European Union*, Cambridge: Cambridge University Press.

Jordan, A.J. and Lorenzoni, I. (2007) 'Is there now a political climate for policy change?' *Political Quarterly* 78(2): 310–319.

Kaminski, I. (2015) 'Rudd admits UK failing to meet 2020 renewable energy target'. *ENDS Report*, 10.11.2015.

Khatri, K. (2007) 'Climate change', in Seldon, A. (ed.) *Blair's Britain, 1997–2007*, Cambridge: Cambridge University Press.

Kerr, A. (2007) 'Serendipity is not a strategy: the impact of national climate programmes on greenhouse-gas emissions', *Area* 39(4): 418–430.

Liefferink, D. and Birkel, K. (2011) 'The Netherlands: a case of cost-free leadership', in Wurzel, R. and Connelly, S. (eds). *The European Union as a Leader in International Climate Change Politics*, London: Routledge, 147–162.

Lockwood, M. (2013) 'The political sustainability of climate policy'. *Global Environmental Change*, 23: 1339–1348.

Mason, R. (2013). 'David Cameron at centre of "get rid of all the green crap" storm'. *Guardian*, 21 November, www.theguardian.com/environment/2013/nov/21/david-cameron-green-crap-comments-storm.

Murray, J. (2009) 'Britain urges European neighbours to adopt UK-style climate law, *BusinessGreen*, 16 November. www.businessgreen.com/business-green/news/2253190/britain-urges-european.
NAO (National Audit Office) (2007) *Cost-Effectiveness Analysis in the 2006 Climate Change Programme Review*, London: National Audit Office.
NAO/DTI (2005) *Renewable Energy*, Report by the Comptroller and Auditor General, HC 210 Session 2004–2005, National Audit Office, London: TSO.
Pétry, F. *et al.* (2015) 'Open letter – action on energy in 2016'. http://news.cbi.org.uk/news/business-calls-for-clear-leadership-and-stable-energy-policy/.
Rayner, T. and Moore, B. (2016) 'Climate policy', in Burns, C. *et al. The EU Referendum and the UK Environment: An Expert Review*. http://environmenteuref.blogspot.co.uk/p/the-report.html.
RCEP (Royal Commission on Environmental Pollution) (2000) *Energy – The Changing Climate*, 22nd report, Cm 4794, London: The Stationery Office.
Ridley, M. (2015) 'My life as a climate lukewarmer'. www.rationaloptimist.com/blog/my-life-as-a-climate-lukewarmer.aspx.
Skovgaard, J. (2014) 'EU climate policy after the crisis', *Environmental Politics*, 23(1): 1–17.
Stern, N. (2006) *The Economics of Climate Change*, Cambridge University Press: Cambridge.
Szarka, J. (2007) *Wind Power in Europe*, Basingstoke: Palgrave Macmillan.
Whiteley, P., Stewart, M., Sanders, D. and Clarke, H. (2005) 'The issue agenda and voting in 2005', in Norris, P. and Wleizen, C. (eds) *Britain Votes*, Oxford: Oxford University Press.
Witte, G. (2015) 'On eve of Paris climate summit, Britain pulls the plug on renewables', *Washington Post*, 20 November, www.washingtonpost.com/world/europe/on-eve-of-paris-climate-summit-britain-pulls-the-plug-on-renewables/2015/11/20/240c5630-8311-11e5-8bd2-680fff868306_story.html?postshare=6891448075871920&tid=ss_mail.
Wurzel, R.K. (2008) *The Politics of Emissions Trading in Britain and Germany*, London: Anglo-German Foundation.

13 Norway

A dissonant cognitive leader?

Elin Lerum Boasson and Bård Lahn

Introduction

Norway is currently responsible for 1/1000 of global greenhouse gas emissions (GHGE). Its domestic emissions in 2015 were slightly higher than in 1990 (Miljøstatus 2015), due mainly to rapid growth in oil and gas extraction. This has also led to a substantial increase in public and private wealth from 1990 and onwards. Norway exports five times more energy than it consumes, ranking as the world's tenth largest exporter of oil and the third largest exporter of natural gas (Boasson 2013: 15; Ministry of Petroleum and Energy 2014: 13).

Despite its increasing dependence on fossil fuels, Norway has persistently sought a leadership role in international climate diplomacy. Together with the USA (see Chapter 16), Norway early on championed ideas such as cost-efficiency and emissions trading. These approaches had a significant imprint on the Kyoto Protocol, and later gained prominence in EU policy-making. More recently, Norway has emerged as a leader in developing new forms of results-based climate finance in developing countries, for example for reducing emissions from deforestation.

Norway's dual role as a major producer of oil and gas and as a country aspiring to climate leadership has been seen as a 'contradiction' (Norgaard 2006: 365), a 'paradox' leading to 'cognitive dissonance' (Eide *et al.* 2014), and a source of 'role-strain' (Eckersley 2013: 395). In this chapter we present a different argument, showing how Norway has adopted a global cost-efficiency approach that serves to reconcile domestic obstacles to climate action with global leadership ambitions. A key element in this approach has been the development of a binding global regime, based on flexible national commitments and emissions trading (Boasson 2015; Lahn 2013; Gullberg 2009: 5). Norway's approach is explicitly based on the view that a country does not necessarily have to achieve domestic GHGE reductions in order to be a global leader – the important thing is to promote solutions that can work well on a global scale. In that sense, Norway's international climate policy leadership has remained remarkably consistent over time.

Therefore, while it may well be argued that the combination of climate leadership and petroleum production presents a paradox, this chapter shows how Norway has championed a climate policy ideal that seeks to overcome this potential contradiction. Instead, we indicate a different and more recent 'dissonance' in Norwegian climate policy: over the past few years, Norway has adopted a series of policies to reduce domestic GHGE that run counter to the ideal of global cost-efficiency. What we see is a country that seeks to provide cognitive and entrepreneurial leadership in promoting global cost-efficiency as an ideal for climate policy internationally, while at the same time adopting a somewhat more proactive domestic policy than warranted by that ideal.

Although Norway is not a full EU member, it is obliged to adopt most EU energy and climate policies under the European Economic Area Agreement (EEAA). When it comes to international climate policy-making, however, Norway has the opportunity to play a different role in multilateral processes than most EU member states. For example, it operates independently from the EU in UN negotiations on climate change, which has made it easier for Norway to provide entrepreneurial and cognitive leadership in this and related processes.

National attitudes to climate change: public support, political conflicts

Norway started to develop a national climate policy already in the late 1980s (Boasson 2013: 21). In 1989, 40 per cent of polled voters stated that they were 'very concerned' about climate change (Tjernshaugen, Aardal and Gullberg 2011: 334–335; Austgulen and Stø 2013: 144). Concern dropped markedly during the 1990s and early 2000s, but increased again towards the end of the decade. Since 2009, some 20 to 30 per cent of those surveyed have consistently ranked climate change among the top challenges facing Norway, rising to 34 per cent in 2015 (TNS Gallup 2015).

Climate issues have been high on the agenda of most political parties in Norway for quite some time. Labour and the Conservatives are opponents on other major issues, but are remarkably similar as regards climate change. Both parties support a global cost-efficiency approach, relying on the development of a binding global regime based on flexible national commitments, global emissions trading and other measures to ensure that least-costly mitigation options are realized first (Boasson 2015; Gullberg 2009: 5; Lahn 2013).

By contrast, a grouping of smaller parties – Christian Democrats, Liberal Party, Centre Party (former Agrarian Party) and Socialist Left Party – has opposed global cost-efficiency and promoted more national measures, such as direct regulation of large emitters and state aid for renewables. The same applies to the Green Party, which won a seat in the Storting (Norwegian parliament) for the first time in the 2013 elections.

Since 2001, government formation in Norway has relied on the ability of either Labour or the Conservatives to create stable agreements with the smaller parties in the Storting (Boasson 2005). This need for alliances with parties that promote domestic-level measures for mitigating climate change has resulted in more national action on the issue. Labour, the Centre Party and the Socialist Left Party formed a majority coalition government from 2005 to 2013.

The right-wing Progress Party is the only party which doubts or denies that climate change is human-induced. This position has been muted, but continues to surface from time to time. However, the party has supported the global cost-efficiency approach. In October 2013, the Conservatives and the Progress Party formed a coalition government. As these parties together still represented a minority in the Storting, the government furthermore required support from the Christian Democrats and the Liberal Party; Norwegian climate policy has therefore continued along much the same path as before.

Norway has a broad range of national environmental NGOs (ENGOs), ranging from traditional nature protection organizations to newer groups focusing on climate change (Boasson 2015; Tjernshaugen 2007). The major international environmental group, Word Wide Fund for Nature (WWF), gained a foothold in Norway after 2000. Environmentalists broadly challenge the global cost-efficiency approach, but are more positively disposed to market measures than are similar groups in other European countries. Various business organizations engage in climate issues, most prominently oil and gas, energy-intensive industry and electricity industry associations. Business primarily supports global cost-efficiency, but there are considerable internal conflicts of interest within this grouping that seldom surface in political discussions (see Boasson 2015). Trade unions have campaigned for more use of gas domestically, but have otherwise not been deeply involved in climate issues.

Phases in Norwegian climate-change policy

1987–1999: global cost-efficiency dominates climate talk and (in)action

In the late 1980s, Norway's international climate involvement was off to a head start with Prime Minister Gro Harlem Brundtland (Labour Party) chairing the UN-appointed World Commission on Environment and Development, which produced the 'Brundtland Report' in 1987 (Andresen and Butenschøn 2001). Moreover, Brundtland was a key figure in the international diplomacy leading up to the adoption of the UN Climate Convention on Climate Change (UNFCCC) in 1992 (Andresen and Boasson 2008).

Norway promptly followed up this international engagement with national action. In 1989, it became the first country to set a national stabilization target for carbon dioxide emissions (CO_2); and it levied a CO_2 tax on about 60 per cent of its national emissions in 1991 (Andresen and Butenschøn 2001; Sæverud and Wettestad 2006). These bold moves were met with hefty protests, not least because it became clear that emissions stabilization would prove very costly. Already in 1993 came a marked shift towards a rhetoric of global cost-efficiency (Boasson 2005; Nilsen 2001; Asdal 2014). The argument was that emission cuts entailing the lowest societal costs should be implemented first – and these were primarily to be found outside Norway. In 1992 a state-appointed commission concluded that Norway should move away from the stabilization target (NOU 1992). After this, the target no longer steered policy development.

During the 1990s, the idea of developing a global price for CO_2 through global emissions trading and joint implementation (JI) gained considerable traction in Norway while less attention was paid to the projected national increase in emissions from petroleum and transport (Boasson 2005: 21–23; Riksrevisjonen 2010: 26, 169). Double regulation was regarded as inefficient: measures introduced in addition to general economic measures would hamper efficiency of the latter; the government should therefore refrain from introducing additional measures (i.e. additional to CO_2-pricing). Hence, the government refrained from applying existing licensing systems for power generation and petroleum exploration to steer national GHGE developments. As Norway had not accepted a reductions commitment in Kyoto in 1997, the Protocol allowed Norway to increase its emissions by 1 per cent between 1990 and 2008–2012.

In the 1980s there had been resistance to high petroleum investments, because this could lead to too much public spending due to high petroleum revenues. But the creation of the Norwegian Sovereign Wealth Fund in 1993 de-coupled annual governmental spending from annual tax income from the petroleum industry, and these worries vanished. This opened the door to steep increases in petroleum exploration and investments, and petroleum policy became largely depoliticized. Oil and gas exploration and investment increased significantly since the mid-1990s, and the CO_2 tax on petroleum production was cut by 40 per cent when the oil prices plunged briefly in 1998 (Boasson 2005: 24).

Plans for the construction of gas-fired power plants mushroomed in the late 1990s (Tjernshaugen 2011). If these proposals were to be realized, the Norwegian stationary electricity provision would no longer be fully based on renewables (large hydro). The environmental movement protested loudly, but the political majority wanted to leave it to the initiators to decide whether to take the risk of high future carbon prices, in line with the Norwegian cost-efficiency approach in global climate politics. Against this backdrop, gas-power construction became one of the most contentious political issues in Norway.

In 1999, the Pollution Control Authority decided that in order to permit these gas-fired plants to be constructed, emissions would have to be reduced by 90 per cent. In effect, this amounted to a requirement to implement carbon capture and storage (CCS) – capturing carbon dioxide and injecting it into geological formations (Tjernshaugen 2011). This requirement led to protests from the majority of parties (Labour, the Conservatives and the Progress Party) in the Storting. Persistent disagreement as to how much Norway should do domestically to mitigate climate change culminated in a stormy gas-power conflict.

2000–2010: global cost-efficiency rhetoric: multiple measures adopted

The rhetoric of global cost-efficiency gained strength throughout the first decade of the twenty-first century. However, in parallel and in contradiction to these arguments, Norway adopted an increasing number of domestic measures.

In March 2000, the majority in the Storting overruled the Pollution Control Authority, ordering the pollution licensing of gas-fired power plants to be altered in order to remove the de facto requirement for CCS (Innst. S. nr. 122 1999–2000). As a result, the minority coalition government resigned. This marked a watershed in the development of Norwegian climate policy. The new Labour government experienced major steering problems and remained in office for less than two years (Boasson 2015). All ensuing governments required a grand compromise that could enable parties with contrasting views on gas power to join coalition governments. CCS was exactly what they needed. It was embraced by the trade unions, which campaigned for more industrial use of Norwegian gas, as well as by the Norwegian environmental movement, which saw CCS as a way of either obstructing power-plant construction or avoiding the associated emissions (Tjernshaugen 2011).

On the same day as the government resigned over the gas-power issue in March 2000, the Storting adopted two important new objectives: (1) the first-ever Norwegian energy-efficiency objective 'to reduce growth in energy demand more than business as usual' and (2) to increase renewable heating by four terrawatt hours (TWh) and wind power by 3 TWh by the year 2010 (St. meld nr. 29 1998–1999). The Storting also agreed to create the state enterprise Enova, mandated to grant state aid to renewable energy and energy efficiency in a cost-efficient manner.

The somewhat fuzzy energy-efficiency objective eventually resulted in the adoption of a broad range of energy efficiency measures (see below). It took longer to reach agreement on renewable electricity. Soon after having agreed to create Enova, the majority in the Storting called for the government to consider a market-based green certificate scheme instead of the Enova scheme (Boasson 2015). However, it was not until September 2009 that Sweden and Norway agreed on the key principles for a common green certificate scheme. The wind power target from 2000 was not met. Norway and Sweden are the only European countries to have created a common renewable energy support scheme.

The Norwegian government had initially planned to introduce a national emissions trading scheme (ETS), while keeping open the option for global emissions trading (Sæverud and Wettestad 2006). But after a very short-lived national pilot scheme, Norway was included in the EU ETS in 2008 (Boasson 2013: 26–27). In the same year, various new climate targets were adopted, with a commitment to a 30 per cent GHGE reduction by 2020 (compared to 1990), and a pledge to overachieve the Kyoto target by an additional 10 per cent (Innst. S. nr. 145 2007–2008). The Kyoto overachievement goal was intended to be fully met through international emissions trading. Further, Norway became the first country to declare that it aimed for carbon neutrality: 'Norway will reduce global GHGE by an equivalent of 100 per cent of its own emissions by 2050' (Innst. S. nr. 145 2007–2008: 2 as translated by the authors; cf. Gullberg 2009; Boasson 2013). Offsetting Norwegian emissions by paying for emissions reductions in other countries was hailed as the key to achieving carbon neutrality, in line with the global cost-efficiency approach. At the same time, as a concession to the smaller parties in the government coalition which emphasized national action, it was agreed that two-thirds of Norway's efforts to reach the 2020 target should be undertaken domestically. The latter target was however more aspirational in form, and the government never developed detailed plans to ensure that it was implemented.

These targets placed Norway among the most ambitious countries prior to the 2009 Copenhagen climate conference. But Norway was also the only country whose high ambitions were premised on the existence of international emissions trading and a global cost-efficiency ideal. In 2007, Norway also launched an international initiative aimed at supporting efforts to reduce emissions from deforestation and forest degradation in developing countries (REDD+) (see below for details).

Domestically, politicians from across the political spectrum in Norway now engaged in promoting CCS. Gas power without CCS had been banned and the government eventually committed itself to investing some €2.8 to 4.5 billion in full-scale CCS facilities (Boasson 2015: 83). Since 2008, an ambitious policy to promote electric vehicles also became an important element in national climate policy (see also below).

2010–2015: Europeanization and increased conflict over petroleum policy

After 2010, the main climate-policy issue in Norway shifted from gas power and CCS to oil exploration in the High North. The partly state-owned Norwegian oil company Statoil remained deeply sceptical of the planned full-scale CCS, presenting increasingly higher cost estimates. Eventually, in 2013, plans for full-scale CCS in Norway were shelved (Gassnova 2013). As Norway primarily relied on purchasing emission allowances through the UN system, this did not influence Norway's ability to reach its climate targets.

Petroleum exploration had now become established in the Norwegian Arctic. This was a hotly contested political issue, but was so far primarily framed as a question of exposing ecologically fragile ecosystems to risks of oil spills and other pollution. It was only after 2010 that the environmental movement succeeded in framing petroleum expansion as a climate-policy issue. After the national elections in 2009, the Socialist Left Party succeeded in ensuring a temporary moratorium on petroleum exploration in large parts of continental shelves off Lofoten, Vesterålen and Senja in Northern Norway, but not in the Barents Sea. The same was achieved by the Liberal Party and the Christian Democrats after the 2013 national elections.

In 2012 a White Paper on climate policy was published (Meld. St. 21 2011–2012). It included various new measures, but no new strategic objectives. The White Paper confirmed that the targets adopted prior to the 2009 Copenhagen climate conference (COP15) were still valid, but lack of specific references to the domestic two-thirds target created uncertainty about the level of domestic ambition. In 2012 Norway signed a second Kyoto commitment period in line with the

2020 target of reducing emissions by 30 per cent compared to 1990 levels (Meld. St. 13 2014–2015: 18). This entails that average emissions for 2013–2020 are to equal 84 per cent of 1990 emissions. Official documents repeatedly declare that Norway is to become a low-carbon society through a 'green transition'. The term seems influenced by the German energy transition (*Energiewende*) (see Chapter 8), although its meaning in the Norwegian context remains unclear.

While many EU member states adopted GHGE targets for 2030, Norwegian discussions continued to centre on 2020. This changed in February 2015, when the government presented the commitment that Norway was to deliver to the climate summit in Paris in December 2015 (Meld. St. 13 2014–2015). At this point, the government declared that Norway intended to be included in the EU effort-sharing decision, not least because this would allow Norway to take part in the flexible mechanisms to be developed by the EU for sectors outside the EU ETS. If a common solution with the EU could not be achieved, the EU's 40 per cent reduction target for 2030 compared with 1990 would be the indicative Norwegian commitment. Surprisingly, the government did not adopt any national objectives or strategy in addition to requesting the EU to include Norway in its effort-sharing agreement.

As it became increasingly clear that an agreement establishing a global price on CO_2 would not be adopted soon, the Norwegian government aimed instead to ensure international cost-efficiency through stronger cooperation with the EU. However, by 2015 Norway already had a rather large basket of climate measures, although the petroleum industry was allowed to continue with a rise of its emissions. The 2015 Paris Agreement was hardly influenced by the Norwegian global cost-efficiency ideal, but the additions on emissions trading and the possible future inclusion of Norway in the EU Effort Sharing Decision (ESD) allowed Norway to continue as before with its main approach to climate mitigation.

Institutional responses, policy instruments and programmes

Public policy and organization of climate issues

Norway created an Environment Ministry in 1972, but it has remained small and rather weak (Boasson 2005). With the exception of emissions trading, most climate measures are administered by other ministries. The Finance Ministry is generally the most powerful ministry in Norway. It has played a very important role in the development of Norway's climate policy (Asdal 2014). It is dominated by economists who steadfastly promote global cost-efficiency (Boasson 2013). For many years, it also had formal responsibility for administering Norwegian participation in the international trade of GHGE allowances. The Ministry of Petroleum and Energy has also played a key role, supporting the general approach of the Finance Ministry (Boasson 2005, 2015).

As a result of the dominant role of the Finance Ministry, Norway largely applies general economic measures to regulate GHGE. The EU ETS applies only to large stationary emitters (onshore and offshore), and allowance prices are low. Hence Norway has continued to tax CO_2 emissions. Since the expansion of the EU ETS in 2013, almost half of Norway's emissions are covered by the ETS which means that in total 80 per cent are included in the ETS and/or covered by a carbon tax (St. meld. 21 2011–2012). Norwegian petroleum activity is covered by a CO_2 tax in addition to ETS. Norway has a high carbon tax (with some sectoral variations) compared with other European countries, but the energy taxation level is relatively low (Boasson 2013: 26–27).

In addition to general economic measures, Norway has adopted a growing number of supplementing measures, including direct regulations such as a CCS requirement for new onshore gas-powered plants (St. meld. St. 21 2011–2012). This regulation is a result of the enduring campaigns against onshore gas power plants in Norway (Tjernshaugen 2011).

Together with Sweden, Norway has a green certificate scheme for renewable electricity. The common objective is to increase renewable electricity production by 26 TWh by 2020 (OED 2012). The scheme is planned to run until 2035, but no plants will be included in Norway after year 2020. Pressure from the electricity industry, and support from the smaller parties in the government coalition can explain why Norway decided to join the Swedish scheme, while implementation of the EU renewables directive contribute to explain the relative high ambitions in the scheme (Boasson 2015). Oversupply of electricity in the Nordic market and falling prices have made the industry argue against including more plants in Norway after 2020.

Considerable energy-efficiency measures have been adopted over the years (Boasson 2013: 42–43). There is a state aid scheme for energy-efficiency measures in industry which is not directly related to carbon emissions (Enova 2013a), and Norway has developed a series of measures directed at reducing energy consumption in buildings. Enova has developed a cost-minimizing state funding scheme for buildings. The energy requirements of the Norwegian building code regulate which techniques and technologies may be applied in building construction. New energy requirements were introduced in 2007, aimed at ensuring that new and renovated buildings used one-fourth less energy than required by the 1997 building code (KRD 2007). The regulations are to be made more stringent every five years, aimed at achieving a passive house standard (or some other demanding holistic standard) in all new buildings from 2020 (Innst. S. nr. 145 2007–2008: 25). Energy certification is required of all large non-residential buildings and other buildings that are rented or sold (NVE 2010; Ot. prp. nr. 24 2008–2009). This significant growth in energy efficiency measures has happened despite low political salience and low electricity prices. This can primarily be explained by entrepreneurial initiatives from civil servants and implementation of the EU Energy Performance of Buildings Directive.

In the transport sector, Norway has received considerable attention for its policies concerning electric vehicles. Zero-emission vehicles have been granted various tax exemptions and other benefits, including free parking in all public inner-city parking zones, and permission to use priority lanes (TØI 2013). This has resulted in a sharp increase in the purchase of electric vehicles, amounting to 17 per cent of new passenger cars registered in 2015 (OFV 2016). Since 2010, the Norwegian electric vehicles market has become one of the largest in Europe – comparable to or even bigger than those of France, Germany or the UK (TØI 2013). The lack of a car manufacturing industry in Norway and an existing high level of taxation on car sales may explain some of the policy entrepreneurship in this area.

Since the mid-1990s, Norway's sovereign wealth fund, known as the 'Government Pension Fund – Global' has received most of the direct government revenues from oil and gas activity. Valued at more than €700 billion, with a combined ownership of 1.3 per cent of listed companies worldwide, it is currently the world's largest sovereign wealth fund (NBIM 2016). This has conferred on Norway a somewhat unexpected structural leadership role as an international investor. Efforts to reconcile this role with the country's climate-policy ambitions have led to noteworthy policy innovations. While the general mandate of the fund is strictly commercial, a Council on Ethics was established in 2004, to oversee the fund's investments in relation to ethical guidelines (Chesterman 2008). Following public campaigns to divest the fund of investments in sectors that contribute disproportionately to climate change (e.g. coal and palm oil industries), the Ministry of Finance proposed in April 2015 supplementing the existing criteria with a separate criterion for exclusion based on company activities that cause GHGE (Meld. St. 21, 2014–2015). The Storting later decided that the fund should divest itself entirely of investments in the coal sector – limited to those companies in which at least 30 per cent of revenues are generated from coal-related activities.

Multi-level climate governance: persistent promoter of a cost-efficient global deal

In the late 1980s, Norway engaged strongly with the 'targets and timetables' approach advocated at the international climate events (Andresen and Butenschøn 2001). During the negotiations leading up to the 1992 adoption of the UNFCCC, however, Norway's approach changed. At the first Conference of the Parties (COP) to the UNFCCC in Berlin, Norway sided with the US (see also Chapter 16), against the EU and G77, in pushing for flexible commitments, joint implementation and tradable credits. In parallel with a rather hard line in the negotiations, Norway supported pilot projects to demonstrate practical approaches to joint implementation (Andresen and Butenschøn 2001). The shift was remarkable. After having been the first country to set a stabilization target, Norway arrived at the Kyoto negotiations without a national target for emissions reductions. This almost complete reversal was criticized by ENGOs and put Norway at odds with the rest of Europe.

Norway performed entrepreneurial as well as cognitive leadership related to Joint Implementation (JI) and global emissions trading more generally; it was an active participant at UNFCCC-related meetings in the 1990s, and Norwegian diplomats and experts travelled extensively, seeking to build alliances (Nilsen 2001). Norway also promoted emissions trading and cost efficiency in the Kyoto protocol. After Kyoto, the EU shifted its position and developed an internal, supranational ETS, much in line with the ideas that Norway had promoted (Sæverud and Wettestad 2006). Norway continued to invest considerable resources in promoting a global cost-efficient approach to tackling climate change after 1997, eventually becoming a large investor in carbon credit-generating projects in developing countries. In order to comply with its commitment under the Kyoto Protocol, and the subsequent pledge to overachieve that commitment, the Government has purchased some 23 million Certified Emission Reductions (CERs) from the Kyoto Protocol's Clean Development Mechanism for the first Kyoto commitment period (2008–2012) (Ministry of Climate and Environment 2015). It has also been authorized by the Storting to procure another 60 million CERs for the second commitment period (2013–2020). Its vast financial resources enabled Norway to act as a structural leader within the field of global emissions trading, despite its small size in terms of population and domestic emissions.

When negotiations on a new international climate agreement were initiated in 2007, Norway again performed entrepreneurial and cognitive leadership. It submitted detailed proposals for a mechanism to finance mitigation in developing countries, as well as proposals for organizing efforts to reduce emissions from deforestation in developing countries (Lahn 2013; Lahn and Wilson Rowe 2015). Government documents and officials frequently stress Norway's ability to act as a bridge builder, aiding the negotiating process by facilitating agreement among other actors (Lahn and Wilson Rowe 2015: 133). Being seen as a small state not beholden to great power interests, as well as being a non-EU member, has provided Norway room for manoeuvre and flexibility in alliance building.

In addition, Norway has performed structural leadership through financial contributions directly underpinning the negotiations. Substantial funding has been channelled through various initiatives within and outside the UNFCCC process, in order to enable outcomes in line with Norwegian priorities. In the period 2008–2012, Norway ranked highest in absolute terms among countries' voluntary contributions to the UNFCCC process (Lahn and Wilson Rowe 2015: 136).

The most visible funding initiative is arguably Norway's International Climate and Forest Initiative, launched by Prime Minister Jens Stoltenberg at the COP13 in Bali in 2007. The ambition was to channel around €370 million annually (in 2007 exchange rates) to reduce emissions from deforestation and forest degradation in developing countries (Hermansen 2015). This mitigation approach, known as REDD+, built on earlier proposals to include tropical forest conservation in carbon markets, and on economic modelling that showed reduced deforestation to be a particularly cost-efficient mitigation option. By committing large amounts of funding to REDD+, Norway underpinned an approach that fits well with its overall approach of global cost-efficiency, while, at the same time, signalling developing countries that they would receive financial support to undertake mitigation action (Lahn and Wilson Rowe 2015: 137; Hermansen 2015).

Norway is the largest donor to REDD+, which has received significant attention within and beyond the UNFCCC process. These efforts have strengthened the international profile of Norwegian political leaders. For instance, former Prime Minister Jens Stoltenberg (Labour) was selected by the UN Secretary-General to chair a high-level panel on climate-change finance, and Norwegian ministers and officials are frequently asked to chair negotiations under the UNFCCC (Lahn and Wilson Rowe 2015).

Norway has performed enduring entrepreneurial leadership in order to persuade others to accept the global cost-efficiency approach, but with far greater success before than after the adoption of the Kyoto Protocol. REDD+ fitted Norway's overarching approach, but here Norway rather offered entrepreneurial and structural assistance in order to support an idea developed by others. Norway's many financial contributions can also be regarded as structural leadership, and partly leadership by example in cases where Norway has been intended to attract followers (i.e. other states also making available significant sums of money for REDD+). However, its efforts did not contribute much to strengthening the global cost-efficiency approach in the UNFCCC negotiations. Instead, the outcome of the 2015 Paris climate conference (COP21) was characterized much less by the ideals promoted by Norway than was the Kyoto Protocol some 20 years earlier.

Developments in international climate politics have influenced domestic Norwegian policies in the sense that new national policies and measures have peaked prior to major international summits, such as Rio 1992, Kyoto 1997 and Copenhagen 2009 (Andresen and Boasson 2008). Interestingly, this did not happen before the 2015 Paris climate conference (COP21). In addition, EU influence has contributed to the increase of national measures, thereby sharpening the contrast between the approach that Norway promotes internationally and its domestic actions (Boasson 2005). Because it is not a full EU member, Norway is primarily a 'taker' of EU climate policies and not a leader in EU policy development. In Norway, EU regulations are not well known and receive scant attention in climate policy White Papers (see St. meld. 21 2011–2012). This may however change with the government's 2015 decision to request participation in the EU ESD.

Conclusions

Ever since the early 1990s, Norway has consistently promoted a global cost-efficiency approach, relying on the development of a binding global regime based on flexible national commitments, global emissions trading and other measures to ensure that least-costly mitigation options will be realized first. This approach has served to reconcile Norway's interests as a major oil and gas exporter with its ambition of acting as an international climate leader. For many years this position constrained the development of domestic policies and measures, with the exception of GHGE pricing measures like CO_2 taxation and emissions trading.

In line with the global cost-efficiency approach, Norway has launched its ambitious REDD+ financing initiative, aimed chiefly at spurring a global regime that can facilitate funding of low-cost mitigation options. In contrast to the global approach, however, Norway has also adopted a comprehensive array of national measures after 2000, mainly as a result of the concessions that its larger political parties have had to make to smaller political actors in order to gain executive power and to implement EU policies (Boasson 2015). This has given rise to a particular 'dissonance' in Norwegian climate policy, where Norway does more in practice to limit domestic emissions than its long-standing rhetoric indicates.

The 2015 Paris Agreement contradicts many elements of the Norwegian approach to international climate politics: It is predominantly non-binding, is based on nationally determined pledges, and provides a weak foundation for global emissions trading. Norway's cognitive and entrepreneurial leadership at the international level has therefore produced only meagre results. Norway experienced significant success in leaving an imprint on the Kyoto Protocol, although this achievement resulted in large part from US leadership (see Chapter 16). Not least because of the many shifts in the US position over time, Norway's international influence has faded. That being said, Norwegian efforts fulfil many of the leadership criteria presented in Chapter 1 of this book.

Despite its small size in terms of population, relative emissions and economy, Norway has acted as a leader. First, it has used its financial strength to act as a structural leader with respect to financial contributions directly underpinning the UNFCCC negotiations on global emissions trading, its REDD+ initiatives and 'shareholder activism' (Chesterman 2008: 581) performed by its sovereign wealth fund. Second, Norway has employed an array of entrepreneurial, diplomatic tools to promote its approach to global climate politics. Third, persistent efforts to re-define the interests of other actors by promoting a global cost-efficiency approach also counts as cognitive leadership. For many years, Norway's lack of an encompassing domestic climate policy fitted perfectly with this approach, which values contributing to solutions that work well on a global scale, instead of focusing on domestic GHGE reductions.

Norway has not gained much international criticism for its approach over the years, partly because of its relative insignificance in terms of global emissions, but probably also due to the constructive role it has taken in global climate talks. Nationally, however, the global cost-efficiency approach has been subject of much debate, and climate policy has been a controversial and salient domestic political issue for more than 20 years. The environmental movement has succeeded in politicizing issues such as gas power, and more recently petroleum production and investment practices. The parliamentary situation where the larger political parties can only form a majority through cooperation with smaller parties that place greater emphasis on domestic emission reductions, has contributed to an increasing number of national climate policy measures being adopted.

Norway's global approach rests on the argument that domestic measures make sense only if these can ensure global cost-efficient GHGE mitigation – hence there is no logical link between adopting comprehensive national measures and performing international leadership efforts. Significant national efforts serve to weaken the consistency of the Norwegian position internationally; and to some extent this undermines, rather than strengthens, Norway's position in global climate diplomacy. The country has maintained surprisingly stable long-term objectives, aimed at bringing about radical political chance at the global level. Nonetheless, given the glaring lack of success, we must conclude that Norway has *not* performed transformational leadership. Given Norway's modest size as a country of five million inhabitants, this lack of global political clout should hardly come as a surprise.

As a highly competent actor outside the EU, Norway has been able to play an active and independent role in the global political arena. However, its relationship to EU climate policy is now set to change, with the 2015 Storting decision that Norway should seek to become fully integrated in all parts of EU climate policies. This gives rise to questions regarding Norway's ability to maintain an international leadership role after the 2015 Paris conference and whether the current dissonance between its domestic climate policies and its cognitive and structural leadership in championing a global cost-efficiency approach will continue, or be reconciled over time.

References

Andresen, S. and E.L. Boasson (2008) 'Internasjonalt klimasamarbeid [International climate cooperation]', in S. Andresen, E.L. Boasson and G. Hønneland (eds) *Internasjonal Miljøpolitikk* [International Environmental Policy]. Bergen: Fagbokforlaget.

Andresen, S. and S.H. Butenschøn (2001) 'Norwegian climate policy: From pusher to laggard?', *International Environmental Agreements*, 1: 337–356.

Asdal, K. (2014) 'From climate issue to oil issue', *Environment and Planning A*, 46: 2110–2124.

Austgulen, M.H. and E. Stø (2013) 'Norsk skepsis og usikkerhet om klimaendringer [Norwegian scepticism and uncertainty on climate change]', *Tidsskrift for samfunnsforskning*, 54(2): 123–152.

Boasson, E.L. (2005) 'Klimaskapte beslutningsendringer?' [A new climate for decision-making?]', *FNI Report* 13/2005, Lysaker: Fridtjof Nansen Institute.

Boasson, E.L. (2013) National climate policy ambitiousness. *CICERO report* 2013:02, Oslo: CICERO.

Boasson, E.L. (2015) *National Climate Policy*, London: Routledge.

Boasson, E.L and J. Wettestad (2013) *EU Climate Policy*, Farnham: Ashgate.

Chesterman, S. (2008) 'The turn to ethics', *American University International Law Review*, 23(3): 577–616.

Eckersley, R. (2013) 'Poles apart?', *Australian Journal of Politics and History*, 59(3): 382–396.

Eide, E., Elgesem, D. Gloppen, S. and Rakner, L. (2014) 'Norske paradokser [Norwegian paradoxes]', in E. Eide *et al.* (eds) *Klima, medier og politikk*, Oslo: Abstrakt.

Gassnova (2013) Mongstad skrinlegges [Mongstad project cancelled]. www.gassnova.no/no/Sider/Mongstad-skrinlegges.aspx, (accessed 18 January 2016).

Gullberg, A.T. (2009) Norsk klimapolitisk debatt og klimaforliket fra 2008 [Norwegian climate policy and the 2008 climate agreement]. *CICERO Working Paper* 2009:3, Oslo: CICERO.

Eurobarometer (2009) Special Eurobarometer 300. Europeans' Attitude Towards Climate Change, Brussels: European Commission.

Hermansen, E. (2015) 'Policy window entrepreneurship', *Environmental Politics*, 24(6): 932–950.

Innst. S. nr. 145 (2007–2008) *Innstilling fra energi- og miljøkomiteen om norsk klimapolitikk* [Recommendation from the Standing Committee on Energy and Environment on Norwegian climate policy], Oslo: The Storting.

Lahn, B. (2013) *Klimaspillet* [The climate game], Oslo: Flamme.

Lahn, B. and E.W. Rowe (2015) 'How to be a 'front-runner', in B. Carvalho and I.B. Neumann (eds), *Small States and Status Seeking*, London: Routledge, 126–145.

Meld. St. 21 (2011–2012) Norsk klimapolitikk [White Paper: Norwegian Climate Policy]. April, 2012, Oslo: Ministry of Environment.

Meld. St. 13 (2014–2015) Ny utslippsforpliktelse for 2030 – en felles løsning med EU [White Paper: New emissions requirements for 2030: common solution with the EU], February, 2015, Oslo: Ministry of Climate and Environment.

Meld. St. 21 (2014–2015) Forvaltningen av Statens pensjonsfond i 2014 [White Paper: Administration of the Sovereign Wealth Fund]. April, 2015, Oslo: Ministry of Finance.

Ministry of Climate and Environment (2015). The background of Norwegian Carbon Credit Program, www.regjeringen.no/no/tema/klima-og-miljo/klima/innsiktsartikler-klima/norwegian-carbon-credit-procurement-program/the-background-of-norwegian-carbon-credit-program/id2415679/ (accessed 6 January 2016).

Miljøstatus (2015) www.miljostatus.no/tema/klima/globale-utslipp-klimagasser/'Globale utslipp av klimagasser' [Global emissions of climate gases], published 30 October 2015 (accessed 12 January 2016).

Ministry of Petroleum and Energy (2014). Fakta 2014: Norsk petroleumsverksemd [Fact sheet 2014: Norwegian petroleum activity], Oslo: Ministry of Petroleum and Energy.

NBIM (2016). Norges Bank Investment Management: Market Value, www.nbim.no/en/the-fund/market-value/ (accessed 20 January 2016).

NOU (1992) *Mot en mer kostnadseffektiv miljøpolitikk i 1990-årene* [Official Norwegian Report: Towards a cost-efficient environmental policy for the 1990s], Oslo: Ministry of Environment.

Nilsen, Y. (2001) *En felles plattform?*. [A common platform?]. PhD thesis, TIK – Centre for Technology, Innovation and Culture, University of Oslo.

Norgaard, K.M. (2006) 'We don't really want to know', *Organization and Environment*, 19(3): 347–370.

OFV (2016) Bilsalget i 2015 [Car sales 2015], Opplysningsrådet for Veitrafikken, www.ofvas.no/bilsalget-i-2015/category679.html (accessed 20 January 2016).

Riksrevisjonen (2010) Riksrevisjonens undersøkelse av måloppnåelse i klimapolitikken [Auditor General's investigation of goal achievement in climate policy]. Document 3:5. Oslo: Riksrevisjonen [Office of the Auditor General].

SFT (1999) Utslippstillatelse for Naturkraft AS-gasskraftverk på Kollsnes 21. januar 1999 [Emissions permit, Naturkraft AS gas plant, Kollsnes, 21 January 1999], www.sft.no/nyheter/dokumenter/gasskraft/kollsnes-utslippstillatelse.html (accessed 1 October 2004). Oslo: Norwegian Pollution Control Authority.

St. Meld nr. 29 (1998–1999), Om energipolitikken [White Paper on energy policy]. Oslo: Ministry of Petroleum and Energy.

Sæverud, I.A. and J. Wettestad (2006) 'Norway and emissions trading', *International Environmental Agreements*, 6: 91–108.

Tjernshaugen, A. (2007) *Gasskraft* [Gas power]. Oslo: Pax.

Tjernshaugen, A. (2011) 'The growth of political support for CO2 capture and storage in Norway', *Environmental Politics*, 20(2): 227–245.

Tjernshaugen, A., B. Aardal and A.T. Gullberg (2011) 'Det første klimavalget?' [The first climate election? Environmental and climate issues in the 2009 elections], in B. Aardal (ed.), *Det politiske landskap*, Oslo: Cappelen Damm Akademisk.

TNS Gallup (2015) TNS Gallup Klimabarometer 2015, www.tns-gallup.no/document-file1906?pid=Native-ContentFile-File (accessed 20 January 2016).

TØI (2013), 85g CO2 per kilometer i 2020. [85g CO2/km by 2020], *TØI report* 1264/2013. Oslo: Institute of Transport Economics., www.toi.no/getfile.php?mmfileid=33029 (accessed 20 January 2016).

Part IV

Civil society

Business and environmental groups

14 Business

Greening at the edges

Wyn Grant

This chapter reviews the response of business to climate change, noting that the range of responses provides opportunities for environmentalists to build coalitions that can promote effective solutions. A key point to note about business is that it does not represent a homogeneous and undifferentiated set of interests and perspectives. Olson's *Logic of Collective Action* 'challenged the view … that a unified business class dominates politics in capitalist society' (Hart 2010: 176). Different sectors of the economy have different interests, while the stance of a particular firm may be influenced by its culture or even the priorities of the chief executive officer. These differences are particularly marked in the case of climate change. For some firms, particularly in fossil fuels, tackling climate change can be seen to represent a major threat to their business. For other firms, for example those firms operating in renewables, mitigating climate change represents a business opportunity. The outcome of the 2015 Paris climate conference (COP21) could be seen as a major encouragement to invest in low-carbon technologies. These divisions mean that environmental non-governmental organizations (ENGOs) find it easier to agree on goals than business as Chapter 15 argues.

One report by consultants Mercer shows a considerable divergence in the impact of climate change on returns by industry sector (although clearly the figures are sensitive to the extent of climate change). Over a 35 year period, returns from renewables could increase by 3.5 per cent a year and from nuclear by 1.8 per cent a year. Utilities could see a 2.5 per cent decline in annual returns, increasing to 4.1 per cent for oil and 4.9 per cent for coal (Mercer 2015).

These divergences provide opportunities for NGOs to build coalitions with some parts of business and to inflict reputational damage on others who seek to resist efforts at climate change mitigation. As Weale (1992: 31) notes, 'a cleavage begins to open up not between business and environmentalists, but between progressive, environmentally aware business and short-term profit takers on the other'. Another distinction that has been made is between 'evangelicals' and 'sustainability capitalists'.

> For the 'evangelicals' such as Unilever, Dutch electronics group Phillips or Britain's Marks and Spencer … sustainability is 'a belief system', often

> driven by a Chief Executive Officer's (CEO) views on long-term trends such as resource scarcity. The 'sustainability capitalists' such as GE or Siemens invest in ventures such as wind energy … because they see short-term growth opportunities.
>
> (Clark 2012: 9)

Leadership

One general qualifying point needs to be made about the exercise of leadership by those in business. The principal function of the chief executive officer of a company is to ensure that the company is competitive and returns profits that satisfy shareholders and the financial markets. Of course, securing those profits in the twenty-first century is not just a matter of an efficient management of the production process combined with effective marketing. That marketing often involves the creation or maintenance of a brand, and it is important to avoid any reputational damage to that brand.

One way in which companies have been suffering reputational damage is through a divestiture movement spearheaded by 350.org. which claims that 'The fossil fuel divestment movement has grown faster than any previous divestment campaign in history'.[1] The intention is to make oil, gas and coal investments as unpopular as tobacco. High profile successes have included Stanford University's endowment fund, and Norway's sovereign wealth fund (see Chapter 13). In May 2015 Axa, one of the world's largest insurers, announced that it would sell €500 million of coal assets between then and the end of 2015.

Insurance companies have been among the most active groups in the financial sector urging governments to take tougher action and show structural leadership to combat climate change because the likelihood of more extreme weather events affects their ability to assess the risk from such events. Financial groups in general have largely kept fossil fuel assets, arguing that they can have a bigger effect on energy companies by engaging with them about climate risks. Given that oil and gas companies have a combined market value of around $2.2 trillion, the amounts divested so far are relatively small in comparison. Nevertheless, the momentum for divestiture concerned ExxonMobil and Shell sufficiently to write to investors arguing that there is little to fear. They claimed that a growing global population means that demand for fossil fuels will remain high for decades and the investment in existing energy infrastructure means that it will take decades for any effect to be felt from government climate policies. To an extent, the oil and coal industries sought to play down the impact of the decisions taken at the 2015 Paris climate conference (COP21), arguing that many developing countries planned to continue burning coal (see Chapters 17 and 18 respectively).

Even so, a chief executive officer has to consider the way in which the company's operations are perceived. Part of her role is therefore to engage in public debate about contemporary issues. Some leeway is given to chief executive officers to take part in think tank or business association activities. However, a commitment by a chief executive officer to climate change mitigation is always vulnerable to the replacement of that individual. The commitment needs to be made part of the culture of a firm in a way that is structurally embedded.

One form in which that can be embedded is to have senior officials within a company whose role is to address climate change and sustainability issues. For example, a letter published in the *Financial Times* on 26 February 2014 called for long-term action to create a low-carbon economy to combat climate change. Among the signatories were the 'Head of Environment and Climate Change' at Aviva, two individuals with the title of 'Sustainability Director' and the Director for 'Responsible Business' at Lloyds Banking Group.

The most usual type of leadership exercised by business is structural (see Chapter 1 for different types and styles of leadership). This reflects the fact that business enjoys hard power deriving from its economic strength (Coen, Grant and Wilson 2010). Business can seek to argue that climate change mitigation measures would damage economic growth and employment or push up input costs. This exercise of structural power seeks to translate power resources into bargaining leverage to bring pressure to bear on others to support the stance taken by business. Necessarily, this often produces a humdrum leadership style that is frequently reactive and incremental in character. Entrepreneurial leadership is not significant in the case of business, as business leaders do not seek to act as mediators producing compromise packages.

There are, however, opportunities for cognitive leadership through the definition or redefinition of interests through ideas. A good example is the Climate-Wise initiative of the insurance industry discussed further below. Business leaders have sought to win acceptance for scientific accounts of climate change and to explore possible solutions and the ways in which they may be implemented. As a consequence of this cognitive leadership, climate change denial as a policy response has been undermined in business circles, thanks in part to the efforts of some business leaders such as Paul Polman chief executive of Unilever and Grant Reid, chief executive of Mars. That is not to say that it has completely disappeared as an argument that is advanced. For example, Greg Boyce, the chief executive of Peabody Energy in the US, one of the world's largest coal companies, tried to claim that the computer models used to predict climate change were 'flawed' (*Financial Times*, 27 May 2015). Other business leaders have sought to go further in portraying climate change as a challenge that business should meet, leading to long-term benefits for business activity.

In April 2015 a group of chief executive officers from 43 companies in 20 sectors published a full-page advert in the *Financial Times* calling for bold action at the 2015 Paris climate conference (COP21), emphasizing the responsibility of the private sector 'to engage actively in global efforts to reduce greenhouse gas emissions, and to help the world move to a low-carbon, carbon-resilient economy'. Among the signatories were the CEOs of BT, Dow Chemical, Marks and Spencer, Toshiba, Unilever and Volvo. Actions such as this at least remind politicians that there is significant business support for effective action on climate change. For companies such as these, the 2015 Paris climate conference represented a good outcome.

In general terms, business leaders are often responding to agenda setting and the framing of issues that has been undertaken by politicians, international organizations and non-governmental organizations. Rather than being proactive, business leaders are seeking to develop an adequate business response to policy initiatives emerging from other actors. One also has to be aware of the possibilities of 'greenwash', a public relations response based around a few token policies and activities, that seek to make companies appear more environmentally friendly than they actually are.

The World Business Council for Sustainable Development (WBCSD) claims to have provided '20 years of leading sustainability'. It was founded on the eve of the 1992 Rio Earth Summit to encourage businesses to become involved in sustainability issues which are framed in terms of the three pillars of environmental, social and economic concerns. Giving a rough equivalence to these three dimensions does mean that environmental needs have to be balanced against considerations of economic viability.

The WBCSD is an organization of just under 200 well-known companies represented at chief executive or board member level. Forty-three per cent (84) of the member companies are in Europe. The number of companies from the US and Canada has dropped since 2007, the date for which membership figures were calculated for Grant (2011: 201): indeed, the number of companies from Canada has dropped from five to one. Companies from Japan have dropped from 28 to 20, but there are eight from India and six from China. There are 15 companies from Latin America whereas there were none before, but the three Russian members have disappeared.

In terms of initiatives, WBCSD places considerable emphasis on its Low Carbon Technology Partnership Initiative (LCTPI), although the scheme is short on detail. It was launched together with SDSN (Sustainable Development Solutions Network) and IEA (International Energy Agency). It aimed to present a series of concrete action plans at the 2015 Paris climate conference for the large-scale development and deployment of low-carbon technologies. The objectives are to 'Accelerate the diffusion of existing technologies by removing technological, market and social barriers and introducing required policy and financial instruments' and 'Develop Public Private Partnerships (PPPs) on the Research, Development, Demonstration and Deployment (RDD&D) of potentially game changing new technologies'.[2]

The sense that one is left with is that WBSCD is very dependent on government and intergovernmental actions to pursue their objectives. Before the 2015 Paris climate conference they stated:

> Business requires a clear international commitment to provide the essential long term signal for decision making on assets, investments and business strategies. COP21 in Paris, where a new climate agreement will be brokered, will be a decisive moment in making the transition to a low-carbon economy a reality.[3]

An alternative model is to build coalitions between business, environmental groups and other stakeholders. This model provides a wider perspective and has more potential for influencing political decision-making. It has been developed in the United States by ceres.org which was originally set up in response to the 1989 Exxon Valdez oil spill. It organized a letter in support of President Obama's plan to curb greenhouse gas emissions from power stations which attracted signatures from more than 220 companies including Kellogg's, Levi-Strauss, Nike and Starbucks. European companies that signed the letter included Adidas, Ikea, Nestlé and Unilever.

The members of Ceres include environmental and social non-profit groups such as NRDC, Union of Concerned Scientists and Oxfam, institutional investors such as the California and New York public pension funds, socially responsible investors (SRIs), labour unions and other key stakeholders. It claims to bring together the investor, business and advocacy communities. The Ceres Company Network includes nearly 70 members from two dozen industries, including technology, footwear and apparel, food and beverage, oil and gas, electric utilities and financial services. More than half of the companies Ceres works with are listed on the S&P 500. Ceres has shown some capacity for leadership in terms of taking initiatives that have an impact. In 1997, Ceres created the Global Reporting Initiative (GRI), now the international standard for sustainability reporting used by thousands of companies worldwide.[4]

By taking its own initiatives and attempting to exercise some forms of leadership, albeit often in a somewhat reactive form, business hopes that it can avoid more onerous or intrusive forms of imposed regulation, using policy instruments that it finds burdensome. Private systems of regulation are likely to be more attuned to the specific needs of business, particularly in terms of protecting product reputation and the integrity of a brand. These considerations serve as the main drivers for business leadership in climate change policies in Europe. Business has evolved specialist organizations to serve as a rallying point for businesses that want to enhance their environmental reputation and hence their standing with consumers.

Multi-level and polycentric climate governance

Elinor Ostrom has sought in her work to explore a space that is neither occupied by the state nor dominated by the market where communities can devise solutions to the management of common pool resources that protect the resource through processes of cooperation that take account of specific local conditions. As Ostrom (2012: 84) has put it, 'what we are trying to think through is ways in which people facing problems on diverse scales can self-organise and cope more effectively with managing those resources over the long run'. Scaling up micro-level solutions to manage inshore fisheries or forest resources to tackle the global challenges presented by climate change is a significant challenge. Localized problems have clear geographical and social boundaries which aid devising solutions and monitoring and enforcing them. Ostrom sought to respond to these challenges through the deployment of the notion of polycentricity.

Her view is that one needs to think in terms of 'a polycentric *system* rather than a monocentric *hierarchy*'. She quotes Vincent Ostrom's definition of a polycentric order as 'one where many elements are capable of making mutual adjustments for ordering their relationships with one another within a general system of rules where each element acts with independence of other elements'. In other words, there is a general framework that guides mutual action, but there is also respect for the autonomy of actors which may appeal to business (Ostrom 2010: 33). In Ostrom's perspective, 'it is these areas of limited autonomy that create, at the local level, a structure of incentives favourable to trust and also create a diverse environment favourable to discovering better solutions to problems' (cited in Tarko 2012: 61).

Ostrom's view is that we cannot wait for an effective global agreement on climate change to be negotiated. There is hence scope for more localized initiatives. Progress needs to be at multiple scales and decision-making units.

> The advantage of a polycentric approach is that it encourages experimental efforts at different levels, as well as the development of methods for assessing the benefits and costs of particular strategies adopted in one type of ecosystem and comparing these with results obtained in other ecosystems.
>
> (Ostrom 2010: 39)

This approach would give business the opportunity to take particular initiatives inspired by policy learning from elsewhere.

The general view taken here is that Ostrom's work is highly stimulating and opens up new perspectives, but in some respects is flawed. In part this arises from her methodology, which is based on the examination of multiple case studies from which an attempt is made to isolate 10 subsystem variables that can guide further work (Ostrom 2009). More generally, there is an issue of her concept of the state which is very much what one would expect from a traditional American liberal. What she describes as 'a very considerable role for large-scale governments' (Ostrom 1998: 17) is actually quite a limited one. Her list includes national defense, internal peace, providing arenas for conflict resolution (i.e. a judicial system), monetary policy, foreign policy, trade policy and moderate redistribution. What she may underestimate is the extent to which common pool resource institutions are underpinned by financial and legal resources provided by the state, although she admits that it is unlikely without some external support that 'reciprocity *alone* completely solves the more challenging common-pool resource institutions' (Ostrom 1998: 17).

'Polycentricity implies that different governance mechanisms are efficiently provided at different scales' (Dolsak *et al.* 2012: 791). However, Ostrom concedes that 'To achieve a complex, multitiered governance system is quite difficult' (Ostrom 1998: 17). A working if imperfect example is the EU. Ostrom's work is relatively silent on the EU and one's sense is that the model of multilevel governance she has in mind is the American federal system with its complex overlapping jurisdictions, particularly at the local level. One comment she did make about the EU, admittedly in the context of fisheries policy is that 'The EU could be providing some very broad overarching ways of conflict resolution, but allowing people to self-organise in ways they had done before we outlawed it' (Ostrom 1998: 85).

The 'letting a hundred flowers bloom' approach to climate change is likely to produce very patchy results. It also needs to take more explicit account of the possibility of initiatives started by or involving business. Ostrom is excited about the fact that 'over a thousand mayors in the USA … have signed an agreement to start working on various ways of reducing greenhouse gases in their cities' (Ostrom 2012: 86–7). But what does this amount to in practice? One suspects that in some cases it may represent little more than rhetorical grandstanding by mayors or token policy efforts such as encouraging citizens to cycle more. In her more general work, Ostrom is alert to the importance of monitoring and enforcement mechanisms. If one is going to give local decision-makers leeway (and resources) one is going to need some means of coordinating their actions, monitoring what they are doing and what the outcomes are. This is very difficult given the complexity of systems below the level of national governments in the EU. Some member states are federal; some are not. Some have devolved subnational governments; others do not. Some have city-regions with considerable resources; others have very small government arrangements at the most local level. Which can deal most effectively with climate change?

In terms of the role that business does or could play in multi-level and polycentric climate governance, the first point to note is that big businesses as production entities operate on an international scale. Their plants or warehouses are, of course, in a particular locality, but the core functions of finance, strategic planning, marketing, and research and development are located elsewhere, quite possibly in another country. Plants or depots may be allowed to make small contributions to local community activities as a goodwill gesture, but they are unlikely to be able to develop their own climate change policies. Smaller businesses are not likely to have much in the way of spare managerial or financial capacity to devote to climate change.

Second, if we look at the business organizations concerned with climate change discussed earlier, or more general business organizations like BusinessEurope and the European Round Table for Industrialists (ERT), their focus is also on an international scale. Their attention is directed to the regular international negotiations on climate change, to international bodies like the UN or the institutions of the EU. They do not focus very much on the national level, and certainly not the subnational or local level. It should be noted, however, that the ERT does favour the development of emission reduction schemes at national, regional and sectoral levels given the lack of global progress. In general, however, business is interested in 'top-down' solutions that produce common global standards of action that do not produce divergences that might lead to competitive advantages or disadvantages. Perhaps it should become more alert to the possible contribution of local initiatives. From a business perspective, diversity in policy or its implementation creates at best uncertainty and, at its worst, incoherence and confusion. Polycentricity would have to be sold to business as a complement to an internationally negotiated framework that made its implementation more effective.

Policy instruments

The limitations of 'command and control' instruments in terms of transaction costs, enforcement problems, etc. are well known. In particular, it imposes

> uniform reduction targets and technologies which ignore the variable pollution abatement costs facing individual firms. In practice, marginal costs of pollution vary widely among industries. Command and control is not only an expensive approach to pollution reduction, but one which, according to many analysts has also reached the limits of its environmental effectiveness.
> (Golub 1998: 3)

This led to a search for new instruments covering a range of environmental taxes and charges, tradeable pollution permit systems, government subsidies, eco-labels, ecoaudits, voluntary environmental agreements and 'altering liability and insurance rules in a manner which benefits the environment' (Golub 1998: 4).

Business tends to favour market-based policy instruments for combatting climate change and uses structural leadership to promote them. These allow firms to make their own judgements about how to respond to policy in terms of their assets, market placement and business plans. The two most common forms of pricing carbon are a tax on carbon pollution and a cap and trade, or emissions trading scheme (ETS). Nearly 40 countries now have some form of carbon pricing. However, while welcoming such approaches in principle, businesses are concerned about their implementation in practice. Energy companies operate on a global scale and they do not want a global patchwork of different, independently operating pricing measures which would create a more complex business environment.

This helps to explain the initiative taken by six of Europe's biggest oil and gas companies in the summer of 2015. This marked a break from their usual strategy of behind-the-scenes lobbying of politicians. It was evident that they felt a more proactive, high-profile approach was needed. They sought direct talks with the UN and governments on devising a global carbon pricing system. The companies involved were Royal Dutch Shell, BP, France's Total, Norway's Statoil, Italy's Eni and Britain's BG Group.

What they had in mind, displaying cognitive leadership, was rolling out a version of EU's ETS and creating a framework linking national or regional schemes. However, the difficulties of creating a global business consensus quickly became apparent. The chief executives of ExxonMobil and Chevron, the two largest US oil producers, made it clear that they would not be joining any European initiative to develop a common position on climate change. The chief executive of Chevron made it clear that he did not favour carbon pricing because it would make fossil fuels more expensive and what customers wanted was low, not high energy prices. Undoubtedly, that is the case, but it is in conflict with increasing demands for effective action on climate change. The initiative was strongly welcomed by the UN. It was admitted that it was unlikely to lead to a single global carbon price, but it was hoped that it would facilitate different carbon pricing schemes working more effectively with each other.

BusinessEurope strongly supports the maintenance of the ETS as the cornerstone of EU climate and energy policy, although it has concerns about carbon leakage. The ETS market-based price signal is seen to offer a means of incentivizing investments to reduce emissions.

> To be effective, limit transition costs and provide a predictable investment framework, it is crucial for the EU ETS to have a stable long-term cap. One hundred per cent of the ETS auctioning revenues should be used to support European businesses in the transition towards a low carbon economy.
>
> (Business Europe 2013: 12)

BusinessEurope is concerned about how measures affect the competitiveness of EU industry. 'For sectors at risk of carbon leakage, full compensation through free allocation based on benchmarks must allow the most efficient companies to be globally competitive without being penalised by direct carbon costs' (BusinessEurope 2013: 12).

It should be noted that business was not unanimously in favour of the adoption of the EU ETS (or national schemes). For example, the German chemical and power industry was initially strongly opposed to emissions trading on any governance level including the EU and national levels (Wurzel 2008). It has since considerably moderated its criticism, partly because it benefits economically from the low allowance prices.

The EU ETS has faced considerable difficulties. A price collapse in 2008–9 gave industry little incentive to diversify away from fossil fuels. The number of allowances available has outstripped demand, making allowances cheaper. There was still a surplus of more than two billion permits in 2014. The EU has, however, been determined to revitalize the ETS. In 2013 a decision was made to backload allowances, delaying the release of 900 million allowances. However, this was a short-term solution that did not permanently reduce the number of allowances and threatened another price crash when they flooded on to the market at a later date.

The longer-term solution is the so-called 'market stability reserve' which would remove or return allowances to the market depending on the size of the surplus. The Commission wanted this to start in 2021, but the European Parliament voted for it to start by the beginning of 2019 (see Chapter 4). This means that the back-loaded allowances, along with unallocated allowances, would be placed into the reserve. In addition, 8 per cent of allowances currently in circulation would be placed into the reserve between January and September 2019. From then on, 12 per cent of surplus allowances would be placed into the reserve every year, at a rate of 1 per cent a month. From a 2015 price of around €7, the price could increase to around €30 by 2020 and to around €40 by 2030.

What was left unresolved was the issue of compensation for carbon leakage and this concerned business. BusinessEurope argued that the post-2020 ETS reform proposal failed to safeguard the competitiveness of European industries and meet the objective of keeping a strong and competitive industrial base in Europe.

> By unnecessarily reducing the volume of free CO_2 emission allowances so drastically, it raises the risk of investment leakage, exposing our industries to unfair competition from countries without comparable climate efforts.... 'Carbon leakage' or in better words 'investment leakage' is already happening and just denying this fact will not help to bring Europe back to [a] growth path.
>
> (BusinessEurope 2015)

Some argue that a carbon tax would be more predictable and effective. Such a proposal was put forward in the EU in the 1990s but encountered substantial resistance from affected industry groups which were able to defeat it (see also Chapter 15). It does not find favour with business interests today. As Zito has pointed out (2000: 101), '[i]ndustrial groups have tended to support voluntary arrangements and even command and control over an environmental tax because industry can negotiate and consult with government during the policy implementation.' The ETS is the only game in town as far as European business is concerned. It is by far the preferred policy instrument.

Green economy

Business responses to the issue of whether the idea of a green or low-carbon economy should be seen as an opportunity or a threat are highly differentiated depending on the sector and firm concerned. This reflects the point made about the heterogeneity of business interests at the beginning of the chapter. 'The most important point of corporative relevance regarding the economics of climate change is that [as] the costs and benefits add up, they will spread unevenly across different sectors of the economy, and even potentially within sectors.' (Dunn 2002: 30). As a general proposition, 'since people are more sensitive to losses than gains, losers are more likely to mobilize politically than winners' (Daugbjerg and Svendsen 2001: 134). Threats are likely to be weighted above opportunities, particularly as a stimulus for political action.

At one end of the spectrum are firms and sectors that produce fossil fuels or are highly dependent on them. They are likely to attract political support from energy-intensive industries that are sensitive to input prices such as steel, glass, aluminium, paper and ceramics. At the other end of the spectrum are those firms that are actively involved in the green economy and are engaged in the development and application of new technologies. It should be emphasized that within any given sector, there will be firms that because of internal leadership, culture or concern about reputation are more active in seeking to move towards low-carbon production than others. It is these firms which generate the more innovative forms of cognitive leadership in terms of providing novel solutions.

The green economy is certainly growing in size, but its political displacement still lags behind its economic displacement. This is not an unusual or surprising phenomenon. Mature industries tend to be more politically effective than newer industries in defending their interests. The mature industries have had a long period of time to develop effective political representation and build networks and connections with political decision-makers. They are likely to be more politically sophisticated than emergent industries. These industries face the challenge of developing an identity, defining their boundaries and creating representative organizations. They may not be able to spare much in the way of management or financial resources for this task. General representative organizations of business are likely to favour established sectors that contribute a considerable proportion of their resources. BusinessEurope has argued for the phasing out of support for energy produced from renewable sources on the grounds of the impact of such support on energy prices and the principle of allowing the market to determine energy choices.

As far as companies that have an incentive to take action on climate change are concerned, '[i]nsurance and reinsurance companies are confronted with enormous liabilities from rising weather-related claims' (Dunn 2002: 31). The Association of British Insurers states, '[a]n increase in the frequency or severity of extreme weather events could mean that insurers have to increase premiums to stay viable' (Association of British Insurers 2008).

In an instance of cognitive leadership, in 2007 the insurance industry set up a global ClimateWise initiative with an international membership of over 30 leading insurance companies and organizations from Europe, North America and South Africa. The initiative is based around six principles: a lead in risk analysis; informing public policy making; supporting climate awareness among customers; incorporating climate change into investment strategies; reducing the environmental impact of insurance businesses; reporting and being accountable. The initiative is run by the University of Cambridge Institute for Sustainability Leadership that also has a knowledge transfer role about sustainability (ClimateWise 2015).

The insurance industry considers that it is particularly well placed to influence consumer behaviour in both retail and commercial markets, given that it is a major source of advice on risk. The insurance industry believes that it faces some real threats from climate change in terms of increased uncertainty, particularly in relation to pricing its products. It considers that if it responds to those challenges in the right way, it can turn a threat into a market opportunity which would also enhance the industry's reputation. The industry wants to be seen as part of the solution rather than a barrier to progress. Unlike, say, the coal industry, climate change will not reduce the total volume of business, but there is a greater risk of making unpredictable losses, so anything that can be done to reduce that risk is beneficial to the industry.

What about industries which are not fossil fuel producers or intensive energy users, but do not face specific threats from climate change? There is increasing evidence that such industries feel an increasing imperative to respond to climate change. In the food and drink industry increased volatility, complexity and uncertainty is seen as presenting new challenges. Climate change is seen as a risk to food security. The EU organization for the sector, FoodDrink Europe (2015) states: 'The food and drink sector is 100% dependent on the climate for the production of the raw material it transforms. Climate change affects the sector, both in terms of production and price stability.' In the UK the Food and Drink Federation (FDF) has set its members a target of securing a 35 per cent reduction in CO_2 emissions by 2020 against a 1990 baseline. The FDF is working closely with DECC on the development of 2050 low-carbon roadmap for the UK food and drink manufacturing sector.

How do these concerns operate at company level? The Hull-based FTSE 350 company Cranswick received the Meat & Poultry Industry Environmental Initiative of the Year Award in 2014 and 2015. It seeks to go beyond compliance with regulatory requirements in its approach to environmental sustainability through what it calls 'green thinking'. 'Targeting all aspects of environmental sustainability Cranswick's dedication to the reduction of its carbon foot print is key to our business strategy and we pride ourselves on being industry leaders in forward planning' (Cranswick plc 2015a). Environmental progress is reported quarterly at board level against performance targets. The company has been measuring its carbon footprint since 2008 using Carbon Trust methodology. Sustainability is seen as a means of improving 'our business model and profitability, while providing the opportunity to be more competitive in an ever tightening market place' (Cranswick plc 2015b: 40).

It should be noted that the farming end of the food chain has made less progress on measures to mitigate climate change. This in spite of the fact that farming is a fossil fuel-intensive industry and is also a generator of a particularly significant greenhouse gas, methane from livestock, as well as nitrogen dioxide from fertilizers. The idea that the Common Agricultural Policy (CAP) might incorporate a third, climate change pillar was swiftly stifled at the beginning of the last round of CAP reform negotiations.

Climate change and energy policies have become more closely interlinked from the perspective of business. BusinessEurope calls for a 'coordinated energy and climate policy for 2030'. It argues that 'Europe has to put cost-competitiveness, security of supply and climate objectives on an equal footing' (BusinessEurope 2013: 4). This leads to a focus on price differentials with competitor countries that has become the subject of increasing business concern in Europe since the 'fracking' boom in the US.

Conclusions

While climate change deniers can still be found in the business community, particularly in the United States (see Chapter 16), most businesses and business associations regard climate change as a 'routine' part of their policy environment which requires a measured and carefully considered response. 'All large oil companies now use a "shadow price of carbon", reflecting expectations of greenhouse gas regulation, when making investment decisions' (Crooks and Clark 2015). The global financial crisis has not distracted business from integrating considerations about climate change into its decision-making procedures, although, as has been emphasized, there has been considerable variation in the nature and effectiveness of the response by business sector and by firm.

The EU has, to some extent, been distracted at a high level from the debate about climate change by the crisis in the Eurozone, and particularly in Greece. That has absorbed a considerable amount of the time and attention of leading decision-makers. In some respects the baton of cognitive political leadership has passed to the United States with President Obama seeking to make climate change a defining theme of the last phase of his presidency. In the business world, it is European companies that have most often taken the initiative. 'In their rhetoric, there is a sharp contrast between European and US oil companies' views on climate policy' (Crooks and Clark 2015).

Business is perhaps no longer the elephant in the room (see Grant 2011). It is part of the process, but not commonly a source of cognitive leadership which often comes from think tanks and NGOs (see Chapter 15). Structural leadership is used to protect business interests, but given the differentiation of business interests, it has been shown that cognitive leadership has emerged from some parts of the business world. Nevertheless, the leadership it offers is more humdrum than heroic, although, given the nature of the challenges faced, it does have to take a long-term perspective. Business does not think it could or should act alone on this issue and an effective partnership with national governments, EU and global organizations is seen as essential to progress. The clear preference is for action at a global level to avoid competitive distortions, and in this sense the progress made at 2015 Paris climate conference is a step forward, but it is accepted that this has to be matched by action at an EU and member state level.

Notes

1 http://350.org/2014-report/ (accessed 10 August 2015).
2 http://lctpi.wbcsdservers.org/the-solution/ (accessed 10 August 2015).
3 http://lctpi.wbcsdservers.org/ (accessed 10 August 2015).
4 www.ceres.org/company-network (accessed 10 August 2015).

References

Association of British Insurers (2008) www.abi.org.uk/Display (accessed 26 June 2008).

BusinessEurope (2013) *A Competitive EU Energy and Climate Policy*, Brussels: BusinessEurope.

BusinessEurope (2015) 'EU ETS reform raises risk of investment leakage', press release, 15 July 2015.

Clark, P. (2012) 'Capitalist conservationists', *Financial Times* 5 June 2012: 9.

ClimateWise (2015) www.climatewise.org.uk/about (accessed 18 August 2015).

Cranswick plc (2015a) 'Caring for the environment', http://cranswick.plc.uk/taking-responsibility/green-thinking (accessed 18 August 2015).

Cranswick plc (2015b) *Annual Reports and Accounts*, Hull: Cranswick plc.

Coen, D., Grant, W. and Wilson, G. (2010) 'Political science perspectives on business and government' in D. Coen, W. Grant and G. Wilson (eds), *The Oxford Handbook of Business and Government*, Oxford: Oxford University Press, 9–34.

Crooks, E. and Clark, P. (2015) 'Energy groups nod to climate of opinion', *Financial Times*, 1 June 2015.

Daugbjerg, C. and Svendsen, G.T. (2001) *Green Taxation in Question*, Basingstoke: Palgrave.

Dolsak, N., Levi, M. and Prakash, A. (2012), 'Elinor Ostrom', *Political Science & Politics*, October 2012: 791–2.

Dunn, S. (2002) 'Down to business on climate change: an overview of corporate strategies', *Greener Management International*, 39(3): 27–41.

FoodDrink Europe (2015) 'A time to act: Europe's food and drink industry shows action to address climate change', www.fooddrinkeurope.eu/S=0/news/press-release/a-time-to-act-europes-food-and-drink-industry-shows-action-to-address-clima/ (accessed 18 August 2015).

Golub, J. (1998) 'New instruments for environmental policy' in J. Golub (ed.), *New Instruments for Environmental Policy in the EU*, London: Routledge, 1–29.

Grant, W. (2011) 'Business: the elephant in the room?', in R.K.W. Wurzel and J. Connelly (eds), *The European Union as a Leader in International Climate Change Politics*, London: Routledge, 197–213.

Hart, D.M. (2010) 'The political theory of the firm', in D. Coen, W. Grant and G. Wilson (eds), *The Oxford Handbook of Business and Government*, Oxford: Oxford University Press, 173–90.

Mercer (2015) *Climate Change,* www.mercer.com/content/mercer/global/all/en/services/investments/investment-opportunities/responsible-investment/investing-in-a-time-of-climate-change-report-2015.html (accessed 18 August 2015).

Ostrom, E. (1998) 'A behavioral approach to the rational choice theory of collective action', *American Political Science Review*, 92(1): 1–22.

Ostrom, E. (2005) 'A general framework for analyzing sustainability of socio-economic systems', *Science*, 325 (5939): 419–22.

Ostrom, E. (2012) *The Future of the Commons*, London: The Institute of Economic Affairs.

Tarko, V. (2012) 'Elinor Ostrom's life and work', in E. Ostrom, *The Future of the Commons*, London: Institute of Economic Affairs, 48–64.

Weale, A. (1992) *The New Politics of Pollution*, Manchester: Manchester University Press.

Wurzel, R. (2008) *The Politics of Emissions Trading in Britain and Germany*, London: Anglo-German Foundation.

Zito, A.R. (2000) *Creating Environmental Policy in the European Union*, Basingstoke: Macmillan.

15 Environmental NGOs

Pushing for leadership

Rüdiger K.W. Wurzel,[1] *James Connelly and Elizabeth Monaghan*

Introduction

Surprisingly little scholarly attention has been paid to the role which environmental non-governmental organisations (ENGOs) have played in EU climate change politics, although there are exceptions (e.g. Long *et al.* 2002; Wurzel and Connelly 2011a; Schoenefeld 2014). This stands in contrast to a reasonably extensive literature on ENGOs in EU environmental policy in general (e.g. Long 1998; Adelle and Anderson 2013). This chapter starts with an overview of the historical development of climate change-related activities by Brussels-based ENGOs. It then assesses the four main themes of this book, namely leadership, multi-level and polycentric governance, policy instruments and the low-carbon economy.

EU-level ENGOs and climate change

The European Environmental Bureau (EEB) which set up an office in Brussels in 1974 was the first EU-level ENGO. In the late 1980s Greenpeace (1988), Friends of the Earth Europe (FoE-Europe)[2] (1989), World Wide Fund for Nature (WWF) (1989) and Climate Action Network Europe (CAN-Europe) (1989) followed. More specialised ENGOs (e.g. European Federation for Transport and Environment (T&E) (1989) and BirdLife International (1993)) opened offices in Brussels in the early 1990s (see, for example, Adelle and Anderson 2013; Long 1998; Wurzel and Connelly 2011). The 2000s saw the arrival of ENGOs specialising in specific climate change issues. CDM Watch (2009) has focused on the Clean Development Mechanism (CDM) which permitted developed countries to sponsor GHGE reduction projects in developing countries for which the former can earn credits (see Chapters 1, 17 and 18). As CDM Watch widened its remit to trying to ensure that the functioning of market-based policy instruments (see below) was not eroded by loopholes it changed its name to Carbon Market Watch in 2012.[3] E3G (2004), set up by former British diplomats, specialises mainly in climate diplomacy issues, while Client Earth, which grew significantly in the early 2010s, offers free legal expertise on climate change issues to ENGOs and Members of the European Parliament (MEPs). The presence of Bellona

Europe, which is an anti-nuclear Norwegian non-profit organisation strongly supporting carbon capture and storage (CCS), within CAN-Europe has occasionally led to tensions with ENGOs strongly opposed to CCS (Interviews 2009 and 2015–16).

For Brussels-based ENGOs, climate change became an important issue when the EU prepared its position for the 1992 Rio 'Earth summit' which adopted the UN framework convention on climate change (UNFCCC) (see Chapter 1). Since the 1990s, CAN-Europe, FoE-Europe, Greenpeace and WWF have established themselves as central players among ENGOs active on EU climate policy issues. Until the early 1990s, the EEB was very active on climate change issues while focusing in particular on the Commission's proposal for an EU-wide carbon dioxide (CO_2)/energy tax (see Chapter 3). However, following the UK's veto of this proposal (see Chapters 3 and 12) and the departure of one of its key climate policy experts, the EEB ceased to be a major player on EU climate policy although it continued work on niche issues (e.g. the EU Directive on fluorinated greenhouse gases or f-gases and GHGE from agriculture and biomass) which are not covered by the four key players. In the 2010s the EEB had only one full-time staff member for climate and energy issues who had to give priority to energy issues (Interviews 2013–16).

Brussels-based think tanks and research institutes undertaking climate policy research often also carry out climate policy advocacy. In the early 2010s the Centre for European Policy Studies (CEPS), Ecologic, Ecofys, European Policy Centre, Öko Institut and the Institute for Sustainable Development and International Relations (IDDR) were most active.

The European Climate Foundation (ECF), Oak Foundation, Kann Rasmussen Foundation and the EU Commission (through its LIFE and LIFE+ programmes) are important external funders for EU-level ENGOs' activities on climate change issues. The ECF, with an annual budget of €5 million in 2014, has developed into the most important external funder for these activities. In 2015 the ECF had a staff of about 100 working in offices in Brussels as well as in Berlin, London, The Hague (headquarters) and Warsaw. Some ENGOs have become concerned about the ECF's transformation from a 'clearing house' for funds from large foundations (e.g. Oak Foundation) to an organisation trying to coordinate climate-related ENGOs activities by creating clusters of NGOs which have to work together on specific projects to receive funding. The ECF itself states that it is guided by 'the co-development of strategy' with its partners and stakeholders which enter into an 'intense dialogue on values, strategies and impact.'[4] Greenpeace (which does not normally accept funding from governments and corporations) has generally welcomed the existence of the ECF as a funder for ENGOs climate change activities, although it took the principled decision not to apply for ECF funding (with the exception of Greenpeace Poland) (Interviews 2013–16). The ECF seems to have responded to the concerns of some ENGOs by emphasising '*now* they [i.e. ENGOs] are always involved from the start of the project deciding the strategy [italics added]' (Interview 2015).

Table 15.1 lists the most important ENGOs, non-environmental NGOs, think tanks and research institutes active in Brussels on EU climate change issues. It also states the main funders for climate change related ENGOs activities and some of the most important alliances between ENGOs and other players.

Alliances and networks

Within the Brussels-based ENGOs community, the EEB, FoE-Europe, Greenpeace and WWF are commonly referred to as the 'Gang of Four' or 'G4' (a parody of G7, the name given to the summits of the then 'great seven powers') (Wurzel and Connelly 2011: 214). The G4 network was set up to achieve better ENGOs cooperation and to counter the influence of business (see Chapter 14). It quickly expanded to the G10 or Green 10,[5] comprising BirdLife Europe, CAN-Europe, CEE Bankwatch Network, EEB, FoE-Europe, Health and Environment Alliance (HEAL), Greenpeace, Naturefriends International (NFI), T&E and WWF. Attempts to expand the Green 10 have been blocked by members fearful of its becoming unmanageable (Interview 2013). Like the G4, the Green 10

Table 15.1 NGOs, research institutes, funders and networks active on EU climate change issues

Most active large ENGOs active on EU climate change policy:
- CAN-Europe, FoE-Europe, Greenpeace and WWF

Small specialised ENGOs active on specific EU climate issues:
- Bellona Europa, Carbon Market Watch, Client Earth and E3G

NGOs other than ENGOs active on EU climate change policy issues:
- ACT Alliance Europe, Christian Aid, CIDSE and Oxfam

Think tanks and/or research institutes active on climate change:
- CEPS, Ecofys, Ecologic, European Policy Centre, IDDRI and Öko Institut

Foundations which fund NGO activities on EU climate change issues:
- ECF, Oak Foundation, Kann Rasmussen Foundation

Examples of networks and ad hoc *coalitions active on climate change:*
- *General network periodically active on climate change issues:*
  Green 10 (made up of BirdLife International, CAN-Europe, Central and Eastern Europe (CEE) Bankwatch Network, EEB, HEAL, FoE-Europe, Greenpeace, Friends of Nature International (NFI), T&E and WWF)

  Specific climate change related networks:
  Coalition for Energy Savings

  EU 2030 climate and energy package:
  - Coalition for Higher Ambition,

  2015 Paris climate conference (COP21):
  - Coalition Climate 21

  2009 Copenhagen climate conference (COP15):
  - Global Coalition for Climate Action, Tcktcktck

Source: interviews 2013–16.

focuses only periodically on climate change issues while giving priority to broader horizontal issues such as the environmental implications of the EU's Better Regulation agenda and the Transatlantic Trade and Investment Partnership (TTIP). The Green 10 represents 'more than 20 million people'[6] which provides its actions (e.g. support for EU climate policy measures) with a certain degree of legitimacy in the form of indirect citizens' support.

CAN-Europe, a network for national NGOs active on climate change issues in Europe, describes itself as

> Europe's largest coalition working on climate and energy issues. With over 120 member organisations in more than 30 European countries (representing over 44 million citizens) CAN Europe works to prevent dangerous climate change and promote sustainable climate and energy policy in Europe.[7]

It was originally set up under the name Climate Network Europe (CNE) as part of CAN-International (Garrelts 2014). Because it initially comprised 'an informal, under-staffed and loosely coordinated transnational group' (Rucht 1993: 91) it was excluded from the G4. However, it quickly developed into a well-organised network and lead coordinator for ENGOs activities on EU climate change issues and plays a major role within the Green 10. In 2016 CAN-Europe employed 18 full-time staff in its Brussels office. CAN-Europe is CAN-International's second largest regional network and thus also a formidable federated network (Schoenfeld 2014).

Temporary networks between different ENGOs (and sometimes non-environmental NGOs and actors such as unions) are formed on an ad hoc basis for collective action on specific climate change issues. High profile climate politics events and important (often multi-level) climate governance arenas provide opportunities for ENGOs' climate change activities. It is for this reason that ad hoc alliances are often formed prior to major Environmental Council and/or European Council meetings (see Chapter 4) and international climate conferences such as the 2009 Copenhagen climate conference (COP15) and the 2015 Paris climate conference (COP21). Such ad hoc alliances were formed prior to both the adoption of the 2020 EU climate and energy package and the 2030 climate and energy package (see Chapter 1) (Interviews 2013–16). The Coalition for Higher Ambition, which included ENGOs, non-environmental NGOs (such as development NGOs), businesses, unions and cities, constitutes one important example. It was formed after the 2015 Paris climate conference (COP21) agreed the Paris Agreement (see Chapter 1) and before the EU adopted its Effort Sharing Decision which will translate the collective GHGE reduction target into national targets.

The 2009 Copenhagen climate conference (COP15) and in particular the 2015 Paris climate conference (COP21) also triggered the setting up of much larger ad hoc alliances, including development NGOs, unions and societal movements such as ATTAC (*Association pour la Taxation des Transactions financière et l'Aide aux Citoyens*) (Interviews 2009 and 2015–16). In the run up to the 2009 UN Copenhagen conference all of the large, and some of the smaller Brussels-based ENGOs and development NGOs, pooled resources with other NGOs (e.g. Amnesty International) for a common campaign 'tcktcktck',[8] while more grass-roots oriented and/or radical groups teamed up for the 'Climate Justice Action'[9] campaign. Shortly after Copenhagen one ENGO representative claimed that the unprecedented level of NGO cooperation could herald 'the beginning of a global civil society movement on climate change' (Interview 2010). Faith-based groups (e.g. ACT Alliance Europe and CIDSE) and, although to a lesser degree, health groups (e.g. European Public Health Alliance (EPHA) and HEAL) also started to participate in informal multi-stakeholder networks to lobby both at EU and international level around the time of Copenhagen. This participation triggered an alteration in the discourse on climate change within the climate network, with climate justice, food security and food prices, climate finance and climate adaptation being more strongly emphasised.

Adelle and Anderson's (2013: 160) assertion that 'many coalitions are quite opportunistic' applies more strongly to temporary ad hoc coalitions than to semi-permanent coalitions like the Green 10 or long-established federated networks such as CAN-Europe. In the run-up to the 2015 Paris climate conference ENGOs representatives were unsurprisingly keen to emphasise the need to avoid 'another Copenhagen' and to 'manage expectations' (Interviews 2015–16). Despite a more cautious attitude an even larger temporary informal network called Coalition Climate 21 emerged. It comprised 'trade unions, international solidarity and green NGOs, social and feminist movements, faith and youth groups'.[10] The main aims of this network were to lobby climate policy decision makers and to mobilise the public by attracting media attention. However, while ENGOs and unions pleaded for a major demonstration at the beginning of the 2015 Paris conference with the aim of influencing the climate negotiations, more radical groups like ATTAC favoured a large demonstration on the final day of the conference to advocate fundamental reforms to the capitalist system (Interviews 2015–16). The terrorist attrocities on 13 November 2015, which killed 130 people and left hundreds wounded, halted both demonstrations because the French government outlawed all large public demonstrations for security reasons. Unusually for a COP likely to adopt a new international climate agreement, only small-scale demonstrations took place including the launch of an air balloon which Greenpeace was allowed to fly (but only a few metres above ground) near the Eiffel Tower, the painting of some streets and a 100-metre-long red ribbon intended to symbolise the emergency of global climate change.

Alliance building, as a defining characteristic of ENGOs working on climate issues, is helped by the fact that coordination among Brussels-based ENGOs is close and working relations are very good for six main reasons. First, most European offices pool resources because they have small staff and budgets, certainly compared with large companies (see Chapter 14). Building alliances, coalitions and networks with non-environmental NGOs, trade unions and other actors improves the chances of success for 'bread and butter lobbying' activities (Long and Lörinczi 2009: 168–9). As the ENGOs community in Brussels grew it developed 'an implicit division of labour' (Rucht 1993: 86) and became professionalised (Long and Lörinczi 2009; Wurzel and Connelly 2011; Adelle and Andersen 2013). This development is in line with the institutionalisation and professionalisation of environmental activism occurring worldwide (Dalton 2015). Second, major funders of ENGOs activities such as ECF have encouraged close cooperation (Interviews 2013–16). Third, competition between ENGOs for members/supporters as well as media and public attention is less intense at EU (rather than national) level (Rucht 1993: 90). Fourth, forming alliances enables ENGOs to claim representation of a larger number of EU citizens. Unlike businesses ENGOs can raise 'input legitimacy' (e.g. public support) *and* 'output legitimacy' (e.g. technical expertise) (Beyers and Kerremans 2005: 114). The Green 10 and CAN-Europe represent a large number of members and supporters: because of this they have perceived legitimacy and can exhibit a considerable degree of structural leadership (see below). The same applies to less well-institutionalised temporary ad hoc climate alliances between ENGOs, non-environmental NGOs and/or other actors such as unions. Adelle and Anderson (2013: 54) argue that NGOs can 'contribute "functional representation" to improve the legitimacy of EU policies'; and Caniglia, Brulle and Szasa (2015: 241–2) state that CAN-International 'is a massive network of NGOs from around the world and carries a great deal of scientific and moral authority as a representative of civil society interests at the climate change talks'. However, whether European ENGOs form part of an emerging EU civil society, which could reduce the EU's democratic deficit, is contested (see CEC 2001; della Porta 2003 and Rucht 1993 for an optimistic view and Greenwood 2003; Monaghan 2013 and Warleigh 2001 for a sceptical view). Fifth, the EU Commission has long encouraged the formation of European-wide interest groups by providing funding, information and privileged access for European umbrella organisations (Mazey and Richardson 1993; Long 1998). Finally, although different ENGOs in Brussels have different organisational structures and strategies (see below), they usually have similar goals, at least on climate change issues (Wurzel and Connelly 2011).

Strategies

Although the main Brussels-based ENGOs have similar goals on climate change issues and frequently join alliances, differences in strategy exist which can be explained primarily by different organisational histories and structures as well as dominant worldviews of core activists and members/supporters.

The interest group literature typically contrasts 'insider' with 'outsider' strategies (e.g. Grant 2000). Like Grant, most European ENGOs argue that these two strategies are not mutually exclusive but can be used simultaneously or sequentially (Interviews 2009 and 2015–16). One ENGO representative explained that 'most politicians realise that it takes pressure from outside which gives them more room for manoeuvre. An MEP or MP can then say: "I am under pressure from an NGO" ' (Interview 2010).

Some ENGOs exhibit stronger preferences for insider strategies (e.g. traditional lobbying) while others show a greater disposition towards outsider strategies (e.g. media stunts). WWF is widely seen as having a preference for insider strategies while Greenpeace is the most prominent ENGO making use of outsider strategies (e.g. Adelle and Anderson 2013). However, neither WWF nor Greenpeace use either strategy exclusively. For example, in 2013, WWF's Brussels office organised a media stunt with one of its activists wearing a face mask depicting the German Chancellor Merkel (Interview, 2013) while WWF Germany attracted huge media attention when it placed 1,000 ice figures on the steps of the German parliament where they quickly melted in the sun.[11] The offices of ENGOs in Brussels are staffed primarily with policy experts rather than campaigners; this includes even the more campaign minded groups such as Greenpeace. For example, climbers who might ascend tall buildings to enfold Greenpeace banners are not part of Greenpeace's EU office staff (Interviews 2015–16).

Another widely found classification in the academic literature differentiates 'mobilisation from above' and 'mobilisation from below' (e.g. Caniglia, Brulle and Szasa 2015: 238). For Greenpeace and WWF, which are both hierarchically structured organisations, campaigns are usually decided on and carried out by professional staff, while involvement from members/supporters is negligible (Rucht 1993). In contrast FoE, which is a decentralised, grassroots-centred organisation that grants considerable autonomy to its national, regional and local organisations, tends to mobilise its members (and the public) from below. The differences in strategy between the Brussels office of ENGOs are not as distinct as on the national level for two main reasons. First, the EU decision-making process, characterised by a dispersal of power across different EU institutional actors does not easily lend itself to mobilisation from below and/or outsider strategies. Second, the absence of a genuine EU/European media (newspapers and television are organised primarily along national lines and cater primarily for national or subnational audiences) makes the adoption of outsider and/or mobilisation from below strategies significantly less attractive for Brussels-based ENGOs. The increased importance of the World Wide Web has changed little in this regard although the internet and social media are of increasing importance for Brussels-based ENGOs. (However, social media arguably play a more important role for national ENGOs which are more oriented towards campaigning rather than lobbying). However, this does not mean that there are no climate change demonstrations or unconventional actions in Brussels, although they are usually dwarfed by similar activities on both the member state level and the international level, at least for the main UN climate change conferences (such as COPs). Brussels-based ENGOs recognise the rise and importance of the internet and social media, although there is a view that it cannot supplant more active campaigning: 'social media are more the means of mobilisation and communication ... the real pressure is from civil society and traditional demonstrations still needs people to move their body. ... the real activism the social media cannot really replace' (Interview 2016).

Because the 2009 Copenhagen climate conference had ended in huge disappointment, participation by development NGOs and faith groups within these networks began to fade. After Copenhagen, ENGOs underwent a period of soul searching, resulting in reduction of staff and time resources devoted to international climate negotiations (Interviews 2013–16).

Many Brussels-based ENGOs (e.g. FoE-Europe) reduced their campaign efforts aimed at the international climate negotiations while putting staff resources into 'disinvestment' campaigns, which demand that large institutional investors (e.g. pension funds) withdraw their investments in fossil fuels and companies which exploit 'dirty fuels' (Interviews 2015). In 2010 a European Coal Network, which received funding from the ECF, was established. Approximately 80 per cent of CAN-Europe's member groups have joined this network which was inspired by disinvestment campaigns from actors such as 350.org, a US-based organisation.

Leadership

All Brussels-based ENGOs have consistently advocated a strong leadership role for the EU in international climate change politics (see Chapter 1). Importantly, ENGOs themselves have also tried to provide different types and styles of leadership. CAN Europe, FoE-Europe, Greenpeace and WWF have been consistently taking the lead on a range of EU climate change issues while offering structural, entrepreneurial and cognitive leadership. For the reasons explained above Brussels-based ENGOs are inclined to offer a humdrum/transactional leadership style with occasional outbursts of heroic/transformational leadership, especially around the time of major international climate conferences (e.g. Paris 2015) and crucial EP, Council and/or European Council meetings. Brussels-based ENGOs have consistently demanded that the EU take on *exemplary leadership* (i.e. a leader-by-example role) while exhibiting a heroic/transformational leadership style (Interviews 2013–16).

Prior to the 2008 Environmental Council and European Council meetings, which agreed the 2020 climate and energy package intended to facilitate the adoption of a follow-up agreement to the Kyoto Protocol at Copenhagen (COP15), CAN-Europe, FoE-Europe, Greenpeace and WWF ran a joint campaign, 'Time to lead', urging the EU to adopt internally ambitious goals and measures and thus to offer exemplary leadership. In the run up to the conference an alternative climate change treaty entitled *A Copenhagen Climate Treaty. A Proposal for a Copenhagen Agreement by Members of the NGO Community* was drafted by a large number of ENGOs (IndyACT *et al.* 2009). For the 2015 Paris climate conference ENGOs did not produce an alternative draft climate treaty. This can partly be explained by the fact that ENGOs were keen to manage expectations and partly by the new complex bottom-up approach, which allowed countries to put forward voluntary (GHGE reduction) pledges (i.e. INDCs), that had emerged in the international climate change negotiations. Initially most Brussels-based ENGOs opposed the bottom-up approach while demanding that the EU should stick to its long-favoured top-down approach (i.e. legally binding GHGE emission reductions enshrined in an international treaty). Eventually ENGOs came to recognise that only the acceptance of a bottom-up approach would pave the way for an international climate change agreement in Paris with development NGOs accepting the new international approach earlier than most ENGOs. However, ENGOs consistently and vehemently insisted on the need to limit global temperature rise to 1.5°C which was eventually enshrined as a target in the Paris Agreement.

ENGOs active on EU climate change issues expected at best a humdrum/transactional leadership style from the EU and governments attending the 2015 Paris conference. Brussels-based ENGOs themselves also relied strongly on a humdrum/transactional leadership style in their own campaign efforts, aimed at managing expectations following the disappointing outcome of Copenhagen (Interviews 2015–16). For the 2015 Paris conference ENGOs, however, demanded exemplary leadership from the EU in the negotiations.

Cognitive leadership, arguably the most important type of leadership offered by ENGOs, tries to influence the wider political discourse and to frame (or reframe) the main issues at stake so that the advantages of early and decisive action become more easily understandable for decision-makers and in particular the public. ENGOs coined the term 'hot air' (Long and Lörinczi 2009: 172) and have demanded an 'energy *(r)evolution*' (Greenpeace 2008). Cooperation between environmental and development NGOs on climate change has slightly altered the political discourse with more attention being paid to issues such as food security and climate justice (Interviews 2008–10 and 2013–16). Brussels-based ENGOs have consistently argued that both the environment and the economy will benefit from the transition of an economy which is highly dependent on high carbon-intensive fossil fuels towards a 'green' or low-carbon economy which relies exclusively (or almost exclusively) on renewable energy. ENGOs have demanded that the EU should adopt a heroic/transformational leadership style when embarking on the route towards a low-carbon economy

The ability of ENGOs to provide *structural leadership* depends largely on their size, capability to get involved in networks, and ability to influence public opinion. Clearly, ENGOs do not have the structural powers of business in resources and jobs (see Chapter 14). ENGOs can, however, put public pressure on laggards among businesses, states and EU institutions. Moreover, large ENGOs (especially) and alliances involving a high number of ENGOs can claim that they represent the views of a large number of EU citizens: this can be seen as increasing the political legitimacy of their goals, strategies and demands. ENGOs which endorse unconventional direct action occasionally offer heroic leadership but this is hampered by the EU's policy-making system, which is generally geared towards a humdrum leadership style. Moreover, forming large alliances can help to increase media attention and to raise further public awareness about climate change as well as support for ambitious EU climate change policies (e.g. Eurobarometer 2015).

ENGOs' entrepreneurial leadership capabilities are arguably more limited than their structural and in particular their cognitive leadership capabilities. The above mentioned alternative climate treaty (see IndyACT *et al.* 2009) for the 2009 Copenhagen climate conference constitutes one important example of entrepreneurial leadership. Some ENGOs have managed to get activists embedded in national delegations which participate in the international climate change negotiations. However, the higher the negotiation level the less likely that ENGOs representatives (albeit 'dressed up' as national representatives) will sit at the negotiation table (Interview 2015).

Most ENGOs acknowledge that the EU has managed to reduce the glaring gap between its heroic rhetoric and humdrum implementation measures (Interviews, 2013–16). However, many acknowledge that it has become more challenging for the EU to provide an international climate leadership role because global politics has changed, with countries such as China and India (see Chapters 17 and 18) playing a more important role.

Multi-level and polycentric governance

The large international ENGOs are generally well adapted to influencing multi-level environmental governance structures. This is particularly true for federated networks such as CAN-Europe. Venue shopping to maximise influence has been widely practiced by large ENGOs (Mazey and Richardson 1993). Smaller ENGOs often find it difficult to engage in multi-level games (Rucht 1993; Interviews 2008–10 and 2013–16). However, even the large Brussels-based ENGOs focus primarily on the EU level in their daily work (Interviews 2008–10 and 2013–16). One important reason is that EU decision-making is 'a mix of inter-governmental and supranational bargaining.... In such a differentiated institutional setting, the problem of interest groups is not a shortage but an over-supply of potential routes to influence between which they must allocate scarce resources' (Long and Lörinczi 2009: 163).

CAN-Europe and WWF's Brussels office are well geared towards tracking the EU, international and member state levels, although their main task is to lobby the EU; regional or city governance level has traditionally not been on their radar. There is awareness among Brussels-based ENGOs of climate change activities by cities (e.g. the Covenant of Mayors). However, a lack of staff and time resources prevents Brussels-based ENGOs from closely following these or participating. As one ENGO representative explained (Interview 2015):

> There were events from the Covenant of Mayors. I didn't attend because when it comes down to my daily work, the kind of documents and consultation responses I have to do, the Covenant of Mayors is low down to what I could deliver information and to ... what I could get information out of. So it is great but, sorry, I have no time.

Others are doubtful whether climate change activities of cities will really make a significant difference because 'the difficulty is that what's inside their border is outside their control', that is, the most GHGE-intensive industries are governed by national, not local, laws (Interview 2013). Joint activities between Brussels-based ENGOs and cities are therefore limited to attending events to which both representatives have separately been invited (Interview 2015). Resource limits prevent any sustained self-coordinating efforts which might contribute to the bottom-up creation of polycentric governance structures (see Chapter 1).

Of the large ENGOs in Brussels, FoE is the most grassroots-oriented in its organisational structure. Its internal organisation structure is multi-level, allowing a considerable degree of decision-making power (e.g. on goals and campaign strategies) for subnational actors. This has advantages (e.g. close contacts with subnational members) and disadvantages (e.g. differences in views among different subnational groups can lead to slower decision-making processes).

European ENGOs lobby all the main EU institutions (with the exception of the European Council and of the European External Action Service (EEAS), although development NGOs have some links with the latter) involved in the EU climate change policy. They pay particular attention to the Commission and EP and, to a lesser degree, Council.

In 2015 an amalgamation of DG Climate Change (established in 2010) and DG Energy was carried out by the Juncker Commission (see Chapter 3). The appointment of Miguel Arias Cañete as the Commissioner in charge of the merged Energy and Climate departments was initially strongly criticised by ENGOs because of Cañete's former links to the oil industry and his lack of expertise on climate change issues. Subsequently, however, most Brussels-based ENGOs seem to have changed their views (Interviews 2015–16). One ENGO representative even stated (Interview 2015):

> [Cañete] definitely doesn't have the commitment of Connie Hedegaard [i.e. the former Commissioner in charge of DG Clima] but what we note is that he is a better political player in that he is better in building alliances inside the Commission. Hedegaard had a lot of enemies inside the Commission and also within the UNFCCC negotiations. She was not always the most diplomatic person.... Cañete is completely different.

Lobbying of national governments is usually undertaken by national ENGOs, although Brussels-based ENGOs assist with information and occasionally with resources, especially in member states (e.g. the Central and Eastern European States (CEES)) with weak civil society activism and/or under-resourced ENGOs. During the international climate change negotiations ENGOs like CAN-Europe activate their respective national member groups when they detect that certain EU member states are not supporting the EU's official position as actively as they should, although this is difficult in the CEES and especially in Poland (Interview 2015).

Because of resource constraints, even the large Brussels-based ENGOs find it difficult to monitor what is going on at different climate governance levels, let alone trying to simultaneously influence the international, EU, national and sub-national levels. One ENGO representative explained: 'I find it quite difficult to handle the multi-level governance structures. I must prioritise every day. I have more to do than what I can handle', while another stated:

> For the moment we are struggling with the international [level]. We believe that it is important to work at an international level, but for the moment we do not have the resources. For the moment we are focusing on the EU and key member states.
>
> (Interviews 2009)

Policy instruments

As explained in Chapter 1, policy instruments can be grouped into three main categories: (1) regulations (which stipulate legally binding targets and deadlines), (2) market-based instruments (e.g. eco-taxes and emission trading) and, (3) voluntary agreements and informational devices (e.g. eco-labels) (e.g. Wurzel, Zito and Jordan 2013). ENGOs have traditionally favoured ambitious regulation, although they were also early supporters of EU-wide eco-taxes.

All European ENGOs have campaigned for ambitious GHGE reduction targets and deadlines to be enshrined in legally binding national/EU regulations and international treaties. For ENGOs '[t]here are a number of problems with market-based approaches which we think can be better dealt with a regulation approach' (Interview 2016). All Brussels-based ENGOs favour an EU-wide CO_2/energy tax, although they realise that it is unlikely ever to be adopted. European ENGOs were initially strongly opposed to emissions trading – a policy which they once had compared to the sale of indulgences. After the EU ETS became operational in 2005 all Brussels-based ENGOs lobbied for improvements. However, since the early 2010s, FoE groups (especially in developing countries) raised concerns, leading FoE-Europe once more to oppose the EU ETS.

ENGOs felt that their scepticism toward voluntary agreements had been vindicated when the voluntary agreement between the automobile industry and the Commission failed. In 1999, the European automobile industry promised to reduce CO_2 emissions from passenger cars by approximately 25 per cent (i.e. 140g/km) by 2008. However, the Commission soon expressed its dissatisfaction with the automobile manufacturers' lack of progress. To the delight of ENGOs, the Commission's proposal for mandatory regulation was adopted by the Council and EP in 2008.

In the run-up to the 2015 Paris climate change conference differences between some ENGOs and development NGOs became apparent over the need for developing countries to accept legally binding GHGE reduction targets. The voluntary pledges (or Intended Nationally Determined Contributions – INDCs) for the Paris conference were initially regarded highly sceptically by most ENGOs. However, after the Paris Agreement had been adopted there was widespread acknowledgement among Brussels-based ENGOs that the voluntary pledges and bottom-up approach had introduced a new dynamic into the international climate change negotiations, although many remain highly sceptical of all countries fulfilling their pledges.

Green economy and low-carbon economy

All Brussels-based ENGOs active on climate change strongly believe that the EU and its member states will benefit, in environmental *and* economic terms, from a move towards a green or low-carbon economy. However, development NGOs have emphasised more strongly the concept of sustainable development, placing equal emphasis on economic, social and environmental concerns, while many ENGOs stress the importance of ecological modernisation, resting on the assumption that ambitious environmental measures are beneficial for both environment and economy (Interviews 2009–10 and 2013–15; Wurzel and Connelly 2011). As mentioned above, climate justice, food security and climate finance as well as climate adaptation are often emphasised more strongly by development NGOs than ENGOs although they have been on convergence course in terms of endorsing the low-carbon economy. Most ENGOs have largely accepted that developing countries will need to be allowed to grow economically (and thus to emit GHGE at least until fossil fuels have been replaced by renewable energy sources) while development NGOs, which often work with local communities rather than with national governments, are increasingly recognising both the economic and environmental benefits of renewable energy for local communities in developing countries.

Development NGOs do not perceive climate change as an issue likely to affect developing countries only in the long term: rather they emphasise that food shortages and environmental refugees are already being triggered by the effects of climate change in some of the poorest countries in the world (Interviews 2013 and 2016). This is clearly expressed in the following Oxfam anti-coal slogan: 'Let them eat coal. Why the G7 must stop burning coal to tackle climate change and fight hunger'.[12]

The 2008 financial crisis and subsequent worldwide economic recession have been identified by ENGOs as posing a danger to their ability to frame climate change as an opportunity for developing a low-carbon economy and promoting ecological modernisation (Interviews 2009 and 2015–16; see also Chapter 4). Some ENGOs have argued that '[i]ndustry has used [the economic recession] to water down the climate and energy package' while raising concerns that '[c]harity spending declines fast in credit crunch times. It might take ten years [to recover]' (Interview 2009). However, as the following statement shows (Interview 2015), some ENGOs representatives are less pessimistic:

> I think overall we are quite sheltered from [the economic crises and Euro crises] with the exception of the discussions on climate finance because it is hard to have a discussion with Finance Ministries. I mean with Environment

Ministers you can have a conversation.... But we all know that they are not the ones who decide on this. So I think that the Finance Ministers need to be more involved.

Conclusion

Climate change is an important issue for all European ENGOs. By the early 2000s, CAN Europe, FoE-Europe, Greenpeace and WWF had developed into the most active European ENGOs on EU climate change issues, with CAN-Europe assuming a coordination role. For major climate change campaigns ENGOs have formed coalitions with non-environmental NGOs (e.g. development and health NGOs) and other actors (e.g. unions). Many of these alliances and networks are supported by funding bodies (e.g. the ECF). Much of the coordination of ENGOs' day-to-day work on EU climate change takes place informally between a few activists with CAN-Europe, Greenpeace, FoE-Europe and WWF having developed into the most important players on EU and international climate change issues.

ENGOs have tried to offer structural, entrepreneurial and cognitive leadership while demanding that the EU show heroic leadership. However, NGOs are aware that the EU usually offers little more than humdrum leadership which is also the leadership style that is used most often by Brussels-based ENGOs, although they engage in spurts of heroic leadership especially around the time of major international climate change conferences.

Brussels-based ENGOs take unconventional direct action only infrequently. Why? First, the EU exhibits complex multi-level governance structures with actors sharing powers varying according to different phases of the policy-making process. Hence singling out a particular environmental villain to target is difficult. Second, there is neither a European-wide media nor a genuine European public: national newspapers, TV, radio and internet news-sites cater primarily for different national publics. Finally, the EU's consensus-seeking decision-making style leaves little room for confrontational direct action. Despite this, Greenpeace, FoE-Europe and (to a lesser degree) WWF have staged some media and public attention-seeking actions.

Even the large European ENGOs find it difficult to stay engaged with the various multi-level climate change governance levels. It stretches them to the full, both in respect of staff and time resources. For these reasons, Brussels-based ENGOs focus primarily on the EU level in their daily work while seeking to take full advantage of the alliances and networks with other NGOs, thinks tanks and actors such as unions.

Notes

1 Rudi Wurzel would like to thank the British Academy (SG46048) and Hull University's Faculty of Arts and Social Sciences for funding research on which parts of this chapter are based. The authors are grateful to representatives from CAN-Europe, The Centre, Client Earth, ECF, EEB, FoE-Europe, Greenpeace, Oxfam, T&E and WWF

who gave up their precious time to be interviewed. Twenty-eight interviews were conducted between 2008–16. All errors and normative judgements remain the responsibility of the authors.

2 FoE Europe was initially called Coordination Européenne des Amis de la Terre (CEAT). See www.foeeurope.org/about/history (accessed 9 September 2015).
3 http://carbonmarketwatch.org/about/ (accessed 16 May 2016).
4 https://europeanclimate.org/mission/strategy/ (accessed 16 May 2016).
5 www.green10.org/ (accessed 10 June 2016).
6 www.green10.org/ (accessed 3 May 2016).
7 www.caneurope.org/about-us (accessed 16 May 2016). The website of the Green 10, which includes CAN Europe, claims the Green 10 represent more than 20 million people while CAN-Europe's website states that it represents 44 million citizens.
8 www.timetolead.eu/ (accessed 15 September 2015).
9 https://climatejusticeaction.net/en/ (accessed 3 May 2016).
10 http://coalitionclimat21.org/en/contenu/about-us (accessed 14 May 2015).
11 http://wwf.panda.org/homepage.cfm?173563/Tiny-ice-figurines-draw-attention-to-big-problem (accessed 14 September 2015).
12 http://policy-practice.oxfam.org.uk/publications/let-them-eat-coal-why-the-g7-must-stop-burning-coal-to-tackle-climate-change-an-556110 (accessed 14 September 2015).

References

Adelle, C. and Anderson, J. (2013) 'Lobby groups', in A. Jordan and C. Adelle (eds), *Environmental Policy in the EU*, London: Earthscan, 152–69.

Beyers, J. and Kerremans, B. (2005) 'Bürokraten, Politiker und gesellschaftliche Interessen: Ist die Europäische Union entpolitisiert?', in R. Eising and B. Kohler-Koch (eds), *Interessenpolitik in Europa*, Baden-Baden: Nomos, 123–50.

Caniglia, B.S., Brulle, R. and Szasz, A. (2015) 'Civil social movements, and climate change', in R.E. Dunlap and R.J. Brulle (eds), *Climate Change and Society. Sociological Perspectives*, Oxford: Oxford University Press, 235–68.

CEC (2001) *European Governance. A White Paper, COM(2001)428 final,25.07.2001*, Brussels: Commission of the European Communities.

Dalton, R. (2015) 'Waxing or waning? The changing patterns of environmental activism' *Environmental Politics*, 24(4): 530–52.

Della Porta, D. (2003) *The Europeanisation of Protest: A Typology and Some Empirical Evidence*, EUI Working Paper 2003/18, Florence: European University Institute.

Della Porta, D. and Parks, L. (2015) 'Democratizing the climate negotiations system through improved opportunities for participation', in M. Dietz and H. Garrelts (eds), *Routledge Handbook of the Climate Change Movement*, London: Routledge, 19–30.

Eurobarometer (2015) *Climate Change. Special Eurobarometer 435*, Brussels: European Commission.

Garrelts, H. (2014) 'Organisational profile – Climate Action Network International', in M. Dietz and H. Garrelts (eds), *Routledge Handbook of the Climate Change Movement*, London: Routledge, 237–39.

Grant, W. (2000) *Pressure Groups and British Politics*, Basingstoke: Macmillan.

Greenpeace (2008) *Energy [R]evolution*, Amsterdam: Greenpeace International.

Greenwood, J. (2003) *Representing Interests in the European Union*, Basingstoke: Palgrave/Macmillan.

IndyACT *et al.* (2009) *A Copenhagen Climate Treaty. Version 1.0*, http://germanwatch.org/klima/treaty1nar.pdf (accessed 14 September 2015).

Long, T. (1998) 'The Environmental Lobby', in P. Lowe and S. Ward (eds), *British Environmental Policy and Europe*, London: Routledge, 105–18.

Long, T. and Lörinczi, L. (2009) 'NGOs as Gatekeepers: A Green Vision', in D. Coen and J. Richardson (eds), *Lobbying the European Union*, Oxford: Oxford University Press, 162–79.

Long, T., Slater, L. and Singer, S. (2002) 'WWF: European and Global Climate Policy', in R. Pedler (ed.) *European Union Lobbying: Challenges in the Arena*, Basingstoke: Palgrave, 87–103.

Mazey, S. and Richardson, J. (eds) (1993) *Lobbying in the European Community*, Oxford: Oxford University Press.

Monaghan, E. (2013) 'A nascent transnational civil society', in J. Hayward and R. Wurzel (eds), *European Disunion: Between Sovereignty and Solidarity*, Basingstoke: Palgrave/Macmillan, 32–47.

Rucht, D. (1993) 'Think globally, act locally? needs, forms and problems of cross-national cooperation among environmental groups', in J. Low *et al.* (eds), *European Integration and Environmental Policy*, London: Belhaven, 75–96.

Schoenefeld, J. (2014) *Furthering Deliberative Democracy? The Role of Environmental Groups in EU Climate Change Policy-Making*, Paper, 15th UACES Student Forum Conference, Aston University: Birmingham.

Warleigh, A. (2001) '"Europeanizing" civil society: NGOs as agents of political socialisation', *Journal of Common Market Studies*, 39(4): 619–39.

Wurzel, R.K.W. and Connelly, J. (2011) 'Environmental NGOs: taking a lead?', in R.K.W. Wurzel and J. Connelly (eds), *The European Union as a Leader in International Climate Change Politics*, London: Routledge, 214–31.

Wurzel, R.K.W., Zito, A. and Jordan, A (2013) *Environmental Governance in Europe*, Cheltenham: Edward Elgar.

Part V

Europe and the wider world

16 The United States

The challenge of global climate leadership in a politically divided state

Guri Bang and Miranda A. Schreurs

Introduction

In 2001, the George W. Bush administration pulled the United States (US) out of the Kyoto Protocol, signalling an end to the climate policies pursued by the Bill Clinton and Al Gore administration. Domestically, the US Congress failed to pass national climate legislation and actively prevented the US Environmental Protection Agency (EPA) from pursuing various climate related activities. As a result, it was not the United States but the European Union that emerged as the global champion of progressive climate action (Schreurs *et al.* 2009).

Almost 15 years later, in Paris in December 2015, the Barack Obama administration announced that the United States would once again pursue a global leadership role on climate change. Obama committed the United States to reduce greenhouse gas emissions (GHGE) by 26 to 28 per cent of 2005 levels by 2025 premised on numerous domestic policy measures that had been or were to be implemented.

Many Congressional representatives, states, and industries, however, objected to the president's plans and have launched initiatives to slow or block their implementation, including through the courts. There have even been calls to pull the US out of the 2015 Paris Agreement, for example, by the Republican presidential candidate, Donald Trump. The multi-level and polycentric nature of the US political system has made it possible for supporters of action at various levels to act as climate pioneers while at the same time opponents of change have been successful in slowing and at times, blocking policy action. This chapter explores efforts to promote US leadership on climate change at the local, state, national, and international levels, as well as the counter-initiatives of industrial and political actors seeking to prevent the introduction and implementation of domestic climate policies.

Vacillating between international climate leadership and international obstructionism

The role of the United States internationally in climate policy making is difficult to characterize as it has vacillated considerably over time. There have been times when the United States has exhibited climate leadership – cognitive, entrepreneurial, and structural – but also periods when the United States has had only symbolic leadership (where there has been more talk than action) and even many periods where the United States has obstructed international climate progress.

In terms of international scientific contributions to the understanding of the causes and possible consequences of climate change, the United States has exhibited cognitive leadership. It has been a consistent and major contributor to the Intergovernmental Panel on Climate Change (IPCC) as well as to capacity building efforts in developing countries. The US has also been an entrepreneurial leader. It was largely due to US demands that the Kyoto Protocol incorporated flexibility mechanisms, including emissions trading and joint implementation (JI), which helped in bringing the Kyoto Protocol to a successful conclusion.

Yet the United States subsequently pulled out of the Kyoto Protocol, thereby threatening the agreement's viability and weakening its effectiveness; the United States became one of the agreement's harshest critics even though the EU had made it clear it planned to stand by and implement the agreement.

At the 2009 Copenhagen climate conference (COP15), and with a new president in office, the United States once again sought to exert climate leadership – cognitive, structural, and entrepreneurial. It was actively engaged in formulating a post-Kyoto agreement. Its entrepreneurial leadership helped prevent a complete breakdown of the negotiations. Its cognitive leadership influenced the development of a compromise strategy – the inclusion in the Copenhagen agreement of voluntary climate action pledges, which were later to become known as intended nationally determined contributions (INDCs) in the 2015 Paris climate agreement, despite the EU's clear preference for binding targets. The United States also played a critical role in reaching formal and informal bilateral climate agreements with major developing countries which helped to bring them into the Paris agreement.

The fact that the United States is the world's second largest emitter of greenhouse gases, but also the world's strongest economy, means that no global agreement can be successful in the long term without US participation. This makes the United States an extremely powerful player in international climate negotiations. Thus, the US has the ability to be a structural leader or a structural blocker of global climate action.

Whether the United States will be more than just symbolically committed to the commitments it made in Paris will depend greatly on domestic political and judicial developments. Understanding the US relationship to the global climate negotiations requires looking at domestic policy developments.

The federal structure of the United States and bottom-up climate leadership

Much like several of the more climate progressive countries in the European Union such as Austria, Denmark (see Chapter 6), Germany (see Chapter 8), and Sweden, some regions in the United States have attracted worldwide attention for their climate leadership. Under the Clean Air Act, California has a special status which allows it to be an initiator of air pollution policies which go beyond those set by the federal government. Other states can choose to follow either California or federal air pollution laws. This is a power no other state has and is a key reason why California is often both a structural and entrepreneurial leader, as can be seen with its global warming law, strong support for renewable energies, and comprehensive GHGE trading scheme (Karapin 2016). Ten Northeastern states were earlier pioneers with carbon emissions trading in the Regional Greenhouse Gas Initiative. In 2005, Seattle launched the US Mayors Climate Initiative, a group of hundreds of US cities which pledged to pursue the goals embedded in the Kyoto Protocol (Bulkeley 2013).

Yet, the United States is well known for the strength of its climate sceptics, who have vocally opposed what they argue to be expensive, job-killing climate initiatives based on uncertain and questionable scientific evidence (Jacques 2009). Climate sceptics, in coalition with supporters of the status quo, have mustered sufficient support in the US Congress to prevent the passage of domestic climate legislation. They also succeeded in persuading the George W. Bush White House to work actively against the Kyoto Protocol. The Bush White House labelled the Kyoto Protocol as an unworkable, ineffective, and unjust regime and criticized the impact it would have on jobs. The Kyoto Protocol was also criticized for not obligating developing countries to take action to curb their greenhouse gas emissions. Behind these voices of scepticism, a variety of interests can be identified: fossil fuel-producing industries, states heavily dependent on fossil fuel extraction and export, evangelicals, libertarians, supporters of the Tea Party, and other individuals and groups.

The polycentric climate landscape in the United States, with its many actors with diverse ideologies and interests has made the efforts of the Barack Obama administration to reassert US climate leadership, especially cognitive and entrepreneurial leadership, challenging. The idiosyncrasies of the US political system have pushed the Obama administration to pursue less traditional political strategies in progressing its climate policy aims in the hope of not being accused of pursuing little more than symbolic leadership.

Ratification of international agreements requires a two-thirds majority in the US Senate; with a highly polarized and ideologically divided Senate this has become an almost impossible hurdle to overcome. Passage of federal climate legislation requires a simple majority in both Houses of Congress; this too has proved an increasingly insurmountable challenge in the ideologically strongly divided Congress where cross-party coalitions are less and less likely to form. Thus, the Obama administration has looked for support from more progressive state and local actors, has broadly interpreted the authority of the US EPA to issue pollution control standards and regulations, and has made full use of its executive authority to pursue its GHGE mitigation goals (Shafie 2014). These moves have in turn invited strong opposition and a turn to the courts by opponents of the Obama administration's actions. Hence the question arises as to whether the US can offer structural and entrepreneurial leadership in the international climate negotiations when the domestic basis upon which its climate policies have been constructed remains shaky and contested.

Public opinion on climate change and US climate leadership

The Obama administration's policies were based on scientific findings and advice from the US National Academy of Sciences and the IPCC, recommendations that resonate strongly among a large majority of Democrats. In 2014, 88 per cent of surveyed Democrats stated they believe that global warming is happening, and 82 per cent expressed willingness to support strict CO_2 emission limits on power plants (Leiserowitz *et al.* 2014). Obama's push for climate change policy is clearly connected to the expressed interests of his party's supporters. However, climate change is an increasingly partisan issue. With the growth of Tea Party power within the Republican Party, congressional Republicans increasingly tended to vote against any proposed climate policy programmes (Skocpol 2013).

The general public's attitudes toward federal energy and climate policy are influenced by their levels of concern about energy security, energy prices, and environmental protection. While slightly more than half of Americans (56 per cent) in 2016 believed that human actions are mostly to blame for climate change, their willingness to endure higher energy prices (for instance as a result of carbon pricing) was much higher, with 70 per cent of respondents supporting setting strict limits to carbon emissions from power plants (Leiserowitz *et al.* 2016). Polling trends spanning two decades show that when the US economy is suffering and unemployment is high, Americans tend to discourage policies that could increase energy costs. Conversely, in a booming economy, their willingness to pay for environmental protection is much higher (Scruggs and Benegal 2012). Hence, public demand for low-carbon energy policy is partly linked to the economy's performance. Public support for federal climate-policy action fell dramatically between 2008 and 2012 during the financial crisis (Scruggs and Benegal 2012).

In 2016, in step with economic recovery, support for climate policy action increased (Leiserowitz *et al.* 2016). Specifically, support for the Obama administration's proposed regulations in the power sector was linked to the positive effect power plant emissions reductions have in improving health conditions. Taking political party alignment into account, a Yale University survey showed that 91 per cent of liberal Democrats and 37 per cent of conservative Republicans supported such limits to power plant emissions (Leiserowitz *et al.* 2016). The survey found a clear difference between moderate Republicans and conservative Republicans. Moderate Republicans conveyed many of the same views as moderate and conservative Democrats did concerning the climate change issue, indicating that Republicans did not uniformly oppose climate policy. Liberal and moderate Republicans' views were close to the mainstream of American public opinion on climate change, while conservative Republicans' views stood out as distinctly different from those of the rest of the American public (Leiserowitz *et al.* 2016). About 30 per cent of Americans describe themselves as evangelical Christians. Compared to non-evangelicals, evangelicals are less likely to believe that climate change is caused by human activity or that global warming is occurring. Among evangelicals, right-leaning evangelicals are more likely to doubt that humans are having an impact on the environment (Smith and Leiserowitz 2013). In general, public opinion also varies significantly between states with high fossil-fuel-energy dependence in the state economy, and states with weaker ties to fossil-fuel energy (Bang and Skodvin 2014). Democrats representing states with fossil-fuel intensive economies have opposed Obama's climate policies.

Pursuing climate leadership in an ideologically divided political environment

Barack Obama's first election campaign was replete with plans to combat climate change through the development of a green economy (Obama's green new deal): investments in clean and renewable energies were to be increased, energy efficiency was to be improved, and the United States was to lead internationally on climate change. Soon after entering office, and with oil and gas prices soaring, the president, together with Vice President Joe Biden, launched the New Energy for America plan, calling for the expansion of renewable energy and clean energy technology to address both energy security and climate change (Obama and Biden 2009).

For many years attempts to change US climate policy were focused on legislative action. The many climate bills introduced into Congress, however, met with persistent and fierce opposition, even when they were initiated by representatives of both parties. Legislation proposing a cap-and-trade system was voted down repeatedly in the Senate in 2003, 2005 and 2008. When the Democrats took over the majority in both chambers of Congress in 2006, they initiated a campaign to put climate change on the Congressional agenda. Hundreds of congressional hearings were held on the subject of federal action on climate policy, and many bills were proposed outlining various strategies for how federal policy could be designed cost effectively (Bang and Schreurs 2011). Climate legislation was introduced and debated in the 111th Congress (2009–2010), and the House of Representatives passed the American Clean Energy and Security Act in June 2009 by a slim majority. However, the bill died in the Senate in 2010 without a floor debate (Wallach 2012). No climate bills have been debated in Congress since then. After the 2010 mid-term election when a Republican majority took hold in the House of Representatives and the Democratic majority in the Senate was reduced, climate legislation had little chance of success.

Pursuing policy change through executive orders

With the constant and seemingly insurmountable opposition to international climate leadership and federal climate action in the US Congress, the Obama administration has sought alternative avenues to pursue action. The US Constitution and various

Acts of Congress delegate to the president and the executive branch of government certain powers to manage operations within the federal government and in developing policies linked to the implementation of broad statutes. Executive orders are subject to judicial review of their constitutionality. Opponents of Obama's plans have attempted to block his executive orders (with limited success) in the courts.

An extremely important existing piece of legislation for Obama's climate program is the U.S. Clean Air Act of 1970, which regulates air emissions from stationary and mobile sources and authorizes the EPA to establish National Ambient Air Quality Standards to protect human health and welfare and to regulate the emissions of hazardous air pollutants.[1] In 2007, a group of 12 states and numerous cities frustrated with the Bush administration's lack of substantial action on climate change brought suit against the US EPA for its failure to regulate greenhouse gas emissions from new motor vehicles. In a 5–4 decision, the Supreme Court in Massachusetts v. Environmental Protection Agency ruled in favour of the states and mandated the EPA to regulate carbon dioxide and other GHGE as pollutants under the Clean Air Act. The Obama administration subsequently issued a series of new regulations and standards for automobiles, light truck, and eventually heavy-duty vehicle fuel efficiency and greenhouse gas emission standards (EPA 2012).

In 2009, the President issued an Executive Order on Federal Leadership on Environmental, Energy and Economic Performance, mandating that federal agencies reduce their greenhouse gas emissions by 17 per cent and enhance their use of renewable energies to 9 per cent. With these actions as an initial basis, and with US emissions already falling as a result of the fracking boom – and the shift away from coal to natural gas that this led to, Obama was able to go to the Copenhagen climate negotiations in December 2009 and pledge to cut US GHGE by 17 per cent of 2005 levels by 2020. Although the 2009 Copenhagen climate conference was widely seen as a failure, given the high and unrealistic hopes that had been set for the conclusion of a binding global climate agreement, the Obama administration sent an important signal to the international community that it was once again planning to exert cognitive and structural leadership – the United States would commit to action, even if only on a voluntary basis given the domestic constraints limiting the administration's room of manoeuvre.

Climate action plan and the clean power plan

Obama was elected to a second term of office in the fall of 2012. In his State of the Union speech in January 2013, Obama stated that he could wait no longer for gridlocked lawmakers:

> I urge this Congress to get together, pursue a bipartisan, market-based solution to climate change.... But if Congress won't act soon to protect future generations, I will. I will direct my Cabinet to come up with executive

actions we can take, now and in the future, to reduce pollution, prepare our communities for the consequences of climate change, and speed the transition to more sustainable sources of energy.

(White House 2013a)

A few months later, in June 2013, the administration presented the Climate Action Plan, outlining its plans to regulate the largest GHG-emitters in the United States by making use of the power of executive order (White House 2013b). The plan was to become the centre piece of the administration's climate policy in the run-up to the Paris climate negotiations. The plan has three main elements: (1) to reduce domestic GHG emissions, (2) to prepare the country for the impacts of climate change, and (3) to be a leader in international climate-change cooperation.

The Climate Action Plan incorporates what is known as the Clean Power Plan (CPP), which aims to cut emissions from US power plants by 32 per cent by 2030 compared to 2005 levels (equivalent of 12 to 19 per cent below 1990 levels (Climate Action Tracker 2015). The plan has been heralded as one of the most important climate steps of the last several decades (*Guardian* 2015). The CPP assigns individual goals for emission cuts to each state, which are to add up to the aggregate national target (EPA 2015). These regulations for new and existing power plants disadvantage coal plants significantly more than natural gas plants because of the higher CO_2 content in coal.

The CPP has two main elements: (1) state-specific goals for emission rate-based CO_2 reductions, and (2) requirements for states to develop, submit and implement plans to reduce CO_2 emissions in their state. The current fuel mix, electricity market, and other factors were considered when the EPA evaluated and decided upon a specific emission rate goal for each state that reflected each state's unique conditions. As a result, state emission rate goals vary from around 0 to above 50 per cent (C2ES 2016). State plans to meet the goals must be consistent with EPA guidelines. States can meet the targets alone, or can collaborate with other states on multi-state plans that could be more cost effective and flexible, for instance cap-and-trade systems (EPA 2015).

In February 2016 the Supreme Court in a 5–4 decision put a stay on the implementation of the CPP until the U.S. Court of Appeals for the District of Columbia, rules on the legality of the new regulations. The case was brought by a group of 29, primarily Republican states, many of which are heavily dependent on coal, which argued that the Obama administration had exceeded its authority with what they considered to be an excessively burdensome plan. The plan, moreover, was seen as a 'power grab' by the federal government over the state's electricity systems (Liptak and Davenport 2016).

The decision of the court illustrates the thin ice on which the Obama administration's efforts to build a climate policy are built and the remarkable power of judicial review in cases where executive orders are used to build policy. Even more striking is what can happen when a member of the Supreme Court dies, thereby shifting the conservative-liberal balance on the court. The death of conservative justice Antonin Scalia just one week after the ruling could influence the future of the CPP and of the commitment the United States made in the 2015 Paris climate negotiations.

Climate change as a threat to the status quo and opportunity for a green economy

Implementation of the CPP requires close interaction between the state and federal levels of government because states are given a key role in developing implementation plans, while the federal government is to perform oversight. However, the individual interests of states as well as the multiple branches of government involved create ample opportunities for actors opposed to policy change to put up barriers. To be effective, governance at multiple levels should be of a reinforcing nature (Schreurs and Tiberghien 2010). In a federal system, however, the strong role of the states can hinder policy implementation, especially since states are granted 'special solicitude' from the federal courts, a critical advantage that helps confer legal standing and means that a matter is less likely to be dismissed by the court. For the energy industry, therefore, states opposed to the CPP are extremely valued partners in trying to stop the Obama administration's regulatory agenda (*New York Times* 2014).

Stakeholder and interest group attitudes to the CPP differ from sector to sector, often depending on whether their own economic activity would be negatively affected. The strongest negative reactions, as in previous rounds of climate policy debates, have come from fossil-fuel-dependent energy industries that fear higher costs. For instance, Southern Company and American Electric Power, two of the biggest coal consumers in the United States, formed alliances with other power energy firms and lobbied politicians to oppose Obama's policy proposals (Greenwire 2014).

Opposition to the CPP was organized in the legislative and judicial branches at both the federal and state levels of governance. In the US Congress, Republicans and lawmakers representing fossil-fuel-intensive states raised harsh criticism of the CPP because it exposed the energy sector in their states. Many legislators were provoked by Obama's use of executive action because it represented a circumvention of Congress, and they expressed a clear request for involvement of the legislative branch in climate policy decisions since this policy field is of high national importance and new policy instruments would affect constituents directly (Greenwire 2014).

State-level politicians' reactions to the CPP varied from state to state, depending on the energy resource mix and which party had the majority. States with a clear fossil-fuel energy profile and with a Republican governor or Republican majority in the state legislature tended to protest sharply against the CPP. For example, Texas governor Greg Abbott (Republican) said the federal government seems 'hell-bent on threatening' the principles of a free market. Oklahoma governor Mary Fallin (Republican) called the plan 'one of the most expansive and expensive regulatory burdens ever imposed on U.S. families and businesses.' Louisiana attorney general James Caldwell (Republican) said the CPP 'will lead to fewer jobs and higher utility bills' (Energy Wire 2015). The Republican Attorneys General Association was very actively engaged in organizing states' judicial efforts to stop Obama's regulatory agenda in the power sector (*New York Times* 2014).

In contrast, in states on the West Coast and in the Northeast, where energy resources are less fossil-fuel intensive, the CPP was met with more enthusiasm. The plan links well to their existing efforts to promote a transition towards a green economy. Most state governors and regulators have had more cautious reactions, reflecting the complex energy landscape of the country. Some states are already implementing various policies to upgrade their energy production systems, and many states have some form of incentive policy for renewable energy phase-in, for example, the 27 states that have Renewable Energy Portfolio Standards. In presenting the rule, the EPA emphasized that state plans of compliance should build on existing policies and fortify existing programs (EPA 2015).

Climate leaders at the state and local levels

There are strong differences in the economic and energy interests of states. At the same time that many states are attempting to block the implementation of the CPP and other air pollution regulations affecting power plants, other states have set ambitious greenhouse gas emission reduction targets, not dissimilar to the targets of more progressive European states. Twenty states and the District of Columbia had (as of 2015) established greenhouse gas emission reduction targets. These include both long-term reduction targets as well as interim targets. For instance, California has a target to return its greenhouse gas emissions to 1990 levels by 2020, a 17 per cent reduction from business as usual emissions, and to reduce emissions to 40 per cent below 1990 levels by 2030. Massachusetts has committed to a 10–25 per cent reduction by 2020 (1990 base line) and Michigan envisions a 25 per cent cut (2005 baseline) (C2ES, n.d.).

Many states are working on phasing more low-carbon energy sources into their electricity generation mix, in part to become green economy leaders. In the Northeast, nine states (Rhode Island, New York, Vermont, Connecticut, Maine, New Hampshire, Delaware, Maryland and Massachusetts) have joined forces in the Regional Greenhouse Gas Initiative (RGGI). The aim is to cap and reduce CO_2 emissions from the power sector through an emissions trading system, which has operated since 2010.

California implemented a comprehensive emissions trading system in 2012. It entered its second commitment period (2015–2017), with new system design features. The scope was extended to include fossil transportation fuels and retail sales of natural gas. Almost 85 per cent of Californian GHGE are covered by the cap-and-trade program.

Moreover, as many as 27 states have enacted renewable portfolio standards of different strength and ambition levels, and states such as Texas and Colorado had a significant growth in their share of renewable energy, in particular wind energy (C2ES 2015).

The pressure for climate policy change from the sub-national level has been a central factor pushing for a reframing of climate change at the federal level. The several state and regional emission trading initiatives that developed in part in response to the US early support for emissions trading during the Kyoto Protocol negotiations as well as a response to the European carbon emissions trading scheme (ETS) have reinforced tendencies to put emission trading at the centre of US climate policy discussions. Similarly, the climate policies and programs of individual states and cities have clearly been influenced by international and especially European examples. In this way, the EU's belated leadership role on ETS and its structural, entrepreneurial and cognitive leadership in the international climate negotiations has impacted both US state and federal policy developments.

International climate leadership

An important part of President Obama's Climate Action Plan concerns US ambitions to lead international efforts to address climate change. Exerting cognitive, entrepreneurial, and even structural leadership depends on two conditions: a direction and design of international climate cooperation that overlaps with US interests, and an ability to make credible domestic policy commitments. For many years, the direction and design of international climate cooperation were focused on specific national emissions cuts targets for developed countries, leaving developing countries without any commitments in accordance with the UNFCCC's principle of 'common but differentiated responsibilities and capabilities'. The United States had reservations about participating in a treaty that did not include commitments for all major emitters and thus sought to shift the debate at the global level making use of both cognitive and entrepreneurial leadership (Hovi *et al.* 2012).

However, this situation changed significantly in 2014 when China and the United States for the first time announced that they had mutually agreed to commit to GHGE cuts within a set period. With China accepting a commitment to cap and reduce its emissions from 2030 onwards, conditions for US participation in international climate cooperation have improved. Since China and the United States together account for over 40 per cent of global GHGE such cooperation is of critical importance. In particular, one argument frequently made for justifying the US's cool reaction to the Kyoto Protocol was that it did not obligate major emerging economies, and especially China and India, to set any target dates for emission reduction cuts.

Importantly, the Obama administration pursued entrepreneurial leadership through bilateral and regional initiatives, some of which had already been initiated under the George W. Bush administration, to engage China on climate change. Examples include the now defunct Asia-Pacific Partnership on Clean Development and Climate, which ran from 2006–2011; various technology cooperation agreements on energy saving buildings, smart grids, and energy efficient cars; and the agreement to phase out hydrofluorocarbons (HFCs) (Lewis 2010).

Linked to Obama's domestic climate change plans and as a central aspect of its entrepreneurial leadership, US Secretary of State John Kerry travelled to China to pursue secret bilateral climate negotiations in February 2014. This was followed by an exchange of letters between President Obama and Chinese President Xi Jinpeng regarding moving in tandem on climate change with both states taking domestic actions that would bring them closer together on climate action (Goldenberg 2014; see also Chapter 17). In an effort to pursue transformational leadership, a side agreement was reached in September 2014 at the UN Climate Summit in New York between Obama and Vice President Zhang Gaoli, who is responsible for climate and energy issues. At the Climate Summit, Zhang Gaoli announced that China will 'make greater effort to more effectively address climate change' and said China will announce post-2020 actions soon that would include 'marked progress in reducing carbon intensity' and 'the peaking of total carbon dioxide emissions as early as possible' (Dugan 2014). The culmination of these moves was a joint announcement by President Obama and President Xi on emission reduction targets and clean energy cooperation (The White House, Office of the Press Secretary 2014). At the US–China Climate Leaders Summit in Los Angeles in September 2015, a Climate Leaders' Declaration was reached under which 11 Chinese and 18 US cities announced emissions reduction targets and pledged to strengthen US–China cooperation at the subnational level (Henderson 2015). These developments paved the way for both states to announce their support of the 2015 Paris Agreement (see also Chapter 17).

Making use of cognitive and entrepreneurial leadership, the US was instrumental in pushing for a new direction and design for international climate cooperation at Copenhagen in 2009, and very clear about not wanting a new climate agreement designed like the Kyoto Protocol (Christoff 2010). Since Copenhagen, most developed countries joined the United States in demanding a new agreement that will apply for all countries – even for developing countries with a large and growing share of total global GHGE. The 2009 Copenhagen Accord contained the important agreement on a 2°C warming above pre-industrial levels as the maximum, accepting that further warming would cause dangerous climate change. It also promoted a pledge-and-review approach to reach an emissions level consistent with the 2°C target, which suited the US well. Under this approach, countries pledge the amount of emissions reductions that are feasible for them within a certain period, and then commit to undergo review to ensure that they deliver on the promise (Stavins 2010). The 2015 Paris Agreement was designed according to a pledge-and-review approach in the form of INDCs, where for the first time all countries agreed to nationally determined pledges, monitoring and verification. The pledge-and-review approach is a politically more feasible option for US participation in international climate cooperation given the gridlock that exists in Congress. It was thus critical for Obama to exert cognitive leadership, shaping the direction of the agreement along with entrepreneurial leadership.

Still problematic for US leadership is the realpolitik of international negotiations, dictating that action speaks louder than words. With an uncertain situation regarding US domestic climate policy, it is by no means certain that the United States *can* be a credible leader or if it will simply be a symbolic leader. A touchy question is whether the Obama administration can expect Congress to be able to pass climate legislation in the next decade. With a climate law that can function as enabling legislation, the United States could more credibly engage in the next rounds of international climate negotiations as an entrepreneurial leader. However, signs that Congress will be able to agree on a climate law very soon are lacking (Bang and Skodvin 2014). More realistically, therefore, US engagement at the international level must be built on further executive action, which will inevitably be subjected to judicial review by opponents. The uncertainties this opens up make it difficult for the US to be convincing as a structural, cognitive, and entrepreneurial leader on climate.

Furthermore, there is a danger that a future president of a different political leaning could undo earlier presidential orders. Such action would, however, come at a cost as it would jeopardize US international credibility in future international negotiations.

Conclusion: political leadership in the United States

President Obama has offered – some would argue belatedly – a transformational, some would even say heroic, leadership style by applying his executive powers in a policy field that is very controversial. Pursuing ambitious domestic climate policy despite the lack of bipartisan support that is necessary for a stable, long-term climate policy solution have been crucial to his efforts to exert cognitive and entrepreneurial leadership internationally.

Generally, when US presidents decide to use executive action, they must consider how political opponents might respond and thoroughly evaluate the likely degree of compliance and the costs and benefits of acting unilaterally rather than collaborating with Congress (Mayer 2001; Shull 2006). The CPP is controversial because it introduces an implicit cost on CO_2 emissions for the first time in the United States, a decision that the US Congress has opposed repeatedly. Furthermore, the risk of delayed or failed implementation at the state level grows when decisions with impacts for a state's economy are made at the federal level without support from political representatives in the state. Judicial challenges have become commonplace.

Although the basis of the Obama administration's policies is contested, substantial shifts in domestic air pollution regulations and energy efficiency standards are slowly putting the country on a new course. The more deeply these changes become institutionalized, the harder they will be for subsequent administrations or for Congress to overturn.

Building on the substantial steps taken by various cities and states to address climate change as well as through the changes achieved through a broader interpretation of the US Clean Air Act, the Obama administration has proven itself one of the most active presidencies in history on the environment. The changing domestic regulatory landscape have made it possible for the administration to pursue a more progressive international climate policy and a stronger international leadership role. However, the credibility of US pledges to be a leader in international climate mitigation will depend heavily on the full implementation of the CPP and on additional domestic emissions cuts.

Note

1 See EPAs website: www.epa.gov/laws-regulations/summary-clean-air-act.

References

Bang, G. and T. Skodvin (2014) 'U.S. climate policy and the shale gas revolution', in J. Hovi, T. Cherry and D. McEvoy (eds), *Toward a New Climate Agreement: Conflict, Resolution and Governance*, London: Routledge.

Bang, G. and Schreurs, M.A. (2011) 'A Green New Deal: framing U.S. climate leadership', in R.K.W. Wurzel and J. Connelly (eds), *The European Union as a Leader in International Climate Change Politics*, London: Routledge, 235–251.

Bulkeley, H.T. (2013) *Cities and Climate Change*, Oxon and New York: Routledge.

C2ES, Center for Climate and Energy Solutions (n.d.), www.c2es.org/us-states-regions/policy-maps/emissions-targets (accessed 31 April 2016).

C2ES (2016) 'Climate Action in US States and Regions', available at www.c2es.org/us-states-regions (accessed 31 April 2016).

Climate Action Tracker (2015) 'USA', 4 September, available at http://climateaction-tracker.org/countries/usa.html (accessed 23 May 2016).

Energy Wire (2014) 'Behind the noise, central states study EPA rule cooperation', available at www.eenews.net/energywire/2014/12/03/stories/1060009833 (accessed 26 January 2015).

Environmental Protection Agency (EPA) Office of Transportation and Air Quality (2012) 'EPA and NHTSA Set Standards to Reduce Greenhouse Gases and Improve Fuel Economy for Model Years 2017–2025 Cars and Light Trucks,' EPA-420-F-12–051.

Environmental Protection Agency (EPA) (2015) 'Clean Power Plan', available at www2.epa.gov/cleanpowerplan (accessed 9 September 2015).

Goldenberg, S. (2014) 'Secret talks and a private letter: how the US–China climate deal was done', *Guardian* www.theguardian.com/environment/2014/nov/12/how-us-china-climate-deal-was-done-secret-talks-personal-letter (accessed 15 September 2015).

Greenwire (2014), 'McConnell, Inhofe vow to strike EPA power proposal', 3 June, www.eenews.net/greenwire/stories/1060000633/search?keyword=epa+power+plant+rule (accessed 26 January 2015).

Guardian (2015) 'Obama's clean power plan hailed as US's strongest ever climate action,' 3 August 2015, www.theguardian.com/environment/2015/aug/03/obamas-clean-power-plan-hailed-as-strongest-ever-climate-action-by-a-us-president (accessed 7 December 2014).

Henderson, G. (2015) 'Chinese and U.S. cities, states and provinces announce climate targets and extensive cooperation,' China FAQs: The Network for Climate and Energy Information, 15 September, www.chinafaqs.org/blog-posts/chinese-and-us-cities-states-and-provinces-announce-climate-targets-and-extensive-coopera (accessed 26 January 2015).

Hovi, J., Sprinz, D. and Bang, G. (2012) 'Why the United States did not become a party to the Kyoto Protocol: German, Norwegian and U.S. perspectives', *European Journal of International Relations*, 18(1): 129–150.

IPCC (2013) *Climate Change 2013: The Physical Science Basis*, Cambridge: Cambridge University Press.

Jacques, P. (2009) *Environmental Skepticism: Ecology, Power and Public Life*, Burlington, VT and Surrey: Ashgate.

Karapin, R. (2016) *Political Opportunities for Climate Policy: California, New York and the Federal Government*, New York: Cambridge University Press.

Leiserowitz, A. *et al.* (2014) 'Climate change in the American mind', *Report*, Yale University Project on Climate Change Communication, October, http://environment.yale.edu/climate-communication/files/Climate-Change-American-Mind-October-2014.pdf (accessed 26 January 2015).

Leiserowitz, A., Maibach, E., Roser-Renouf, C., Feinberg, G., and Rosenthal, S. (2016) *Politics and Global Warming*, New Haven, CT: Yale University and George Mason University.

Lewis, J. (2010) The State of U.S.–China Relations on Climate Change: Examining the Bilateral and Multilateral Relationship. *Woodrow Wilson International Center for Scholars, China Environment* 11: 7–39.

Liptak, A. and C. Davenport (2016) 'Supreme Court deals blow to Obama's efforts to regulate coal emissions', *New York Times*, 16 February, www.nytimes.com/2016/02/10/us/politics/supreme-court-blocks-obama-epa-coal-emissions-regulations.html?_r=0 (accessed 26 May 2016).

Mayer, K. (2001) *With the Stroke of a Pen: Executive Orders and Presidential Power*, Princeton, NJ: Princeton University Press.

New York Times (2014) 'Energy firms in secretive alliance with attorneys general', 6 December, www.nytimes.com/2014/12/07/us/politics/energy-firms-in-secretive-alliance-with-attorneys-general.html?hp&action=click&pgtype=Homepage&module=first-column-region&region=top-news&WT.nav=top-news&_r=1 (accessed 7 December 2014).

Obama, B. and Biden, J. (2009) New Energy for America, http://energy.gov/sites/prod/files/edg/media/Obama_New_Energy_0804.pdf (accessed 27 September 2015).

Schreurs, M. and Tiberghien, Y. (2010) 'European Union leadership in climate change: mitigation through multilevel reinforcement', in K. Harrison and L. McIntosh Sundstrom (eds), *Global Commons, Domestic Decisions: The Comparative Politics of Climate Change*, Cambridge, MA: MIT Press, 23–66.

Schreurs, M., Selin, H. and VanDeveer, S.D. (eds) (2009) *Transatlantic Environment and Energy Politics*, Farnham, UK and Burlington, VT: Ashgate.

Scruggs, L., and Benegal, S. (2012) 'Declining public concern about climate change: can we blame the great recession?', *Global Environmental Change* 22(2): 505–515.

Shafie, D.M. (2014) *Presidential Administration and the Environment: Executive Leadership in the Age of Gridlock*, Oxon, UK and New York: Routledge.

Shull, S.A. (2006) *Policy by Other Means: Alternative Adoption by Presidents*, College Station, TX: Texas A&M University Press.

Skocpol, T. (2013) 'Naming the problem: what it will take to counter extremism and engage Americans in the fight against global warming', *Paper* presented at the symposium *The politics of America's fight against global warming*, Harvard University, 14 February.

Smith, N. and Leiserowitz, A. 'American evangelicals and global warming', *Global Environmental Change*, 23(5): 1009–1017.

Stavins, R. (2010) 'Why Cancun trumped Copenhagen: warmer relations on rising temperatures', *Christian Science Monitor*, 20 December, http://belfercenter.ksg.harvard.edu/publication/20625/why_cancun_trumped_copenhagen.html (accessed 9 September 2015).

Tichner, S. (2015) 'WV Attorney General Patrick Morrisey asks court to intervene in Clean Power Plan,' *The State Journal*, 13 August, updated 12 September 2015, www.statejournal.com/story/29782357/wv-attorney-general-patrick-morrisey-asks-court-to-intervene-in-clean-power-plan (accessed 7 December 2014).

US Department of State (2009) U.S.-China Memorandum of Understanding to Enhance Cooperation on Climate Change, Energy and the Environment, 28 July, www.state.gov/r/pa/prs/ps/2009/july/126592.htm (accessed 3 May 2015).

Wallach, P.A. (2012) 'U.S. Regulation of Greenhouse Gas Emissions', *Brookings Institution Report*, October.

White House (2014) 'U.S.-China Joint Announcement on Climate Change and Clean Energy Cooperation', Fact sheet, 11 November, www.whitehouse.gov/the-press-office/2014/11/11/fact-sheet-us-china-joint-announcement-climate-change-and-clean-energy-c (accessed 26 January 2015).

White House (2013a) 'Remarks by the President in the State of the Union Address', U.S. Capitol, Washington, D.C., 12 February, www.whitehouse.gov/the-press-office/2013/02/12/remarks-president-state-union-address (accessed 14 December 2014).

White House (2013b) 'Climate Change and President Obama's Action Plan', June, available at www.whitehouse.gov/climate-change#section-carbon-pollution (accessed 25 January 2015).

17 China

From a marginalized follower to an emerging leader in climate politics

Xinlei Li

Introduction

Having joined the international climate change regime in the early 1990s, China has been confronted with great challenges and opportunities. The country has been identified as a key actor in global climate talks, especially since becoming the largest emitter of greenhouse gases in 2007 (Liang 2010: 61), and has undergone increasing international reputational pressure to make a legally binding commitment to reductions in greenhouse gas emissions (GHGE). China is a key player in international climate negotiations for two reasons: first, considering that it is responsible for the largest share of GHGE, any global GHGE reduction agreement cannot really achieve its goal without China's contribution; second, as the largest (rapidly) developing country in the world with an influential voice in the UN Security Council, 'its status and influence in the G-77 of Third World states give China prominence in climate negotiations' (Heggelund 2007: 159). For the international climate change regime, China has become a system relevant actor with considerable structural power.

After the 2008 global financial crisis, climate change negotiations have undergone a great shift. First, due to the rapid economic development and high energy consumption of major emerging countries (such as the BASIC countries – China, India, South Africa and Brazil), negotiation interaction shifted from a 'North–South' conflict, to opposition between large and small GHG emitters. The split among the G77 Group has grown since the 2009 Copenhagen climate conference (COP15). Pressures to reduce GHGE in developing countries are not only coming from developed countries but also from within the group of developing countries (including the least developed countries). While under great international pressure, China was surprised to find itself being blamed for the failure of the 2009 Copenhagen climate conference. This has had a profound impact on its subsequent approach to the international climate change talks (Yu 2015). Second, along with the increased role of large transitional countries in global governance and the relative decline of the EU's climate leadership, the BASIC countries, led by China, emerged at the Copenhagen conference as central players in the international climate negotiations. China's rising entrepreneurial leadership attracted special attention (see below).

After the Copenhagen climate conference, China's leaders began to admit that a more flexible stance needed to be taken. They recognized that if they failed to take action, the country's national reputation and credibility would be greatly damaged. China began to pay more attention to entrepreneurial leadership and cognitive leadership (see Chapter 1) especially since the 2011 Durban climate conference (COP17) at which it showed, for the first time, its willingness to accept legally binding obligations by 2030 if they matched China's economic development and capabilities. Furthermore, in order to improve its cognitive leadership, China hosted its own pavilion at the UN climate conferences and also made its negotiators more visible during the talks, including more regular meetings with NGOs and journalists (Geall *et al.* 2016). It is remarkable that along with China becoming a leading country in renewable energy (RE) installation capacity and investment since 2010, it began to promote 'clean energy' South–South cooperation. Through leadership by example, China shared its RE best practices and project experience with other developing countries coupled with relevant RE financial aid, technology transfer and personnel training.

Since President Xi Jinping came to power in 2013, there has been a significant change in Chinese climate diplomacy: from the doctrine of hide one's capacities and bide one's time to the doctrine of diligent and ambitious diplomacy (Ren 2014). The new diplomatic style symbolized that China began to perceive itself as a rising big power which could play a more central role in the international arena while having to take on its international responsibilities. China's climate diplomacy became more demonstrative and confident in showing its considerable determination in pursuing structural and entrepreneurial leadership. Before the 2015 Paris climate conference, China promoted several bilateral climate agreements with big emitters like the US, the EU, India and Brazil (Xinhuanet 2015). The *EU–China Joint Declaration on Climate Change* was issued when Chinese Premier Li Keqiang arrived in Belgium for the 17th EU–China Leaders' Meeting in June 2015, which coordinated bilateral climate positions in advance and contributed a lot to the success of the Paris climate conference. In Paris, China's efforts to construct a new type of big power diplomatic coordination with major developed and developing countries helped ensure that the landmark Paris Agreement was adopted on time. Of central importance for the outcome in Paris was China's cooperation with the US to push through the bottom-up approach in international climate governance, which greatly facilitated the voluntary GHGE pledges of intended nationally determined contributions (INDCs). This bottom-up model replaced the former top-down approach promoted by the EU and enshrined in the 1997 Kyoto Protocol (see Chapters 1, 2 and 5).

This chapter tries to answer the following three main questions: How can we understand China's gradually changing position and its leadership attempts in the international climate change negotiations? What are the (long-term) opportunities and challenges for China's low carbon governance and energy transformation? What are the patterns of China's multi-level and polycentric climate governance?

China's climate debate and its changing attitude to climate change

China's climate change policy is not only a domestic issue, but has significant repercussions for the rest of the world. How to understand China's position and interest in climate politics has become a hot research topic at the centre of which is the question whether its climate policy has changed significantly or remained constant (e.g. Harris 2011; Lewis 2008; Liang 2010; Zhang 2006).

Some scholars have pointed out that China's stance in the climate negotiations has remained constant in its refusal to accept legally binding GHGE. China has refused to set a cap for total GHGE reductions until it reaches the average level of a developed country by 2030. Although China has adopted some voluntary carbon intensity reduction initiatives, the global climate target will not be achievable without further efforts (e.g. Harris 2009 2011; Moore 2011). China's priority for economic development, which remains the dominant domestic goal, has been perceived as the main driver for its unchanged stance in the international climate negotiations (Liang 2010: 70–71). Although the attention to climate change has recently increased among China's leadership, 'it has not surpassed economic development as a policy priority' (Lewis 2008: 154).

Although it is true that economic growth is essential for regime survival and social stability (Harris 2011: 8), others contend that this 'does not mean that China's original stance on climate change remains unchanged' (Chen 2009: 116). It has been argued that a gradual change in China's stance has taken place. In contrast to its previous scepticism about global warming, China has gradually become involved in a green transformation, as illustrated by the inclusion of a strategic target in the 12th and 13th Five Year Plans (FYP).[1] Examining China's historical positions in the climate talks since the 1990s, Zhang (2006) argues that there are both continuities and changes. As Table 17.1 shows, there is an accelerating trend in China's attitude change, especially after China becoming the largest GHG emitter. For instance, it changed its ambivalent attitude towards a domestic carbon emissions trading scheme (ETS) around 2009, launching several ETS trials beginning in 2011. There is a clear tendency that, with its increasing involvement in the regime, China's stance on climate change has shifted from being conservative and suspicious to being active as well as more flexible and open.

China's changing attitude on climate change, from suspicious and ambivalent to active and assertive, has reflected a shift in national preference. Initially China took part in the international climate negotiations mainly as a foreign policy requirement, to enable it to break out of its diplomatic isolation after the 1989 Tiananmen incident. It allied itself with developing countries in dealing with international issues. China has evolved from an observer in the Montreal Protocol, a marginalized participant in the UNFCCC, an active leader of the G77 in the Kyoto Protocol, to a major player in the Copenhagen Accord (Hodgson 2011: 32–35), and an emerging leader in the 2015 Paris climate conference.

Climate change has become a politically more salient issue together with related domestic environmental issues. In particular, the frequency of extreme

Table 17.1 Diachronic comparison of China's stance in climate politics

Issues	*1991*	*1997*	*2001*	*2005*	*2007*	*2009*	*2011*	*2012*	*2013*	*2014*
Financial and technology assistance from developed countries?	Yes	Yes	Yes	Yes	Yes	Yes	Yes	Yes	Yes	Yes
Mention of promotion of renewables development	No	Unclear	Yes	Yes	Yes	Yes	Yes	Yes	Yes	Yes
Support for CDM?	No	Hesitated	Yes	Yes	Yes	Yes	Yes	Yes	Yes	Yes
(Voluntary) commitment to GHGE reductions?	No	No	No	No	Yes	Yes	Yes	Yes	Yes	Yes
National climate change plan?	No	No	No	Planning	Yes	Yes	Yes	Yes	Yes	Yes
Quantitative indicator for carbon-intensity reduction?	No	No	No	NO	No	Yes	Yes	Yes	Yes	Yes
Climate change legislative framework at the domestic level?	No	No	No	No	No	Planning	Preparing	Yes	Yes	Yes
Acceptance of a legally binding commitment?	No	No	No	No	No	No	Planning	Yes	Yes	Yes
Softened stance on one-track negotiations?	No	No	No	No	Unclear	Yes	Yes	Yes	Yes	Yes
Adoption of carbon tax or/and carbon trading system?	No	No	No	No	No	Discussion	Designing	Yes	Yes	Yes
Official deadline to reach GHGE peak (by 2030 or as earlier as 2025)?	No	No	No	No	No	No	No	Unclear	Ambiguous	Yes
Willingness to show global climate leadership	No	No	No	No	No	No	No	Preparing	Yes	Yes

Sources: own data from 2007–2014; 1991–2005 data partly adapted from Zhang (2006).

environmental events in recent years, such as the large-scale particulate matter (PM) 2.5 smog pollution in the northern parts of China, further enhanced the convergence of domestic environmental problems with international climate governance. The government has recognized that incorporating climate change into its national sustainable development strategy and overall socio-economic development planning is conducive to the establishment of a resource-saving and environment-friendly society.

China's attitude to climate leadership also evolved from never taking the lead (due to the principle of 'hide one's capacities and bide one's time') to taking a leading role as a responsible great power. The so-called G-2 model developed by the economist C. Fred Bergsten in 2005 emphasized the influential roles of the US and China in global affairs. Many foreign media used the model to label China as one of the 'two key players' along with the US in the Copenhagen climate negotiation (Harrabin 2009). However, China was very cautious about the G-2 model and still regarded itself as the leader in the bloc of developing countries and placing emphasis on economic development rights and differentiated responsibilities for emissions reductions. Since 2013, the Chinese mainstream official media (e.g. XinhuaNet) has begun to focus on China's rising leadership at the 2015 Paris climate conference. According to XinhuaNet report, 'China's climate endeavours have been an indispensable part of global efforts leading to the making of a final breakthrough at the climate talks' (Xinhuanet 2015).

Three phases of China's climate change policy and institutional responses

This part distinguishes three stages of China's participation in the climate change regime.

Phase 1 (1990–1997): joining the climate change regime without sufficient preparation

China's participation in the climate change regime served as an ideal channel for the country to make up the diplomatic rift with Western countries and regain its international position in the early 1990s (Chen 2009: 106–107). Due to the lack of a comprehensive and in-depth understanding of the climate change regime, China treated it as a basic international environmental agreement. This limited recognition of the climate change issue was reflected in the institutional composition of the National Group of Co-ordination on Climate Change established in September 1990. It aimed to coordinate diplomatic negotiations[2] but failed to include officials in charge of economic development (Yan and Xiao 2010: 83).

Although China did not initially realize the impact of climate change on its energy regulation and economic development, the country treated joining the UNFCCC as an opportunity to reinforce its diplomatic relations with other developing countries. These efforts promoted the formation of the 'bloc of G-77 plus China'. Since 1991, China has aligned itself closely with India and strongly opposed legally binding emission reduction targets for developing countries and insisted on industrialized countries providing technology and financial aid to developing countries (Chen 2009: 104; see also Chapter 18).

Phase 2 (1998–2008): climate change issues arrive on the national political agenda and trigger significant domestic institutional changes

The 1997 Kyoto Protocol became the first international treaty to set legally binding obligations for industrialized countries to reduce GHGE. Along with increased involvement, China has gradually realized that the Kyoto Protocol not only relates to environmental concerns, but also has great influence on national economic growth. In this context, climate change began to gain more attention at the domestic level, which led to institutional changes in the bureaucratic system. The main responsibility for climate change negotiations shifted to the most powerful agency, the State Development and Planning Commission (SDPC) in 1998, which symbolized that climate change had entered the political mainstream in China. In the same year, SDPC representatives signed the Kyoto Protocol and promoted an attitudinal change in relation to the three flexible mechanisms in the Kyoto Protocol. China gradually changed its skeptical attitude towards the Clean Development Mechanism (CDM) around 2001 and this external impetus has stimulated the initial development of large-scale RE (Hatch 2003; Heggelund 2007).

After it surpassed the US to become the largest GHGs emitter in 2007, China has been subject to increasing pressure. The 2007 Bali climate conference (COP13) marked the formal beginning of designing a post-Kyoto climate regime, with the emergence of the Bali Roadmap helping to pave the way for the post-Kyoto negotiations. The watershed in the Bali conference was that developing countries showed a willingness to discuss voluntary reduction commitments for the first time (Liang 2010: 68). In June 2007, along with the announcement of China's first global warming policy initiative (*China's National Climate Change Programme*), the Climate Change Coordination Group under the National Development and Reform Commission (NDRC – Successor of SDPC) was elevated to the National Leading Group on Climate Change (NLGCC) directly under the State Council and was led by former Premier Wen Jiabao (Hallding *et al.* 2009: 125–126).

Phase 3 (since 2009): heading for a diligent and ambitious climate diplomacy

The biggest change between the 2009 Copenhagen climate conference and the preceding climate change negotiations was the greater pressure placed on large developing countries. At the conference China for the first time announced its voluntary national climate mitigation action targets, and pledged to cut carbon intensity by 40 to 45 per cent per unit of GDP by 2020 compared with 2005 levels, and to aim for 15 per cent energy from non-fossil fuels by 2020.

However, this voluntary target did not meet the high expectations of the Copenhagen climate conference. Also for the first time China, instead of the US, was blamed for dragging its feet on international climate negotiations and was accused of contributing to the conference's 'disappointing outcome' (Zhang 2010: 239–240).

Following the frustrations of the Copenhagen conference and the struggles to save the multilateral climate regime at the 2010 Cancun Conference (COP16), the 2011 Durban conference (COP17) was an important turning point that helped resuscitate the Kyoto Protocol and paved the way for the post-Kyoto framework (IISD 2011). The establishment of the Ad hoc Working Group on the Durban Platform for Enhanced Action (ADP) demonstrated the strong pressure for designing a legally binding framework to cover all COP member countries by 2015. Xie Zhenhua[3] declared that China should negotiate a legally binding document after 2020 and would accept a legally binding arrangement, although with conditions. This declaration was interpreted as taking an active stance in global negotiations by promising to put in place legally binding carbon reduction plans (Chinafaqs 2011). Responding to the commitment made at the 2012 Doha conference (COP18), China established its top climate change think-tank, the National Center for Climate Change Strategy and International Cooperation (NCSC) under the NDRC to lead a comprehensive research on its GHGE peak.

In 2013, China introduced its first initiatives to cap the use of coal, aiming to restrict its share in the national energy mix to 65 per cent by 2017. In *China Energy Outlook 2030* issued by Chinese Energy Research Institute, it is anticipated that China's GHGE may peak by 2025, five years ahead of the date it has pledged to the UN (Chinapower 2016). Especially after the 2014 Lima climate conference (COP20), China became 'less defensive and more inviting, trying to take on a leadership role' (Soutar 2015). China submitted its Intended Nationally Determined Contribution (INDC) according to which it planned to reduce the carbon intensity of its economy by 60–65 per cent per unit of GDP by 2030 (compared to 2005), and repeated a previously announced aim that non-fossil fuel should make up 20 per cent of its primary energy supply. When President Xi visited the US in June 2015, he repeated China's pledge that its carbon emissions would peak by 2030, or even as early as 2025. In November 2015, in the run-up to the Paris talks, China's lead negotiator Xie Zhenhua stated that China hoped to see an ambitious and binding agreement being reached in Paris. This ran counter to longstanding international opinion, which expected China to take a conservative stance (Yu 2015).

Climate change: from threat to opportunity

Due to its involvement in the climate change regime, China's stance has undergone a gradual change, becoming more flexible and proactive. Three main driving factors are responsible for this change.

First, China tends to play a more pragmatic role in achieving material incentives from the climate change regime due to a rational calculation. China's position on climate change has consistently revolved around the themes of 'common but differentiated responsibilities' (Zhang 2003: 82). China's involvement in the climate change regime has been motivated by the availability of external financial assistance, technologies transfer, foreign investments, and know-how diffusion. And the move to a less carbon-intensive economy has also been bolstered by its active participation in international climate change cooperation (Zhang 2003: 73). For instance, CDM, the China Renewable Energy Development Project (REDP), and the China Renewable Energy Scale-up Program (CRESP) supported by the Global Environment Facility (GEF) and World Bank have provided a significant impetus for large-scale RE development in China. Through international climate cooperation, China can also boost its image as an environmentally responsible nation and use climate change to further its foreign policy goals.

Second, green growth concerns are being integrated into China's core interest of economic development. This deeply rooted 'development first' mindset is hard to change and plays a significant role in shaping the national position in the international climate change negotiations (Chen 2009: 115). Climate policy 'must be weighed against the grand strategic objectives of stability, security, and prosperity laid out over the past 30 years of reform and opening' (Moore 2011: 147). Any climate change initiative that harms the economy or hampers economic growth is likely to be vigorously resisted by China (Zhang 2003: 82). Despite this, China's understanding of economic development has undergone a gradual transformation. The new interests behind a transformation from an economic growth model to a low-carbon economy have also been a driver for climate change policy (Kang 2010: 66). China changed its conservative attitude on the carbon tax and ETS system after the Copenhagen conference: since then it has tended to rely on market-based policy instruments combining green growth with climate change mitigation while using climate change to promote economic structural transformation. Global competition in low-carbon technology also influences the climate change negotiations. Boyd (2012) points out that 'the idea of low-carbon leadership has reframed China's global role and supported China's emergence at the forefront of low-carbon markets' and that the Chinese government has 'strong interests in seeing China successfully compete and lead in global low-carbon energy markets' (Boyd 2012: 2).

Third, enhancing China's international image and prestige in the international community are important goals of Chinese foreign policy (Zhang 2003: 78). Alastair Johnson points out that compared with most countries, China is particularly sensitive to external criticisms, and would go out of its way to avoid diplomatic isolation and international censure. Chinese leaders 'are concerned to project an image of a responsible major power' (1998: 560). When China surpassed the US to become the largest GHGs emitter in 2007, it was confronted by increasing pressure from the international community. The rapid growth of China's GHGE stimulated heated and continuing arguments about the country's responsibility in the post-Kyoto framework's design. In order to save its national reputation and minimize external pressure, national leaders for the first time agreed to set quantitative indicators for carbon intensity reduction target before the Copenhagen conference and began to abandon their hard-line opposition to carbon pricing policy (Carraro and Tavoni 2010).

Low carbon governance and energy transformation: achievements vs. challenges

For China the essence of climate change mitigation and adaptation is to optimize the national energy structure. The energy supply mix in China is dominated by coal, which constitutes over 70 per cent of energy consumption. Despite this carbon locked-in energy structure, China took what could be called a late development advantage to emulate innovative RE policy instruments from pioneering countries since the beginning of the twenty-first century, such as the tendering policy of the UK model and feed-in tariffs policy of the German model. Especially with the introduction of the Renewable Energy Law in 2005 (amended in 2009), it set out core regulations mandating grid connection, determining electricity pricing, cost allocation and preferential funding. After one decade of rapid growth, China ranked top in RE investment and became the world leader in installed wind power capacity, solar panels/cells manufacturing, and solar water heating utilization in 2010 (Zhang *et al.* 2012: 1). In 2014 China's RE investment rose to $89.5 billion (around £58.4 billion), and China invested nearly 73 per cent more than the US, the world's second largest market (Geall *et al.* 2016). According to the report *China 2050 High Renewable Energy Penetration Scenario and Roadmap*, RE will account for over 60 per cent of China's total energy consumption and more than 85 per cent of its total electricity consumption by 2050. These targets are comparable with all but the most ambitious European countries.

However, the road to successful energy transformation is strewn with many hindrances. First of all, grid-connection problems for RE has plagued China's green transformation. According to China's Renewable Energy Industries Association (CREIA), the country's average wind curtailment rate stood at a record high of 15 per cent in 2015. In wind-rich Northern provinces such as Gansu, Ningxia or Heilongjiang, rates soared to 60 per cent. However up to 33.9 billion KWh of wind electricity was unconnected to the grid, leading to 18 billion yuan ($2.8 billion) of losses (Li 2016). Even though the feed-in prices for different RE were agreed gradually, the implementation of FIT policy has still encountered strong resistance from conventional energy forces. The 2009 Renewable Energy Law Amendment aimed to combine a quota system (the US model) with feed-in tariffs to set mandatory renewables quota for energy and grid enterprises (see Chapter 16). In 2012, China released a discussion draft of its Climate Change Law aiming to provide a strong dynamic for the implementation of the quota system. However, this draft was not adopted and integrated into a formal legal system until 2016. The efforts made since 2009 have shown that introducing the quota system to ensure the grid-connection for RE power was met with considerable resistance.

Furthermore, China's economic slowdown is likely to have a short-term beneficial effect in terms of reduced GHGE but a medium- to long-term negative effect in terms of reduced investment in the low-carbon transformation. China's GDP growth was officially 7.4 per cent in 2014. Growth is expected to be 7.1 per cent and 6.9 per cent in 2016 and 2017 respectively (World Bank 2015). This is a much slower pace than the annual average of 10 per cent attained in the past three decades. China's economic slowdown could lead to lower emissions and help to realize its commitments made under the 2015 Paris Agreement. However, in the context of acute economic slowdown, stimulus measures usually turn to energy-intensive industries, since it would be a high-risk endeavour for the leaders to base attempts primarily on newly emerging RE industries (Wuebbeke and Conrad 2016). This compromise to maintain political stability may hinder a shift to lower-carbon growth.

Domestic top-down climate governance and low-carbon pilots at the local level

At the domestic level, the signalling mechanism and 'learning by doing' pattern are the two main approaches for local climate governance. The top-down signalling mechanism means that once the leaders make (voluntary) commitments to the international climate talks in order to raise national reputation and reduce the international pressure, they will send positive signals to the domestic audience and make efforts to force the policy implementation at the local level. China's state-over-society structure enhanced the effect of signalling mechanisms, especially after national elites reached a general consensus on a climate mitigation commitment. China has a top-down approach to designing and implementing climate policies, such as setting long term, consistent climate targets (e.g. the FYPs), spurred by the political and economic resource reallocation at the domestic level (Schuman and Lin 2012: 102). In order to realize the national voluntary carbon-intensity reduction and RE development targets by 2020, the State Council issued the *Energy Conservation and Emissions Reduction Comprehensive Work Plan for the 12th FYP (2011–2015)*. The plan devolves the national green target onto the provincial level, through the rational index allocation of energy conservation and emission reduction targets. Furthermore, international commitments have promoted the greening of standards in the local official performance assessment system. Emissions reduction and RE development requirements have become key evaluation standards to measure local officials' political performance.

The learning by doing pattern means that, in order to avoid unnecessary mistakes and setbacks, national leaders typically conduct policy trials on a small scale in order to gain experience before introducing change on a large scale. Pilot programmes and demonstration projects introduced on a small scale were seen as a better approach to carry out innovative low-carbon policies, to correct errors in a timely manner, and to accumulate experience through learning by doing. The successful performance achieved in the pilot programmes boosted national confidence in nationwide expansion. Low-Carbon Pilot Cities and Carbon-Trading Pilot Cities are two important examples.

The NDRC launched the Low-Carbon Pilot Cities and Provinces Project in August 2010. Five provinces (Guangdong, Liaoning, Hubei, Shaanxi, Yunnan) and eight cities (Tianjin, Chongqing, Shenzhen, Xiamen, Hangzhou, Nanchang, Guiyang, Baoding) have acted as low-carbon pilots to demonstrate the feasibility of low-carbon development plans and RE application in urban life. A further 16 cities were named as the second batch of pilot cities in February 2012 (NDRC 2012).[4] The prosperous development of the trial cities might subsequently form a basis for the promotion of large-scale low-carbon reforms and integration of innovative policies instruments into the policy system. Having drawn lessons from European ETS experience, China issued the *Notice on Carbon Trading Trials Scheme* in October 2011, initiating a group of city- and provincial-level ETS trials (Wang 2012). In this scheme, five cities and two provinces were chosen as pilots: the cities of Beijing, Tianjin, Shanghai, Chongqing and Shenzhen as well as the provinces of Hubei and Guangdong. Based on the successful experience of these ETS pilot cities and provinces, Chinese leaders have held on to a relatively optimistic plan to launch a national-level ETS by 2017 (Fialka 2016), with a view of inter-connecting it with ETSs in other parts of the world after 2020.

With increasing public awareness of air pollution and the salience of climate change issues at the local level, there is a move to a 'bottom-up' approach of low-carbon initiatives promoted by multiple actors, such as municipal authorities, local think tanks, environmental NGOs and RE industry associations. Even though they are not included in the national-level pilots programme, small cities like Rizhao, Dezhou and Weihai, coupled with support from civil society, put forward their own low-carbon and ETS development blueprint.

Heading for an all-around international climate leadership

Although China benefited from its increased influence in the BASIC group since the 2009 Copenhagen conference, it also learned the lessons from its current diplomatic strategy and began to show a more flexible and pragmatic position in international climate politics. China has gradually started to exhibit entrepreneurial, cognitive and structural leadership while relying primarily on a transactional leadership style in international climate governance. It has engaged in four kinds of climate diplomatic efforts to pursue a leading role in international climate politics.

First, diplomatic coordination with major powers has been enhanced, especially to construct new kinds of 'win-win' relations to avoid the appearance of negotiation deadlocks in COPs, such as with the EU and the US. In 2010, the Europe–China Clean Energy Center (EC2) and the China–EU Institute for Clean and Renewable Energy (ICARE) were established to improve the institutionalization of Sino-EU climate cooperation. When Premier Li Keqiang visited the EU in June 2015 to attend the 40th anniversary of Sino-EU diplomatic relations, the two sides agreed to develop a sustainable low-carbon economy through technological innovation and enhance their cooperation in ETS under China's 'Belt and Road' development framework.[5] The Asian Infrastructure Investment Bank (AIIB) and the Silk Road Fund will both offer funding for Chinese and European joint ventures on 'Belt and Road' low-carbon designing projects, such as linking the interregional Euro-Asian carbon-trading market. At the Asian-Pacific Economic Cooperation (APEC) Summit in Beijing in November 2014, President Xi Jinping and his US counterpart, Barack Obama, made a historic joint announcement, vowing to jointly push the international climate negotiations for a new agreement to be reached as planned at Paris in 2015. In September 2015, the two countries issued a joint statement in Washington to reaffirm the commitment of the two sides, which laid a foundation for reaching an ambitious agreement in 2015.

Second, China endeavoured to promote two aspects of 'South-South' cooperation: (1) a new kind of cooperation with other major developing countries and (2) traditional 'South–South' cooperation with small developing countries and least developed countries. China took an initiative to ally with India, Brazil, and South Africa to form the group of BASIC, in order to hold their ground in the two-track negotiation framework. In 2012, China first offered to support South-South climate cooperation with $10 million in funding. To show its great commitment, Beijing also announced the establishment of an independent South-South cooperation fund of Renminbi 20 billion ($3.1 billion) in 2015 to help developing countries affected by global warming. Through providing more international public goods (such as aid), China tries to bridge the differences between separate negotiation groups, boost its profile and enhance its structural leadership in international climate politics.

Third, China engaged in establishing some low-carbon institutions, aiming to promote a leading role in international multi-level cooperative platforms. The APEC Sustainable Energy Centre (APSEC), established at the 2014 APEC Energy Ministers Meeting, is expected to strengthen exchanges on energy research and technological innovation, as well as optimize the energy structure in the Asia-Pacific region. In early 2016, China used its G20 Presidency and its leadership role in new international financial institutions (such as AIIB and the New Development Bank – 'BRICS Bank') to take a decisive step forward by mainstreaming low-carbon financing into the colossal new financial institutions in Asia. Indeed, this task is perhaps the largest priority for climate diplomacy in 2016 (Ha *et al.* 2016).

Fourth, China has taken efforts to enhance entrepreneurial leadership by promoting public climate diplomacy. At the 2015 Paris conference, the Chinese delegation seemed more confident and increasingly open to the international community. As mentioned above, the Chinese pavilion hosted an active programme of events that included business, civil society, expert and the presence of a number of non-Chinese contributors (Hilton 2015). Increasingly, Chinese NGOs have operated across borders and initiated some bottom-up international climate cooperative projects. The Chinese environmental NGO Global Environmental Institute (GEI) aims to promote the development of clean energy and energy efficiency industries through exploring market-oriented solutions in China and abroad. GEI has launched RE Demonstrations projects to accelerate the RE technologies diffusion and utility in ASEAN countries such as Myanmar (GEI2015).

However, mainly due to calculations of short-turn interest, China's leadership intention sometimes shows degradation in certain issue. For instance, China strongly opposed the EU's attempt to extend its ETS to aviation, including foreign (e.g. Chinese) aircraft carriers. In addition to lobbying the Commission, the Chinese government also targeted the German, French and British governments in what could be considered a 'divide and rule' strategy which resulted in the EU halting the extension of its ETS to foreign aircraft. A global ETS for the aviation sector has been discussed by the International Civil Aviation Organization which, however, has made only very slow progress partly due to the indifference of China, India and the US. Maybe a turning point will arrive if and when a truly global ETS finally emerges which may link the EU ETS with an Asia interregional ETS in the future.

Conclusion: political leadership in China

China's stance has undergone a three-step gradual change, making it more flexible, tangible and self-motivated. Meanwhile, more and more agencies have been embedded in the domestic bureaucratic system as institutional responses. China has gradually come to treat the climate change issue as an opportunity to speed up its domestic green transformation rather than merely a threat to its economic development. This kind of evolution has resulted from China's pragmatic self-adaptation to the international context, the increasing integration of green growth concerns into its core national interest and the increasing international pressure due to the country's rapid GHGE growth.

China's attitude to leadership in climate politics has also undergone considerable changes, initially from a marginalized follower to an active key player, resulting eventually in an emerging leader which exhibits considerable entrepreneurial, cognitive, exemplary and structural leadership. In the 2009 Copenhagen conference, China began to assume entrepreneurial leadership through using the BASIC group to hold together the major large developing countries on the crucial climate change issues negotiated at this conference. Having been blamed for the Copenhagen failure, China took efforts to boost its international reputation by providing cognitive leadership which resulted in an enhanced reputation of its public climate diplomacy. Having become a top leading country in renewables installation capacity and green investment since 2010, China has become more confident to show its leading performance in international climate politics. The promotion of 'clean energy diplomacy' in 'South–South' cooperation has shown China's willingness to lead by example through sharing its successful RE experience with other developing countries. The doctrine of 'diligent and ambitious diplomacy' proposed by the new leader President Xi has stimulated a significant improvement of China's structural leadership in climate politics since 2013. This has led to an emerging leadership role for China in the international climate change negotiations where it has made efforts to bridge differences among various groups. China has also promoted South–South climate finance by establishing new Southern-led international organizations such as the BRICS bank, the AIIB and the Silk Road Fund. Besides, China's 'Belt and Road' strategic framework has offered opportunities for the diffusion of know-how and the transfer of green technology to its neighbouring countries, promoting their energy transformation while also helping China's economy.

In summary, China's efforts to distinguish itself in climate politics have shown the country's intention to adopt a transformational leadership style. However, if constraints such as the domestic energy transformation dilemma hold back China's climate diplomatic progress, then, in practice, China will end up with a transactional leadership style.

Notes

1 Five-Year Plan (FYP) refers to a series of social and economic strategic development initiatives to map the strategies for economic development, set the growth targets, and launch the structural reforms, which provide a significant blueprint for the immediate future. Since China launched its first FYP in 1953, China's five year plans have played a pivotal role in guiding the general economical and social development.
2 The main leading institutions of this delegation were from the China Meteorological Administration (CMA) and the State Environmental Protection Administration (SEPA).
3 Xie Zhenhua is the vice chairman of the NDRC and the leader of the Chinese delegation to the UN Climate Change Conference.
4 They are Beijing, Kunming, Xi'an, Ningbo, Guangzhou, Shenyang, Harbin, Huai'an, Yantai, Haikou, Chengdu, Qingdao, Zhuzhou, Bengbu, Shiyan and Jiyuan.
5 The Belt and Road Initiative is a development strategy and framework proposed by Chinese leader Xi Jinping primarily between China and the rest of Eurasia.

References

Boyd, O. (2012) *China's Energy Reform and Climate Policy: The Ideas Motivating Change*, CCEP Working Paper, Centre for Climate Economics & Policy, Crawford School of Public Policy, the Australian National University.

Carraro, C. and Tavoni, M. (2010) 'Looking ahead from Copenhagen: how challenging is the Chinese carbon intensity target?', available at www.voxeu.org/article/china-s-copenhagen-commitment-business-usual-or-climate-leadership (accessed 17 July 2014).

Chen, G. (2009) *Politics of China's Environmental Protection: Problems and Progress*, Singapore: World Scientific Publishing Co.

Chinafaqs (2011) 'Propelling the Durban Climate Talks', available at www.chinafaqs.org/blog-posts/propelling-durban-climate-talks-china-announces-willingness-consider-legally-binding-comm (accessed 17 July 2014).

Chinapower (2016) 'The issuing of China Energy Outlook 2030', www.chinapower.com.cn/finance/20160322/21414.html (accessed 27.05.2016).

Fialka, J. (2016) 'China will start the world's largest carbon trading market', www.scientificamerican.com/article/china-will-start-the-world-s-largest-carbon-trading-market/ (accessed 24 May 2016).

Geall, S. and Hui, L.N. (2016) *China's Low Carbon Future Offers Global Opportunities*, Chinadialogue Brief in 2016.

GEI (2015) 'Programs of the Global Environmental Institute', www.geichina.org/index.php?controller=Default&action=index(accessed (27 May 2016).

Ha, S., Hale, T. and Ogden, P. (2016) 'Ahead of the Paris Treaty signing ceremony, it's "all hands on deck" for climate finance', www.chinadialogue.net/article/show/single/en/8792-Ahead-of-the-Paris-Treaty-signing-ceremony-it-s-all-hands-on-deck-for-climate-finance (accessed 24 May 2016).

Hallding, K., Han, G. and Olsson, M. (2009) 'China's Climate- and Energy-Security Dilemma: Shaping a New Path of Economic Growth', *Journal of Current Chinese Affairs*, 38(3): 119–134.

Harris, P.G. (ed.) (2009) *China and Global Climate Change*, Hong Kong: Proceedings of the conference held at Lingnan University, 18–19 June 2009.

Harris, P.G. (2011) 'Diplomacy, responsibility and China's climate change policy', in P.G. Harris, (ed.), *China's Responsibility for Climate Change: Ethics, Fairness and Environmental Policy*, Bristol: The Policy Press, 1–25.

Hatch, M.T. (2003) 'Chinese politics, energy policy, and the international climate change negotiations', in P.G. Harris, (ed.), *Global Warming and East Asia: The Domestic and Inernational Politics of Climate Change*, London and New York: Taylor & Francis Group, 43–67.

Heggelund, G. (2007) 'China's climate change policy: domestic and international developments', *Asian Perspective*, 31(2): 155–191.

Hilton, I. (2015) 'China adopts greater openness at UN climate talks', 06.12.2015, available at www.chinadialogue.org.cn/article/show/single/en/8405-China-adopts-greater-openness-at-UN-climate-talks (accessed 27 June 2015).

IISD (2011) 'Summary of the Durban Climate Change Conference', available at www.iisd.ca/vol. 12/enb12534e.html (accessed 23 March 2014).

Johnston, A.I. (1998) 'China and international environmental institutions: a decision rule analysis', in M.B. McElroy, (ed.), *Energizing China: Reconciling Environmental Protection and Economic Growth*, Cambridge, MA: Harvard University Press.

Kang, X. (2010) 'Perception of interests and internalization of international norms: a case study on China's internalization of norms in international climate cooperation', *World Economics and Politics*, 1: 66–83.

Lewis, J.I. (2008) 'China's strategic priorities in international climate change negotiations', *Washington Quarterly*, 31(1): 155–174.

Li, Y.(2016) 'Blowing in the wind', 31 May 2016, www.chinadialogue.net/article/show/single/en/8965-Blowing-in-the-wind (accessed 23 May 2016).

Liang, W. (2010) 'Changing climate? China's new interest in global climate change negotiations', in J.J. Kassiola and S. Guo, (eds), *China's Environmental Crisis*, Basingstoke and New York: Palgrave Macmillan, 61–84.

Moore, S. (2011) 'Strategic imperative? reading China's climate policy in terms of core interests', *Global Change, Peace & Security*, 23(2): 147–157.

Ren, W. (2014) 'The 3.0 version of China's diplomacy', 15 January 2014, available at http://mil.news.sina.com.cn/2014-01-15/1430760216.html (accessed 27 May 2015).

Schuman, S. and Lin, A. (2012) 'China's renewable energy law and its impact on renewable power in China: progress, challenges and recommendations for improving implementation', *Energy Policy*, 51: 89–109.

Soutar, R. (2015) 'A "less defensive" China can help spur global climate deal', www.chinadialogue.net/article/show/single/en/8248-A-less-defensive-China-can-help-spur-global-climate-deal (accessed 16 October 2015).

Wang, T. (2012) 'China's carbon market challenge', China Dialogue, available at www.chinadialogue.net/article/show/single/en/4936-China-s-carbon-market-challenge (accessed 26 July 2014).

Wuebbeke, J. and Conrad, B. (2016) 'How China's economic downturn could delay promised CO2 peak', 26 February 2016, www.chinadialogue.net/article/show/single/en/8652-How-China-s-economic-downturn-could-delay-promised-CO2-peak (accessed 16 March 2016).

World Bank(2015) World Bank Report 2015, www.worldbank.org/content/dam/Worldbank/document/EAP/China/ceu_06_15_en.pdf (accessed 16 March 2016).

Xinhuanet (2015) 'Spotlight: China makes active contribution for breakthrough at Paris climate talks', http://news.xinhuanet.com/english/2015-12/13/c_134912237.htm (accessed on 16 March 2016).

Yan, S. and Xiao, L. (2010) 'Evolution of China's position in international climate talks (in Chinese)', *Journal of Contemporary Asia-Pacific Studies*, 1: 80–90.

Yu, J. (2015) 'Entering the mainstream: an evolution in China's climate diplomacy, 01.12.2015, www.chinadialogue.org.cn/article/show/single/en/8369-Entering-the-mainstream-an-evolution-in-China-s-climate-diplomacy (accessed 16 March 2016).

Zhang, Z. (2003) 'The forces behind China's climate change policy: interests, sovereignty and prestige', in P.G. Harris, (ed.), *Global Warming and East Asia: The Domestic and Inernational Politics of Climate Change*, London and New York: Taylor & Francis Group, 66–86.

Zhang, H. (2006) 'China's position in the negotiations on international climate change: continuities and changes (in Chinese)', *World Economics and Politics*, 10:36–43.

Zhang, X., Chang, S. and Eric, M. (2012) 'Renewable energy in China: an integrated technology and policy perspective', *Energy Policy*, 51: 1–6.

Zhang, Z. (2010) 'Copenhagen and beyond: reflections on China's stance and responses', in E. Cerda and X. Labandeira, (eds), *Climate Change Policies: Global Challenges and Future Prospects*, Celtenham and Northampton: Edward Elgar.

18 India

The global climate power torn between 'growth-first' and 'green growth'

Kirsten Jörgensen

Introduction

In the mid-2010s, rapidly industrializing India with its fast growing economy is, after the US and China (see Chapters 16 and 17 respectively), the world's third largest greenhouse gas emissions (GHGE) producer, emitting 2,407 million tons carbon dioxide (CO_2) in 2013 alone (Global Carbon Atlas 2013). India is likely to overtake the fastest growing economy in the world, China, in the next decade 'as the primary source of growth in global energy demand' (Bloomberg Business 2015).

However, India is both an emerging economy and a developing country with widespread poverty. According to India's Planning Commission, 21.9 per cent of its population live in poverty (Government of India 2013), 59.2 per cent of the population in 2011 lived on less than $2.00 per day and 300 million people do not have access to electricity (World Bank 2014).

The reason why India ranks so highly in terms of GHGE is not due to its high per capita emission levels; India represents 17 per cent of the world's population but produces only 6 per cent of worldwide CO_2 emissions. In fact, due to energy poverty, India still has very low levels of per capita CO_2 emissions. Rather India contributes substantially to global CO_2 emissions because of the sheer size of its population combined with the rapid pace of its economic development. India's gross domestic product (GDP) has risen rapidly since economic liberalization in 1991. The majority of the country's CO_2 emissions are generated by the energy sector (70 per cent) (Dubash *et al.* 2015). In 2013, India and the US both contributed about one-fifth of the net growth in global coal consumption. India's energy sector, and in particular its coal consumption, which amounts to 59 per cent of its total energy share, will increase by a further 25 per cent between 2016–2017 according to India's Annual Reports and the 12th Five-Year Plan (Netherlands Environmental Assessment Agency PBL 2013). With regard to the emissions intensity of the national economy, India's energy emissions per GDP are higher than those of the EU but lower than those of the US.

India has been a key player in international climate negotiations since the beginning of the 1990s. This chapter will explore the dynamics of climate policy-making in India. Considering the four main themes – (1) leadership,

(2) multi-level and polycentric climate governance, (3) policy instruments, and (4) the green and low-carbon economy – proposed in the Introduction of this edition (Chapter 1), this chapter will shed light on the prospects for the development of a climate policy framework which would allow India to develop in a more climate-compatible manner (Dubash *et al.* 2015) and to promote an effective global climate regime. The first section of the chapter investigates national attitudes towards climate change in India, while considering public opinion and political salience among parties and other important stakeholders. The next section outlines benchmarks of India's climate debate and considers whether climate policy is perceived as a threat or opportunity in India's shifting climate change discourse. The section will also explore the way India's domestic and international climate policies have developed in three phases since the adoption of the United Nations Framework Convention on Climate Change (UNFCCC) in 1992. India's role in multilateral climate cooperation will be considered in relation to the global system of multi-level climate governance. What role has India, which is often regarded a deal-breaker (Betz 2012), played? Was it able to play a greater role in international climate negotiations, and, if yes, what kind of leadership did India exert? A change of government can open a window of opportunity for policy change. In 2014 India elected a new national ruling party with the Bharatiya Janata Party (BJP) replacing the Indian National Congress Party's (INC) coalition government. Are there signs that a new form of national leadership is emerging, one that is necessary to bring about substantial policy change in India?

In a third section the choice of strategies and policy instruments in India's energy and climate policy are explored against the backdrop of traditional and new modes of governance and low-carbon development. The fourth section explores multi-level and polycentric climate governance structures in India and asks whether India's federal structure allows for leadership and stimuli in the form of bottom-up approaches from various levels of government and the private sector. Finally, wrapping up the previous sections, climate policy leadership in India will be explored.

National attitudes to climate change

It is hard to assess public opinion and the public mood towards climate change in India as data tends to be sparse and/or unverified. Among its population of 1.2 billion people, 288 million people live in poverty (World Bank 2014), two-thirds in rural areas and a large number of people are assumed to have very low levels of education and access to information, since only 42 per cent of all students complete high school (Sahni 2015). For these citizens climate change may still be a topic that they are not yet familiar with. According to a 2010 Gallup report, the 'average Indian was unaware of climate change', and only 32 per cent of Indians stated that 'they knew at least something about climate change' (Gallup 2010). However 40 per cent of those questioned were of the opinion that the Indian government is not doing enough to reduce emissions from motor vehicles and factories.

Another national survey conducted a year later by the Yale Project on Climate Change Communication and GlobeScan (2012) found that 72 per cent of Indians questioned stated that they believed global warming was happening after they had been provided with a definition of climate change. Forty-one per cent of the 4,031 Indian adults questioned in urban and rural areas were of the opinion that the government in India should be doing more to address global warming, 54 per cent agreed that 'India should be making a large or moderate-scale effort to reduce global warming, even if it has large or moderate economic costs.' This survey was however of a small number and does not represent millions of people living in rural areas and along India's long coastline who are already facing the impacts of climate change. In India media coverage of climate change started slowly but rose sharply in 2009 (Boykoff 2010; Thaker and Leiserowitz 2014). According to Jogesh (2012) media attention between 2009–2010 was driven primarily by the international climate negotiation process, particularly during the run-up to the 2009 Copenhagen climate conference (COP15) and immediately thereafter. Media reports covered the achievements and errors in the work of the Intergovernmental Panel on Climate Change (IPCC) which had been chaired by an Indian scientist and which received the Nobel Peace Prize in 2007 (Isaksen and Stokke 2014). In second place, the media covered domestic climate policy and politics. The third most important issue attracting media attention, receiving a coverage of 10 per cent, was the topic of the increasing number of scientific publications about India's vulnerability and environmental problems (Jogesh 2012: 272).

In 2009 climate change also appeared in the election programmes of the two leading parties, the INC and the BJP, with both promising to undertake initiatives to combat climate change, promote low-carbon energy sources, and thereby safeguard India's ecological resources (Worldwatch Institute 2009). In the 2014 election campaign, the INC and BJP, as well as other parties (such as the Dravida Munnetra Kazhagam (DMK), Biju Janata Dal (BJD), Nationalist Congress Party (NCP), All India Trinamool Congress (AITC) and Aam Admi Party (AAP)), have paid attention to climate policy, clean (and affordable) energy and sector approaches to climate protection (Climate Parliament 2014). Yet as Nivela (2014) states 'environment and climate change are not political priorities in India. Unsurprisingly, politicians still focus on what they see as the unsolved development challenges' (Nivela 2014). Green parties do not exist at the national and subnational level. A few party activists from the INC, BJP and various other regional parties have picked up on the topic of climate change though.

NGOs have always played an important role in various policy domains, particularly in development policy but also with regard to environmental protection and water management (Shah 2004). A few environmental NGOs (ENGOs) with close ties to the national and subnational governments have been involved in the domestic and international framing of global climate governance from the outset of international climate policy (Fisher 2012). Over the past decade, climate change has become a more relevant topic for increasing numbers of civil society organizations and transnational organizations working in India (Lele 2012).

India's private sector remains fragmented (Betz 2012) when it comes to climate issues. Yet, stakeholders from the industrial sector showed a strong interest in the introduction of the market-based Clean Development Mechanism (CDM) in the Kyoto Protocol after the COP8 which was held in Delhi in 2002. Therefore India reversed its position, dipping 'half a toe in the water' (Dubash 2013: 192) by agreeing to the introduction of the CDM in the Kyoto Protocol. In the years that followed, entrepreneurship related to CDM boomed in India. Moreover, stakeholders from the private sector paid increasing attention to the problem of climate change, in particular with regard to its impact on business and as a potential hindrance for growth (Das 2012; Pulver 2012).

India's climate debate and three phases of climate change policy

Knowledge about dominant national climate discourses and the way problems, interests, and positions are framed contribute to a better understanding of a country's climate politics, the stability of its existing policy framework as well as the likelihood of policy change. India's climate discourse, which was relatively stable from 1991, started to diversify in 2007 (Isaksen and Stokke 2014; Thaker and Leiserowitz 2014).

The first dominant and durable social construction of the climate change problem in India was the climate equity frame (Dubash 2013: 193) or, to use Isaksen and Stokke's (2014) notion, the Third World discourse according to which industrialized countries should be held to account. Initially put forth by NGO activists in 1991 (Gupta *et al.* 2015), climate change was perceived as a problem created by industrialized countries through (1) historical stocks of GHGE and (2) the annual flows of high per capita emission levels emitted by the global North (Dubash 2013). Thus it was also seen as the sole responsibility of the global North. India only paid attention to international negotiations and not to domestic climate action. The climate equity frame and the development-first paradigm were shared by government officials and NGOs and initially informed India's international climate policy. Accordingly, as a former colony and developing country, India's first priority should be development (Dubash 2013) and poverty reduction while these goals should not be hampered by emission reduction goals. Despite the persistent coalition of 'growth-first stonewallers' from the public and private sector (Dubash 2012) and a broad continuous support for the equity frame in India (Dubash 2013), from 2007 onwards India's climate debate started to include new angles and approaches.

India's vulnerability to the negative impacts of global warming has gained increasing public awareness. The threat posed to development and growth from global warming, including the melting of the Himalaya glaciers, droughts, a rising sea level along India's long coastline and the increase in extreme weather events became more evident. At the same time, the developmental advantages and potential economic and technical benefits of climate action, the development of renewable energy and the economic potential of CDMs (Shukla and Dhar

2011; Fisher 2012; Michaelowa and Michaelowa 2012) became more obvious, culminating in the co-benefit and green growth debate.

Furthermore, the changing international context and the new geopolitical alliance of the four largest newly industrialized BASIC countries (Brazil, South Africa, India and China) led to India being identified as an emerging economy rather than a developing country (Isaksen and Stokke 2014). In line with this there is a perception that India as a rising world power should play a more central role in the international arena and accept its international responsibilities.

Phase 1 of India's climate policy (1990–2007): initiating climate diplomacy

In the 1990s climate policy first appeared on India's governmental agenda under the auspices of the Ministry of External Affairs (Fisher 2012). Agenda setting was however largely catalyzed by international climate negotiations resulting from the passing of the Climate Convention in 1992. Government action therefore focused on the emerging international regime and framed as well as institutionalized the subject accordingly, i.e. as an international issue, leading to the establishment of a negotiation team which was influenced by both governmental actors and NGOs (Michaelowa and Michaelowa 2012). Climate equity and pre-existing concerns regarding energy security and India's development trajectory guided the development of India's negotiating position (Fisher 2012).

In the context of the international negotiations between 1990 and 2015, India was often perceived as obstructionist as it continued to resist binding goals and obligations for developing countries. However, India was able to exert cognitive and entrepreneurial leadership in various ways: in 1991 India submitted a paper to the Intergovernmental Negotiating Committee introducing the equity principle to the international climate negotiations and received high approval from the other developing countries (Gupta *et al.* 2015). India also became a rule-maker in formulating the central principles of the international climate regime. Indian negotiators managed to introduce the 'common but differentiated responsibility' wording into international treaties in 1992. Furthermore India took together with China the lead in the negotiating group of developing countries (G77) (Gupta *et al.* 2015) in the run-up to COP1 in Berlin in 1995 when insisting on the differentiated architecture of the treaties, thus leaving developing countries free from obligations. According to the Kyoto Protocol, which entered into force in 2005, this required that only the industrialized countries (so-called Annex I countries) would have to commit to GHGE reduction obligations and that technology transfer and financial mechanisms should be put in place to offset climate protection costs in developing countries (Michaelowa and Michaelowa 2012).

Domestic climate policy did not exist prior to 2007. However, driven by India's need for energy sufficiency and its aspiration for rapid industrial growth, a renewable energy framework was developed during the 1980s. During the first stage in the 1990s, India's Union Government and the states focused on promoting wind power. In the second stage, attention switched to solar energy

(Chaudhary *et al.* 2014). Linkages between energy, industrial policy und climate mitigation started to appear triggered by the introduction of the CDM. India ratified the Kyoto Protocol in 2002 and created the institutional framework for the implementation of the CDM.

Phase 2 of India's climate policy (2007–2014): reluctance on the international level, cautious engagement on the domestic level

Since 2007, beginning with the run-up to the 2009 Copenhagen climate conference (COP15), India has seen an emerging climate policy towards domestic mitigation and adaptation. In 2007, during the G8 Summit in Heiligendamm (see also Chapter 8) India's incumbent Prime Minister Manmohan Singh announced that per capita GHGE in India will never exceed those of the Organization for Economic Co-operation and Development (OECD) nations. In 2007 Singh also established a National Advisory Panel which was mandated to develop a national climate strategy. The National Action Plan on Climate Change (NAPCC) was launched in 2008. It included national objectives such as the promotion of renewable energies ('Solar Mission') and energy efficiency ('Enhanced Energy Efficiency') (Government of India 2008).

A rapid development of climate policy measures in India occurred and the amount of newly introduced regulations increased. India's domestic climate policy was increasingly driven by India's pressing energy security concerns and a growing scientific consensus about the impacts of climate change (Gupta *et al.* 2015).

This changing policy can also be explained by international factors. In the international arena India's negotiating position was increasingly questioned by both domestic and international actors. Betz (2012: 17) has argued that '[t]he Indian government is committing less in international negotiations than it is willing to push through domestically'.

India's former Environmental Minister, Jairam Ramesh (2009–2011), tried to reposition India in terms of 'style and substance' in the international negotiations (Ramesh 2015: 450), a move that faced strong resistance from India's chief climate negotiators and triggered controversial parliamentary and media debates. Ramesh suggested that India should dissociate itself from the G77 group and take on emission cuts commitments (Gupta *et al.* 2015).

Together with the other BASIC countries, India agreed to voluntary commitments to reduce its emissions intensity and it helped to influence the design of the 2009 Copenhagen Accord including the commitment to limit global warming to 2 degrees Celsius. India pledged to the UNFCCC to reduce the emissions intensity of its GDP by 20 per cent to 25 per cent by 2020 (compared to 2005), hence signaling a modification of its longstanding position of refusing reduction commitments. The commitment made by India was easily achievable, as it was in line with several domestic policies and developments currently in place (Climate Action Tracker 2015). In contrast to China (see Chapter 17), India's ability to exert leadership among the BASIC and the G77 during the 2009

Copenhagen climate conference (COP15) was impaired by the red lines set for India's negotiators (Jairam 2015).

Phase 3 of India's climate policy (May 2014–December 2015): Modi's arrival at power, major environmental regulation rollbacks, and a glint of hope at the Paris Summit

During the BJP's 2014 election campaign in which Narendra Modi emerged as strong political and administrative leader, great expectations started to develop with regard to the breaking of gridlock in various domestic policy domains in India. As incumbent Chief Minister of Gujarat, Narendra Modi was known for his successful industry policy of promoting the expansion of solar energy in his federal state, which in turn propelled economic growth. The question arose as to whether Prime Minister Modi would pursue an aggressive industrialization strategy or whether he would, as his book *Convenient Action* suggested, strive to set up a national climate and environment-compatible growth strategy and develop new approaches to low-carbon development (Modi 2011).

In his first year of government, Modi chose the first option as can be seen by developments such as the dismantling of environmental legislation, relaxing restrictions on the licensing of environmental harmful mining activities, and the banning of environmental organizations (*New York Times* 2014).

Yet, shortly before the 2015 Paris climate conference (COP21), India submitted the country's Intended Nationally Determined Contributions (INDCs) which included three goals for 2030. First, a rather moderate new goal for the reduction of emissions intensity of GDP by 33–35 per cent by 2030 (compared to 2005). Second, and even more importantly, a highly ambitious goal for renewable energy, of a 40 per cent share of non-fossil fuels in the installed electricity mix by 2030. Third, additional carbon reductions of 2.5 to three billion tonnes through additional forest cover. Particularly the goal to enhance the solar power capacity to 100 GW by 2022 is outstanding. It is linked to a remarkable international initiative taken by India at the 2015 Paris climate conference, namely the creation of an International Solar Alliance involving more than 120 countries, including African countries (Ministry of Environment, Forestry and Climate Change 2015).

Low carbon governance in India – strategic approaches and instruments

During the first phase of India's climate debate, when climate mitigation centered on the idea of equity of responsibility, domestic climate mitigation and multilateral mitigation obligations were regarded as a greater threat to domestic development than India's vulnerability to the impacts of climate change (Dubash 2012). However, after the country's climate debate started to diversify in 2007, the co-benefit concept became increasingly popular, opening up a window of opportunity for domestic climate mitigation. India's National Action Plan

Climate Change (2008) introduced the co-benefit principle which focused on measures that promote development objectives 'while also yielding co-benefits for addressing climate change effectively' (Government of India 2008: 2).

Some of India's strategic approaches launched as part of the domestic NAPCC are industrial policies which also benefit climate policy. The *Mission for Energy Efficiency* and the *Solar Mission* are tailored around the principles of ecological modernization and offer great potential for the achievement of environmental improvements through technological eco-innovation and the development of new green markets (Jänicke 2008). India labels these policies low-carbon development. Due to India's overwhelming energy crisis, low-carbon development is a familiar concept. Of particular importance therein is the growing market for renewable energy producers in India (IPCC 2011). Renewable energy and climate change were also discussed in the 11th and the 12th National Five Year Plan. India's low-carbon development strategy includes sector-based mitigation policies (Sant and Gambhir 2012), building on long-established institutions in the energy sector, and is particularly detailed when it comes to the development of renewable energy and energy efficiency. Sector strategies are promoted and administrated by the national Bureau of Energy Efficiency (BEE) India and the Ministry of New and Renewable Energy. A 2014 report prepared by the Expert Group on Low Carbon Strategies for Inclusive Growth (which had been set up by India's now abolished Planning Commission) suggested low-carbon developments would be consistent with inclusive growth (Planning Commission 2014). Twelve low-carbon targets to be achieved by 2030 were set in the fields of coal and renewable energy technologies, urban transportation, vehicles, industries and buildings.

India's National Action Plan on Climate Change deals specifically with the deployment of appropriate technology as well as new and innovative forms of governance, such as market, regulatory and voluntary mechanisms plus efficient and cost-effective strategies (Government of India 2008).

The types of policy instruments currently largely belong to new modes of governance. India's renewable and energy efficiency policy mix includes targets and builds on a variety of market-based instruments such as generation-based incentives (GBI), feed-in tariffs, wheeling of powers, the third-party sale of power, cap and trade, Renewable Energy Purchase Obligations (RPO), energy pricing, carbon tax and the Perform Achieve and Trade (PAT) Scheme) both at the national (Chaudhary *et al.* 2014) and state levels (Jörgensen *et al.* 2015).

India pioneered the PAT system in the context of the developing countries (Vaidyula 2015). In international negotiations India has been more open to the idea of market mechanisms than many developing countries (Michaelowa and Michaelowa 2012). Although India initially resisted the introduction of market mechanisms and emission trading into international treaties, the CDM became the dominant form of carbon market activity in India. India is in fact the world's second largest supplier of Certified Emissions Reductions (CERs), after China (Vaidyula 2015; see also Chapter 17). The country's share of CDM projects reached 50 per cent in 2012, making India the second largest host country

(Michaelowa and Michaelowa 2012: 578). In the mid-2010s India ran various pilot projects on emission trading related to air pollutants.

One of India's flagship climate change mitigation undertakings is the Jawaharlal Nehru National Solar Mission (JNNSM) (Chaudhary *et al.* 2014). The JNNSM is an industrial policy which links the promotion of solar energy to climate mitigation efforts. It was initiated in 2009 and developed jointly by the Union Government, various states, national and internationally funded NGOs and international organizations as well as solar manufacturers. Driven by the motives of economic competitiveness, energy security and demands from the international climate community to take action, the National Solar Mission is pursuing the goal of installing 22,000 MW of grid-connected and off-grid solar power using photovoltaic and concentrated solar power. The JNNSM promotes the development of a renewable energy market by making Renewable Purchase Obligations mandatory at the state level and introducing tradable Renewable Energy Certificates (Chaudhary *et al.* 2014).

Multi-level and polycentric climate governance

Different from the quasi-federal system of the EU, policy-making in India's federal system, which consists of 29 federal states, six union territories and 4,000 cities, is more centralized in spite of amendments to India's constitution in the 1990s which aimed to effectively enhance the sharing of power between the levels of government.

The EU system of climate governance is characterized by fluid actor and institutional constellations. It combines elements of intergovernmental decision-making between the member states, complex networks influencing regulation and technological standards and supranational institutions which exert influence in agenda setting, political decision-making and jurisprudence such as the EU Commission, European Parliament and the European Court of Justice (Lenschow 2013: 66; see also Chapters 3–5). Polycentric climate governance in the EU stimulates innovation, learning and diffusion at various governance levels but also puts limits on centralized steering (Rayner and Jordan 2013). The EU's multi-level governance structure provides opportunity structures and various access points for entrepreneurs to stimulate policy change and move their policy proposal higher up the political agenda of the EU (Schreurs and Tiberghien, 2010) but also enables veto players.

In comparison climate governance in India has been rather centralized since the 1990s and shaped by national institutions and elites. The 2008 National Action Plan Climate Change was developed in a top-down process, involving exclusive policy networks and selected actors from the public and private sector. It was not subject to a wide consultation process. Following this pattern, India's climate negotiating position is developed within rather exclusive circles. According to some academics, 'an effective forum where state and central government, private companies, municipalities, NGOs and citizens can … drive meaningful and measurable action' on climate change is missing (Dutta *et al.* 2015: 7).

However, there are indications that multi-level and polycentric climate governance structures are emerging in India which are more dynamic and changeable than India's centralized federal structures would suggest. The influence of subnational levels is noticeable in the context of domestic and transnational networks involving ENGOs, the corporate sector and donor organizations, all of which are influential at all levels of policy-making (Fisher 2012).

Energy, which is a highly important policy field for the mainstreaming of climate mitigation goals, falls under shared competencies between the central government and the states. Aside from implementing national climate and energy policy, India's states also partially promote their own priorities. A number of states such as Andhra Pradesh, Gujarat, Kerala, Maharashtra, Rajasthan and Tamil Nadu have recognized the economic co-benefits of renewable energy policy and some of them have developed ambitious independent approaches. Instruments employed at the state level include green energy funds, heavy industry taxes and preferential tariffs (Chaudhary *et al.* 2014; Jörgensen *et al.* 2015). In 2011 Narendra Modi, as incumbent Chief Minister (2002–2012) of India's federal state Gujarat, published the above mentioned-book *Convenient Action* which focused on adaption to and mitigation of climate change. Under his leadership, Gujarat introduced a Department of Climate Change and set up measures for the promotion of solar energy. India's states often link renewable energy to climate action in the context of their climate action plans (Jogesh and Dubash 2015; Jörgensen *et al.* 2015).

By the mid-2010s, significant forms of polycentric climate governance could not be observed in India's cities (Beermann *et al.* 2016). One significant constraint of subnational action in general is that the states and, even more so, the cities depend on financial transfers from the national level.

Apart from the multilateral context, bilateral relations between key players from the global North (such as the US and the EU) and the rapidly developing BASIC countries have become alternate avenues for the enforcing of global climate mitigation governance (Keukeleire and Bruyninckx 2011). In 2004 the EU and India agreed on a strategic partnership that was aimed at strengthening their ties in areas of mutual interest, including energy and climate change as well as multilateral relations. The potential for leadership by the EU in the form of regulatory models is high in energy-related green growth approaches, yet common ground with India is more limited with respect to multilateral approaches to international climate agreements. Although the EU-India partnership proved to be more fertile in the energy sector and the development of environmental standards (Torney 2015), it did not result in a better collaboration between the two parties in global climate negotiations.

Conclusion: political leadership in India

Despite the enormous challenges posed by India's rapid growth (i.e. the increasing demand for energy and the need to lift roughly 40 per cent of the population out of poverty) the transition towards climate-compatible development in India is not without hope. Starting in 2007, India's domestic climate policy output has grown, driven by India's vulnerability and international factors. In the context of India's centralized federal system, the 2008 National Action Plan on Climate Change emerged in a top-down process under the structural political leadership of India's Union Government. India's regulatory output, consisting predominantly of new market-based instruments of governance, is remarkable. Energy security and a strong focus on development remain the predominant motives in the Indian climate narrative. In this narrative, mitigation is merely regarded as a co-benefit of low-carbon strategies to inclusive growth.

India's low-carbon inclusive growth strategy solidly couples renewable energy and energy efficiency policies with climate mitigation. Long-established institutions such as the Bureau of Energy Efficiency and the Ministry of New and Renewable Energy and the elaborate policy mix for renewable energy and energy conservation ensure a reliable path-dependent continuation of innovative low-carbon approaches. India already has a strong position in the world market for renewable energy. The target included in India's INDCs to achieve a 40 per cent share of non-fossil fuels in the electricity mix by 2030 is very ambitious. Low-carbon approaches are, however, still outnumbered by examples of high carbon development in India's process of urbanization, specifically in the building, transportation and energy sectors.

Climate change also resonates at other levels of India's multi-level governance structures such as the states, cities, civil society and the private sector. With regard to the actor constellations that drive India's climate governance, NGOs and the private sector deserve a special mention. It was NGOs, not representatives of India's Union government, which initially provided the cognitive leadership and a stable framing of India's international climate policy, introducing the equity principle to India's emerging climate discourse, which gave the sole responsibility for climate mitigation to the developed countries. In the run-up to the Kyoto Protocol, stakeholders from India's corporate and industrial sector lobbied for the introduction of flexible mechanisms which paved the way for the growth of India's CDM-based carbon market.

Multi-level and polycentric climate initiatives are emerging, India's states are beginning to experiment with individual approaches to renewable energy development, tailored to their regional specifics, and their economic and political importance is growing (Jörgensen *et al.* 2015). In spite of this, bottom-up approaches are still limited because India's financial federalism and lacking decentralization do not yet provide sufficient space, capacities and resources, particularly at the local level (Beermann *et al.* 2016).

India's role in the global climate governance system is changing. At certain times it has been quite successful in positively influencing the global climate negotiations while at other times, impaired by the red lines set through India's domestic climate discourse, it was mainly perceived as an obstacle. India's international negotiators exerted entrepreneurial leadership in the early stage of the international climate negotiations, hence ensuring the inclusion of the Common But Differentiated Responsibilities (CBDR) concept in the UNFCCC. India's international climate policy has changed since 2009, when India – collectively with the BASIC countries Brazil, South Africa and China – advocated the 2009 Copenhagen Accord and committed to a voluntary decrease in its emissions intensity of 20–25 per cent by 2020 (compared to 2005 levels). Surprisingly at the 2015 Paris climate conference (COP21) India even exerted directional leadership by launching the International Solar Alliance involving more than 120 countries.

However, doubts have been expressed in recent political and academic debates as well as by the domestic and international media as to whether India's climate policy will see strong policy coordination and the forceful assertion of political will required for a significant transition to low-carbon development in India in the short term. Since the BJP came to power in 2014, Prime Minister Narendra Modi's strong, top-down style of leadership has failed to promote strategies for low-carbon development in India's coal sector. India's BJP government follows an aggressive industrial growth strategy. In doing so, it caters – notwithstanding the changing climate discourse in India and the forceful directional leadership in the promotion of renewable energy – to the interests of a strong growth/development-first advocacy coalition uniting lobby groups and institutionalized vested interests from the coal sector.

References

Beermann, J., Damodaran, A. and Jörgensen, K. (2016) 'Climate action in Indian cities: an emerging new research area', *Journal of Integrative Environmental Sciences*, 1–12.

Betz, J. (2012), *India's Turn in Climate Policy: Assessing the Interplay of Domestic and International Policy Change* [online], GIGA, www.econstor.eu/bitstream/10419/57188/1/689598971.pdf (accessed 8 September 2015).

Bloomberg Business (2015) *The 20 Fastest-Growing Economies This Year* [online], www.bloomberg.com/news/articles/2015-02-25/the-20-fastest-growing-economies-this-year (accessed 8 September 2015).

Boykoff, M. (2010) 'Indian media representations of climate change in a threatened journalistic ecosystem', *Climatic Change*, 99(1–2): 17–25.

Chaudhary, A., Krishna, C. and Sagar, A. (2014) 'Policy making for renewable energy in India: lessons from wind and solar power sectors', *Climate Policy*, 1–30.

Climate Action Tracker. *India Rating* (2015), http://climateactiontracker.org/countries/india.html (accessed 25 July 2015).

Climate Parliament (2014) *Indian elections connect with climate change – 10 Apr, 2014*, http://climateparl.net/cp/404 (accessed 8 September 2015).

Das, T. (2012) 'Climate change and the private sector' in N.K. Dubash (ed.) *Handbook of Climate Change and India. Development, politics, and governance*. Abingdon, Oxon, New York: Earthscan, 246–253.

Dubash, N. (2012) 'Climate Politics in India: three narratives' in N.K. Dubash (ed.) *Handbook of Climate Change and India. Development, politics, and governance*. Abingdon, Oxon, New York: Earthscan, 197–207.

Dubash, N., Khosla, R. Rao, N.D. and Sharma, R.K. (2015) *Informing India's Energy and Climate Debate: Policy Lessons from Modelling Studies*. Climate Initiative, Research Report, New Delhi: Centre for Policy Research.

Dubash, N.K. (2013) 'The politics of climate change in India: narratives of equity and cobenefits' *Wiley Interdisciplinary Reviews: Climate Change*, 4(3): 191–201.

Dutta, V., Dasgupta, P., Hultman, N. and Gadag, G, (2015) 'Evaluating expert opinion on India's climate policy: opportunities and barriers to low-carbon inclusive growth', *Climate and Development*, 1–15.

Fisher, S. (2012) 'Policy storylines in Indian climate politics: opening new political spaces?', *Environment and Planning C: Government and Policy*, 30(1): 109–127.

Global Carbon Atlas (2013) Emissions, www.globalcarbonatlas.org/?q=en/emissions (accessed 25 July 2015).

Gallup (2010) *Indians Largely Unaware of Climate Change*, www.gallup.com/poll/125267/Indians-Largely-Unaware-Climate-Change.aspx?utm_source=position6&utm_medium=related&utm_campaign=tiles (accessed 8 September 2015).

Government of India (GOI) (2008) *National Action Plan on Climate Change (NAPCC)*, New Delhi: Prime Minister's Council on Climate Change.

Government of India, and Planning Commission (2013) *Report of the expert group to review the methodology for measurement of poverty*, New Delhi.

Gupta, H., Kohli, R.K. and Ahluwalia, A.S. (2015) 'Mapping "consistency" in India's climate change position: Dynamics and dilemmas of science diplomacy', *Ambio* 44(6): 592–599.

IPCC (2011) *IPCC Special report on renewable energy sources and climate change mitigation*, Cambridge: Intergovernmental Panel on Climate Change.

Isaksen, K.-A. and Stokke, K. (2014) 'Changing climate discourse and politics in India. Climate change as challenge and opportunity for diplomacy and development', *Geoforum*, 57: 110–119.

Jänicke, M. (2008) 'Ecological modernisation: new perspectives', *Journal of Cleaner Production*, 16(5): 557–565.

Jogesh, A. (2012) 'A change in climate? Trends in climate change reportage in the Indian print media', in N.K. Dubash (ed.) *A Handbook of Climate Change and India. Development, Politics, and Governance*. New Delhi: Oxford University Press, 266–286.

Jogesh, A. and Dubash, N.K. (2015), 'State-led experimentation or centrally-motivated replication? A study of state action plans on climate change in India', *Journal of Integrative Environmental Sciences*, 12(4): 247–266.

Jörgensen, K., Mishra, A. and Sarangi, G.K. (2015) 'Multi-level climate governance in India: the role of the states in climate action planning and renewable energies', *Journal of Integrative Environmental Sciences*, 12(4): 267–283.

Rayner, T. and Jordan, A. (2013) 'The European Union: the polycentric climate policy leader?', *Wiley Interdisciplinary Reviews: Climate Change*, 4(2): 75–90.

Keukeleire, S. and Bruyninckx, H. (2011) 'The European Union, the BRICs, and the emerging new world order', in C. Hill and M. Smith (eds) *International Relations of the European Union*, Oxford: Oxford University Press, 380–403.

Lele, S. (2012) 'Climate change and the Indian environmental movement' in N.K. Dubash (ed.) *Handbook of Climate Change and India. Development, Politics, and Governance*. Abingdon, Oxon, New York: Earthscan, 208–217.

Lenschow, A. (2013) 'Studying EU environmental policy', in A. Jordan and C. Adelle (eds) *Environmental Policy in the EU*, London: Routledge, 49–72.

Michaelowa, K., and Michaelowa, A. (2012) 'India as an emerging power in international climate negotiations', *Climate Policy*, 12(5): 575–590.

Ministry of Environment, Forestry and Climate Change (2015) India's Intended Nationally Determined Contribution to UNFCC, www.moef.gov.in/sites/default/files/INDIA%20INDC%20TO%20UNFCCC.pdf (accessed 31 January 2016).

Modi, N. (2011) *Convenient Action: Gujarat's Response to Challenges of Climate Change*, Gandhinagar: Macmillan.

Netherlands Environmental Assessment Agency (PBL) (2013) *Trends in global CO2 emissions: 2013 Report*, The Hague: Netherlands Environmental Assessment Agency.

New York Times (2014) 'Narendra Modi, Favoring Growth in India, Pares Back Environmental Rules', www.nytimes.com/2014/12/05/world/indian-leader-favoring-growth-sweeps-away-environmental-rules.html?_r=0 (accessed 1 February 2016).

Nivela (2014) India and Elections: Will the environment and climate ever be a priority?, www.nivela.org/articles/will-the-environment-and-climate-ever-be-a-priority-for-the-world-s-largest-democracy/en.

Planning Commission (2014) *The Final Report of the Expert Group on Low Carbon Strategies for Inclusive Growth*, Delhi: Planning Commission.

Pulver, S. (2012) 'Corporate responses to climate change in India', in N.K. Dubash (ed.) *Handbook of Climate Change and India. Development, Politics, and Governance.* Abingdon, Oxon, New York: Earthscan, 254–265.

Ramesh, J. (2015) *Green Signals. Ecology, Growth, and Democracy in India.* New Delhi: Oxford University Press.

Sahni, U.(2015) *Primary Education in India: Progress and Challenges. The Second Modi-Obama Summit: Building the India-U.S. Partnership.* Brookings Institution, www.brookings.edu/~/media/research/files/reports/2015/01/20-building-the-india-us-partnership/the-second-modi-obama-summit-briefing-book.pdf (accessed 10 September 2015).

Sant, G. and Gambhir, A. (2012) 'Energy, development and climate change', in N.K. Dubash (ed.) *Handbook of Climate Change and India. Development, Politics, and Governance.* Abingdon, Oxon, New York: Earthscan, 289–302.

Schreurs, M., and Tiberghien, Y. (2010) 'European Union leadership in climate change: mitigation through multilevel reinforcement', in K. Harrison and L.M. Sundstrom (eds) *Global Commons, Domestic Decisions. The Comparative Politics of Climate Change.* Cambridge, MA: MIT Press, 23–66.

Shah, G. (2004) *Social Movements in India. A Review of Literature* (2nd edn). New Delhi: Sage.

Shukla, P.R. and Dhar, S. (2011) 'Climate agreements and India: aligning options and opportunities on a new track', *International Environmental Agreements: Politics, Law and Economics*, 11(3): 229–243.

Thaker, J. and Leiserowitz, A. (2014) 'Shifting discourses of climate change in India', *Climatic Change*, 123(2): 107–119.

Torney, D. (2015) *European Climate Leadership in Question. Policies toward China and India.* London: MIT Press.

Vaidyula, M. (2015) *INDIA: An Emissions Trading Case Study*, https://ieta.memberclicks.net/assets/CaseStudy2015/india_case_study_may2015.pdf (accessed 10 September 2015).

World Bank (2014) *Poverty headcount ratio at $2 a day (PPP) (% of population)*, http://data.worldbank.org/indicator/SI.POV.2DAY (accessed 8 September 2015).

World Resources Institute (2015) *6 Graphs Explain the World's Top 10 Emitters*, www.wri.org/blog/2014/11/6-graphs-explain-world%E2%80%99s-top-10-emitters (accessed 8 September 2015).

Worldwatch Institute (2009) *Will India's Elections Sway Climate Policy?*, Available from: www.worldwatch.org/node/6072 (accessed 8 September 2015).

Yale Project on Climate Change Communication, and GlobeScan Incorporated (2012) *Climate change in the Indian mind* [online], http://environment.yale.edu/climate/files/Climate-Change-Indian-Mind.pdf (accessed 8 September 2015).

Part VI
Conclusion

19 Conclusion

Re-assessing European Union climate leadership

Rüdiger K.W. Wurzel, Duncan Liefferink and James Connelly

Introduction

The preceding chapters analysed the *types* and *styles* of leadership which different EU institutional, member state, societal and non-EU actors have offered in international climate change politics. They also considered *when* and *how* those type(s) and style(s) of leadership were exercised. Chapter 1 introduced four leadership types (structural, entrepreneurial, cognitive and exemplary) and two leadership styles (heroic/humdrum and transformational/transactional.

All chapters adopted an actor-centred approach to the analysis of EU climate change politics while focusing on the key themes of: (1) leadership, (2) multilevel and polycentric governance, (3) policy instruments, and (4) green economy and low-carbon economy. Because leadership is the over-arching analytical theme, the preceding chapters have shed light on the paradox that the EU developed into a leader in international climate change politics despite having been set up as a 'leaderless Europe' (Hayward 2008) in which power is shared among a wide range of EU institutional, member state and societal actors, thereby increasing the potential number of veto actors. Schreurs and Tiberghien (2007: 24) argue that the EU's climate policy-making process follows 'a kind of logic that is the reverse of that of veto points or veto players' because it offers an 'open-ended and competitive governance structure … [which] has created multiple and mutually-reinforcing opportunities for leadership'. Jänicke (2011: 142) has observed in Germany's case that a symbolic leader which adopts a heroic leadership style may even find itself caught by a 'kind of "enforced leadership"' through EU institutions (e.g. the Commission) and (although to a lesser degree) international organisations. However, the UK's 2016 Brexit decision (see Chapters 1 and 12) dramatically shows that there are limits to the EU's ability to provide 'enforced leadership'. The insights and evidence offered throughout suggest that resistance from (potential) veto actors to the EU's leadership role can be overcome, but only under certain conditions and with the help of certain types and styles of leadership.

Leadership

The preceding chapters provide fewer empirical examples of *structural* leadership (which relates to an actor's hard power and depends on material resources such as economic strength) compared with the more widespread occurrence of *entrepreneurial* leadership (i.e. diplomatic, negotiating and bargaining skills in facilitating agreement) and *cognitive* leadership (i.e. the definition and/or redefinition of interests through ideas such as the concept of a green or low-carbon economy). Importantly, entrepreneurial and cognitive leadership often seem to work in conjunction. In addition, *exemplary* leadership by individual member states can play an important role in mobilising the EU as a whole. Setting an example for the rest of the world figures prominently in the climate ambitions of many of the EU institutions (see Chapters 3–5). This could be interpreted as support for the claim that the EU is a normative power (e.g. Manners 2002; Chapter 2) which advocates a Kantian world order (Chapter 2; Schmidt 2008). It could, however, also be construed as evidence for claims (from neorealists/intergovernmentalists) that because the EU is not a state it lacks the kind of structural powers which powerful states exhibit.

Structural leadership

In 2016 the EU constituted the world's largest internal market and hosted a population of about 500 million people from 28 member states. The victory of the Leave Campaign in the 2016 Brexit referendum is likely to trigger the UK's exit from the EU. However, given that a Norwegian-type of cooperation agreement between the UK and EU may be one possible future option (see Chapters 12 and 13), it was not clear at the time of writing, whether this would also mean that the UK will lose its access to the EU's internal market. Although the EU's military powers are, compared to most states and military alliances such as NATO, relatively unimportant, it nevertheless possesses considerable economic structural 'presence' (Chapter 2). But the EU often punches below its economic weight because it is either unable or unwilling to translate its putative economic structural powers into bargaining leverage in international negotiations. When the EU tries to make use of its economic weight (or structural powers), as was the case with the half-hearted attempt to extend the EU Emissions Trading Scheme (ETS) to the aviation sector including foreign aircraft, the limitations of its power vis-à-vis important global powers such as the US and China become apparent (Torney 2015). When faced with strong American, Chinese and Indian opposition the EU decided not to make use of its 'market power' (Damro 2012; Chapter 1) and put on hold the implementation of its decision to allow more time for negotiations within the International Civil Aviation Organization (ICAO) on a possible global ETS for the aviation sector (Chapters 4, 16 and 17).

Within the context of the international climate change negotiations, the limitations to the EU's structural power became apparent also at the 1992 UN Rio conference, which adopted the UN Framework Convention on Climate Change

(UNFCCC), and in particular at the 2009 Copenhagen climate conference (COP15) which agreed the Copenhagen Accord. On the other hand, without the EU's determined leadership in the 1990s there would not have been a Kyoto Protocol (Grubb and Gupta 2000). Without the EU it is also unlikely that the 2015 Paris climate conference (COP21) would have been able to adopt a meaningful climate change agreement (Chapters, 1, 2 and 5).

Realist/neorealist international relations theory, including new and liberal intergovernmentalist approaches argue that only the most powerful states possess sufficient hard power to be able to influence 'high politics' decisions (e.g. Bickerton *et al.* 2015; Chapters 2 and 5). At first sight, this seems to be confirmed by the empirical findings reported in the preceding chapters showing that large member states exhibit greater structural powers than medium-sized or small member states. Germany's economic strength and advanced innovation system were enabling factors for its international climate leadership (Chapter 8). However, Germany's exemplary leadership on its domestic energy transition (*Energiewende*), which influenced other countries also to move towards a low-carbon economy, is not merely the result of 'hard power' which Germany translated into structural leadership. Instead it also relied heavily on entrepreneurial and cognitive leadership (Chapter 8), or 'soft power' (Nye 2008).

Although the UK has not displayed the degree of structural leadership exhibited by Germany, it has offered considerable entrepreneurial and cognitive leadership and some exemplary leadership (cf. Chapters 8 and 12). The impending departure of the UK from the EU following the 2016 Brexit referendum is therefore likely to result in a perceptible loss of leadership capacity for the EU (and the UK). At first sight France exhibited mainly entrepreneurial leadership combined with a heroic leadership style which contributed greatly to the success of the 2015 Paris climate conference (COP21). A comparison between the handling – or mishandling – of the 2009 Copenhagen climate conference (COP15) by the Danish Presidency (Chapter 6) and France's skilful climate diplomacy (Chapter 7), reveals that competent entrepreneurial leadership is crucial for the successful outcome of international climate change conferences. However, structural leadership may also be relevant here. The French entrepreneurial efforts were not least aimed at contributing to the country's diplomatic prestige and international *grandeur* – or in short: to its 'presence' in the world – and thus also at enhancing its structural leadership capacities (Chapter 7).

The Commission's structural powers are primarily the result of its formal Treaty competences. The Commission's role as 'initiator' and 'guardian' of EU laws enables it to direct the behaviour of others within the EU and thus provides it with structural leadership capacities (Chapter 3). A good example is the Commission's rejection of Germany's second National Allocation Plan (NAP) under the EU ETS, which forced Germany to resubmit a significantly more ambitious revised NAP.

Similarly, the formal legislative competences of the Council and the EP under the 'ordinary legislative procedure' (formerly codecision procedure) provide these EU institutional actors with structural powers. The adoption of the 2008 climate and energy package and the launch and revision of the EU ETS offered particularly the Commission (Chapter 3) but also the EP and Council (Chapters 4 and 5) an important opportunity to adopt a novel policy instrument (ETS) which was initially resisted by some member states, including Germany (Chapter 8). However, both the Commission's and the EP's structural leadership capacities are contingent on the wider political and economic context, as becomes clear from the fact that the Commission and EP have become less willing to offer climate leadership since the 2008 financial crisis (Chapters 3 and 4).

Business and environmental NGOs (ENGOs) also possess some structural leadership capacities through their ability to create economic wealth and jobs (business) and their large supporter/member base (ENGOs) (Chapters 14 and 15). All member governments are influenced, to some degree, by important societal actors such as business and ENGOs. It is not by chance that Poland and, to a lesser degree, Spain (Chapters 10 and 11) (where only relatively few ENGOs are active on climate change issues) have frequently opposed ambitious EU climate change policies, while Denmark, Germany, Norway, the Netherlands and the UK (Chapters 6, 8, 13, 9 and 12) (where ENGOs' climate change activities are extensive) have often pushed for relatively ambitious environmental policy measures. In the climate leader states, moreover, ambitious climate change goals are often backed up by support from significant industries, e.g. the renewables sector which is still in its infancy in Spain and especially Poland. Nevertheless, even in environmental leader states, business can also effectively oppose or even veto their government's climate leadership ambitions, as illustrated by the German Chancellor Merkel's willingness (in cooperation with France) to water down EU carbon dioxide (CO_2) standards for cars (Chapter 8).

The 2008 financial crisis appears to have reduced the EU's structural leadership capacities at a time when the US, China and India became more active in the international climate change negotiations. In 2014 China and the US jointly committed themselves to greenhouse gas emissions (GHGE) cuts and to work towards an agreement at the 2015 Paris climate conference (Chapters 16 and 17). The creation of the BASIC group (Brazil, South Africa, India and China) reflects the increased importance of transitional and fast developing countries in global politics including international climate change politics (Chapters 2, 17 and 18; Torney 2015). The 2016 Brexit decision is likely to weaken further the EU's structural leadership capacity while also reducing the UK's relative influence in international climate change negotiations. The decline of the EU's structural leadership capacities can, however, not only be explained with reference to its relative economic decline vis-à-vis the US and, especially, fast developing countries such as China and India. Somewhat paradoxically the EU's reduced structural leadership capacities are at least partly also the result of its relatively successful climate change policies. As the EU's global share in GHGE has declined, the relative share of countries such as India and in particular China has increased significantly. As the UK is the EU's second largest GHG emitter, its departure will further decrease the EU's global share in GHGE. In future the EU therefore will arguably need to make even more extensive use of entrepreneurial and cognitive leadership in achieving its climate policy goals.

Entrepreneurial leadership

Grubb and Gupta (2000: 19–20) argued that 'the EU is clearly ill suited to a strictly entrepreneurial style of leadership … [because n]egotiators for the EU … have more limited scope for entrepreneurial action than representatives of equally weighty nation-states'. The chapters in this book at least partly refute this claim.

The six-monthly rotating EU Presidency provides the office holder with entrepreneurial leadership opportunities for bringing about integrative bargains in both internal and external EU climate change policies (e.g. Wurzel 1996). However, its role and influence in the international climate change negotiations has diminished since the introduction of 'lead negotiators' and 'issue leaders'. The new system was introduced in order to accomplish greater continuity and better EU representation in the international climate change negotiations (Chapters 2 and 5).

The Lisbon Treaty, which came into force shortly before the 2009 Copenhagen conference, left the rotating Council Presidency unchanged, but introduced an elected President for the European Council. Around 2005 the European Council took on an important 'initiating' (it cannot formally initiate EU laws) and coordinating role in the field of climate change, partly at the expense of the Commission and Council (Chapters 3 and 5). This may be interpreted as an indication of the member states' growing desire to gain closer control of key areas of European integration and international politics such as climate change. In other words, it could be seen as an expression of the 'new intergovernmentalism' which seems to be gaining ground in the EU (Bickerton *et al.* 2015).

Individual member governments sometimes launch climate diplomacy activities which are not coordinated on the EU level. The inability to present a unified EU position was painfully demonstrated at the 2009 Copenhagen climate conference (Chapters 2 and 6). On the other hand, coordinated national climate diplomacy efforts can help to bring about better outcomes in negotiations. The UK has frequently made use of its special relationship with the US to persuade American climate negotiators to accept more ambitious climate agreements (Chapter 12). Within the EU the UK initiated the 'Green Growth Group' between like-minded countries demanding more ambitious EU climate change policies. Post-Brexit the UK would have to try to continue such activities from outside the EU. The extensive French climate diplomacy prior to the 2015 Paris climate conference facilitated the adoption of the Paris Agreement. Spain has been a driving force behind the Ibero-American Network of Climate Change Offices which includes 21 other countries (Chapter 11). However, Poland's attempts to form alliances with the Visegrad nations (Hungary, Slovakia, the Czech Republic and Poland) were primarily aimed at reducing the EU's climate change ambitions (Chapter 10). Despite Poland having gone through a period of 'ecologic enthusiasm' between the late 1980s and early 1990s (Chapter 10), these activities cannot be classified as climate leadership as defined in Chapter 1: only efforts which help to bring about more ambitious climate policies count as offering leadership.

The above-mentioned bilateral US–China climate agreement and the Intended Nationally Determined Contributions (INDCs) which the US, China and India put forward for the 2015 Paris climate conference can also be perceived as constituting entrepreneurial leadership, although their respective voluntary pledges varied considerably in their ambition (Chapters 16, 17 and 18).

Cognitive leadership

The EU's cognitive climate leadership largely relies on the Commission's ability to draw together research and policy approaches and to weld them into innovative EU policy proposals which are then modified and adopted (or rejected) by the Council and EP. Although the European Council has taken a stronger interest in climate change issues since the mid-2000s, it relies only infrequently on cognitive leadership which, however, is often used by the Environmental Council (Chapter 5) and in particular the EP (Chapter 4).

The Commission has tried to recast climate policy from a purely environmental problem to one which encompasses energy security issues (e.g. dependence on oil from the Middle East and gas from Russia) and the modernisation of Europe's economy into a low-carbon economy. The Council and, to a lesser degree, the European Council have propounded the argument that climate change policies and economic growth are not mutually exclusive as can be seen from their endorsements of the 'green' and low-carbon economy.

The Netherlands tried to compensate for its lack of structural leadership capacities (and increasingly also its lack of domestic action) by providing cognitive (and entrepreneurial) leadership: it sought to persuade by expertise, scientific and experiential knowledge (Haverland and Liefferink 2016; Chapter 9). The UK, through the Stern report (Stern 2007) provided the clearest expression of cognitive leadership which presented ambitious climate change policy measures as a pro-growth strategy (Chapter 12). The pioneering Danish (carbon dioxide) and UK (GHGE) national emissions trading schemes offer additional examples of cognitive leadership (Chapters 6 and 12). Germany has been offering cognitive leadership since the early 1990s by providing a better 'knowledge base' for climate policy and by advocating ecological modernisation. However, the most prominent example of cognitive leadership is arguably its domestic energy transition (*Energiewende*) which has been partly justified with reference to the ecological modernisation of the German economy (Chapter 8).

Since the 1998 burden sharing agreement (followed by the effort sharing decision (Chapters 1, 2 and 5)), the EU's internal GHGE reduction targets have demanded more strenuous efforts from the highly developed richer Northern European member states than from the less highly developed poorer Southern and Eastern European member states.

ENGOs have been highly active in providing cognitive leadership by trying to frame the public debate about climate change in Europe. ENGOs coined the term 'hot air' which crystallised their objections to merely symbolic GHGE reduction rhetoric. Greenpeace has been promoting renewable energy, characterising it as an energy '[r]evolution'. Of all the actors assessed in this book, it is ENGOs which have most vociferously and forcefully argued that both the environment and economy will benefit from a low-carbon economy. ENGOs have demanded that the EU should adopt a heroic/transformational leadership style when embarking towards a low-carbon economy (Chapter 15).

Exemplary leadership

In Chapter 1, a distinction was made between *unintentional* and *intentional* exemplary leadership. In the former, a country develops ambitious climate policies primarily for domestic reasons without the intention of attracting followers. In the latter, setting an example for others to follow is one of the explicit aims of a country's domestic policy.

Several chapters refer to *unintentional* exemplary leadership in the field of energy policy and renewables. Both Denmark and Germany adopted transformational domestic energy transformations with a view to ecologically modernising their economies (Chapters 6 and 8). Acting as an energy transition pioneer resulted in export opportunities for 'green' industries and setting an example for other countries gradually became a more explicit goal. In 2011 a new centre-left government in Denmark claimed that ambitious environmental policies would turn the country into a 'green laboratory', making it possible to 'reap economic first-mover benefits' (Chapter 6). Even Spain can be characterised as an early climate pioneer (at least in renewables) with solar energy being of particular importance (Chapter 11). However, due to political fragmentation, inconsistent policies and the impact of the economic crisis, Spain failed to realise its first mover advantages in the way that Denmark and Germany did (compare Chapter 11 with Chapters 6 and 8).

The UK provides a good example of exemplary leadership that was intentional from the outset. The 2002 UK ETS was explicitly intended to deliver first mover advantages to give UK companies practical experience prior to the anticipated launch of the EU's ETS (Chapter 12). UK officials were sent to Brussels to promote the British ETS example and to assist the Commission in drafting the EU ETS Directive. Although the scheme eventually adopted by the EP and Council differed in important respects from the UK ETS (Wurzel 2008), the UK's role constituted a clear case of exemplary, accompanied by cognitive and entrepreneurial leadership.

All major EU institutions have emphasised the importance of having convincing and effective *internal* policies that can act as an example *externally*. This is about credibility (i.e. to show to the outside world that the EU is able to deliver). But it is also related to the EU's intention to exert exemplary climate leadership (Chapters 2 to 5).

Interaction between different types and styles of leadership

More than one type of leadership is required to achieve integrative institutional bargaining success (Young 1991). Given the limited relevance of military power in international climate change politics, structural climate leadership primarily works through economic power. The sheer size of the internal markets of the US, the EU and increasingly China and India, give these 'big players' considerable leverage in global climate politics (Chapter 2). However, the sole use of structural leadership will not be sufficient when trying to solve a highly complex, long term, cross-border and cross-sectoral policy problem such as climate change; conversely cognitive or entrepreneurial leadership alone will not suffice.

The chapters in this book suggest that cognitive, entrepreneurial and exemplary leadership are often interlinked. Ambitious domestic policies can (intentionally or unintentionally) act as examples for others, but they may also provide a firm knowledge base in the form of scientific expertise and experiential knowledge (Haverland and Liefferink 2012). Entrepreneurial leadership efforts may start at the domestic level and subsequently be used to set the EU and/or international agenda. Examples include the UK pushing of emissions trading (Chapter 12), Denmark, Germany and the Netherlands propagating an EU-wide CO_2 tax (Chapters 6, 7 and 9) and France trying to push nuclear energy as a carbon-free alternative for fossil fuels (Chapter 7). Germany's efforts in setting up the International Renewables Energy Agency (IRENA) provides another example (Chapter 8).

The timing and sequencing of different types of leadership is important because cognitive leadership operates on a different timescale to structural and entrepreneurial leadership. Cognitive leadership

> is a deliberative process; it is difficult to articulate coherent systems of thought in the midst of the fast-paced negotiations associated with institutional bargaining … [and] new ideas generally have to triumph over the entrenched mindsets or worldviews held by policymakers.
>
> (Young 1991: 298)

A good example of the EU failing to achieve the right combination of types of leadership necessary for a successful outcome of international climate change conferences constitutes the 2009 Copenhagen conference. At other critical junctures of the international climate conferences (e.g. 2010 Cancún (COP16), 2011 Durban (COP17), 2014 Lima (COP20) and Paris (COP21)) the EU has, however, skilfully managed to use its leadership capacity. For example, the EU forged alliances with developing countries (entrepreneurial leadership) by emphasising the importance of equity issues and economic advantages in the move towards a low-carbon economy driven by renewable energy (cognitive leadership). It could be argued that while the EU's structural leadership capacities are declining its entrepreneurial and particularly its cognitive leadership capabilities are becoming increasingly more important for achieving its goals in international climate change politics. The

2016 Brexit decision has arguably accelerated the relative decline of the EU's structural powers in international (climate change) politics. The EU may try to compensate for this loss of structural leadership capacity by increasing its entrepreneurial and cognitive leadership capacities, although that will not be easy.

In addition to analysing different leadership types the preceding chapters have also assessed different leadership styles, namely heroic versus humdrum and transformational versus transactional styles. In the early 1990s the EU aspired to a heroic/transformational leadership style, but often defaulted to a humdrum/transactional leadership style because of the wide gap between its rhetorical ambition in international climate change politics and its lack of specific domestic implementation measures (Chapter 5; Grubb and Gupta 2000; Jordan *et al.* 2010). The climate policy measures which the EU adopted to implement its Kyoto Protocol obligations have helped it to close its 'credibility gap' (see Chapter 5), resulting in a more transformational style in the mid-2000s (Chapters 3 and 5). The mix of entrepreneurial and exemplary leadership pursued by the EU has been characterised as 'leadiating', a neologism combining 'leading' and 'mediating' (Bäckstrand and Elgström 2013; Chapter 5).

Green economy and low-carbon economy

The concept of a 'green' or low-carbon economy has been gathering support not only among EU institutional actors (Chapters 3, 5 and 4), member states (especially Chapters 6 and 8) and societal actors (especially ENGOs, see Chapter 15) but also in European non-EU states (Chapter 13). Importantly, although Europe can be identified as the region where support for a low-carbon economy has been strongest, the concept is gaining traction in the US and China and starting to enter political discourse in India (Chapters 16–18). The concept of a low-carbon economy is closely related to the concept of ecological modernisation which assumes that ambitious environmental policy measures can be beneficial for both the environment and economy (e.g. Jänicke 1993; Chapter 1). From this perspective, climate change poses not only an environmental threat but also as an opportunity to modernise the economy.

The EP and ENGOs (Chapters 4 and 15) were the earliest, strongest and most consistent proponents of a rapid move towards a low-carbon economy. But the accession of the poorer Central and Eastern European States in 2004/2007, together with the 2008 financial crisis and subsequent recession, has created doubts over the enduring commitment to move speedily towards a low-carbon economy. The Commission (Chapter 3) and the European Council (Chapter 5) and even the EP (widely regarded as the 'greenest' of the EU's institutions) have become less outspoken in their support for a rapid move towards a low-carbon economy. In other words, their previously transformational leadership style has increasingly given way to a transactional leadership style (Chapter 1). Broadly speaking, members of the EP (MEPs) from the poorer Eastern (and Southern) European member states and/or the political right have been less supportive of a rapid move towards a low-carbon economy than MEPs from more affluent

Northern European member states and/or the political left (including Green Parties). However, Spain has considerably increased its renewable energy and particularly solar energy capacities (Chapter 11).

Since 2005, the European Council has dealt frequently with climate change issues (while often making reference to the low-carbon economy). After a dip in 2012 there was again an increase in how often the European Council referred to climate change, explainable by the upcoming 2015 Paris climate change conference (Chapter 5).

Germany's energy transformation (*Energiewende*) clearly tries to achieve a rapid move towards a low-carbon economy. However, although the German government has tried to use a transformational leadership style to speedily bring about the *Energiewende*, in practice it frequently had to resort to transactional leadership. This is not surprising considering the enormous challenges which policy makers must confront when attempting to implement an ecological industrial policy (*ökologische Industriepolitik*) of such magnitude (Chapter 8). Nevertheless, Germany's adoption of an ecological tax reform (including obligatory feed-in tariffs) in the 1990s arguably prepared the ground for a rapidly growing and internationally successful 'climate protection industry' which created jobs (Chapter 8). This sector has grown considerably since the German government decided to embark on the *Energiewende* following the decision to accelerate the phasing out of nuclear power after the 2011 Fukushima nuclear disaster. Germany could therefore be identified as an exemplary and occasionally also a cognitive leader which is not to deny that Germany also has blind spots (e.g. its reliance on coal) (Chapter 8). A similar pioneering role has been played by Denmark which has tried to foster 'green growth' and a 'green transition' through a green tax reform and a rapid expansion of its wind energy sector. Denmark was ranked as top performer as regards its will and skill to transform into a low-carbon economy by the Climate Change Performance Index (Chapter 6). Norway has officially also endorsed the concept of a low-carbon economy through a 'green transition' which is a term that was influenced by the German energy transition (*Energiewende*) (Chapter 13).

Although there was consensus among the mainstream political parties in the 2015 UK election campaign for the need to accelerate the transition to a competitive, energy-efficient, low-carbon economy, phasing out unabated coal for power generation, its energy transformation has been somewhat slower and more patchy (with the exception of off-shore wind energy) (Chapter 12). Spain has found itself caught in the dilemma of whether to use the EU's climate change policies to boost its fledgling renewable-based energy transition in line with a green economy or to continue with business-as-usual to meet the EU and international climate change commitments (Chapter 11). France arguably has had an ambivalent attitude towards a 'green' economy, which relies primarily on renewable energy, while it has traditionally been a strong supporter of a low-carbon economy due to its high reliance on nuclear power (Chapter 7).

The Dutch economy's relatively low renewable energy potential (with the exception of wind energy) and dependence on high energy intensive industries (e.g. refineries and chemical industry) helps to explain why the Netherlands have made comparatively little progress (at least when compared to Denmark and Germany) on their way towards a low-carbon economy (Chapter 9). However, of the European countries assessed in this book, it is Poland which has been struggling most with the concept of a low-carbon economy. The main reason for this is the central importance which (high carbon intensive) coal plays in the Polish economy in general and electricity generation in particular (Chapter 10).

Clearly the concept of a low-carbon economy is gaining traction also outside Europe as the chapters on the US and China and (to a lesser degree) India show (Chapters 16–18). Although President Obama's proposed 'green new deal' quickly ran into implementation difficulties, many subnational green economy initiatives have been undertaken by US states (Chapter 16). Although China's astonishing economic growth is still largely dependent on fossil fuel use, there has been a rapid extension of the renewable energy sector albeit from a low level. In fact, after one decade of rapid growth, China ranked top in renewable energy investment and became the world leader in installed wind power capacity, solar panels/cells manufacturing, and solar water heating utilization in 2010 (Chapter 17). India, another rapidly developing country still largely dependent on fossil fuels for its economic growth, has experienced a boom in the renewable energy sector and in particular in solar energy. China aims to enhance its solar power capacity to 100GW by 2022. India's *Mission for Energy Efficiency* and *Solar Mission*, which are influenced by the concept of ecological modernisation, help to explain why India initiated the creation of an International Solar Alliance involving more than 120 countries (including African countries) at the 2015 Paris climate conference (Chapter 18).

While all Brussels-based ENGOs have broadly and consistently endorsed the rapid transformation of fossil fuel dependent economies towards low-carbon economies, the views among businesses vary greatly with some businesses (especially in the renewable energy sector) strongly in favour of such a transition while others (especially high energy intensive industries and the coal industry) are strongly opposed to such a move (Chapters 15 and 14 respectively).

Policy instruments

The use of new environmental policy instruments (NEPIs) (such as eco-taxes and emissions trading) is a typical but not a necessary concomitant of ecological modernisation and the transition toward a low-carbon economy. Germany, one of the energy transition pioneers, was initially hostile towards the adoption of the EU ETS and later implemented this novel policy instrument in such a way that its energy intensive industries were granted exemptions (Chapter 8). Denmark, in contrast, embraced emissions trading early and, together with the UK, played a pioneering role in paving the way for the adoption of this policy instrument at the EU level (Chapter 12).

On the EU level in particular it is easier to secure political agreement for some NEPIs than others. The Commission's 1991 proposal for an EU-wide

CO_2/energy tax was vetoed by the UK on sovereignty grounds (Chapter 12), whereas its 2003 proposal for an EU ETS was adopted speedily with only relatively minor changes by the Council and EP (Chapter 8). The EP still largely retains a preference for traditional regulation and eco-taxes, although it has accepted the EU ETS as a useful policy instrument.

Member states vary considerably in their national policy instrument repertoires and preferred EU policy instrument mixes (Wurzel, Zito and Jordan 2013). France's relative unease with ecological modernisation extends to certain NEPIs. Its 2000 Climate Plan included proposals for carbon taxes and emissions trading but after massive protests against fuel duty increases, carbon taxation was no longer deemed desirable by the French government. Because of its heavy reliance on nuclear power, France lobbied (unsuccessfully) for an EU-wide CO_2 tax instead of the Commission's CO_2/energy tax proposal but eventually accepted the EU ETS (Chapter 7). In addition to the adoption of traditional regulatory instruments, the Netherlands has strongly relied on voluntary agreements ('covenants') between industry and the government, in line with its corporatist tradition emphasising negotiation and consensus. The Netherlands was one of the EU's eco-taxes pioneers and strongly supported the wide use of the Kyoto Protocol's flexible instruments (i.e. CDM and Joint Implementation (JI)) (Chapter 9). Spain has been slow in adopting NEPIs on the domestic level and experienced difficulties with the implementation of the EU ETS (Chapter 11). Perhaps surprisingly, Poland has exhibited a preference for market-based instruments, at least in domestic climate change policy, which partly explains Polish policy makers' distrust for 'command-and-control' regulations that are still associated with the Communist past (Chapter 10).

For long Brussels-based ENGOs favoured traditional regulation and environmental taxes; they were initially opposed to emissions trading, suspicious of voluntary agreements and considered informational tools (such as energy consumption labels) merely as supplementary policy instruments. ENGOs pragmatically accepted the EU ETS, while campaigning for the tightening of its rules. However, in the 2010s, Friends of the Earth (FoE) Europe started again to oppose the EU ETS. ENGOs still favour ambitious GHGE reduction targets, deadlines and monitoring requirements enshrined in legally binding EU (and national) regulations and international treaties (Chapter 13).

Businesses have generally favoured voluntary agreements and market-based instruments (emissions trading in particular, although there are important exceptions) over 'command-and-control' regulations (Chapter 14). Since the late 1990s, the main differences over policy instruments between Brussels-based ENGOs and businesses seem to have shifted from advocating different policy instrument types (voluntary agreements and emissions trading (business) versus stringent regulations and eco-taxes (ENGOs)) to the strictness of the rules for one particular type of policy instrument. For example, most European businesses favoured the free allocation (so-called grandfathering) of allowances under the EU ETS, while most Brussels-based ENGOs campaigned for the full auctioning of allowances.

The US has been a pioneer in the use of emissions trading on a regional level (although not for GHGE). However, compared with Denmark, the UK and EU, American states (e.g. California) have somewhat belatedly set up regional GHGE trading schemes which are beginning to be linked on a 'trans-subnational' level (Chapter 16). The regional ETSs in the US could possibly be linked to the EU ETS. China has started to endorse market-based instruments and in particular emission trading (Chapter 17). Of the countries assessed in this book India has arguably experimented least with NEPIs, preferring instead mainly traditional (command-and-control) regulation (Chapter 18).

In the international climate change negotiations, the EU has for a long time strongly supported traditional regulation in form of legally binding treaties with clear targets and explicit deadlines. It was initially opposed to voluntary pledges (or INDCs) which were enshrined as Nationally Determined Commitments (NDCs) in the 2015 Paris Agreement (Chapters 1 and 2). However, the determination of the US and China (which received strong support from India) for a bottom-up approach (with voluntary national pledges enshrined in an agreement the legal bindingness of which is contested) won the day over the EU's favoured top-down approach (with targets and deadlines enshrined in a legally binding treaty). The EU was nevertheless able to secure regular monitoring of the pledges and successfully insisted on the strengthening of transparency measures enshrined in the 2015 Paris Agreement.

Multi-level and polycentric governance

The EU is itself a multi-level governance system in which powers are shared between EU institutional and member state actors. Societal actors (e.g. business and NGOs) try to influence both EU institutions and member states (Chapters 14 and 15). Chapter 1 argued that multi-level governance (MLG) and polycentric governance share certain core assumptions, but diverge on the relative role they attribute to governmental and societal actors in policy-making or societal rule-making. MLG-inspired EU studies usually emphasise the mutual dependency of EU and member state actors while polycentric studies focus more strongly on societal self-coordination at different governance levels with a strong emphasis on the regional and local level.

This book provides clear evidence of the importance of MLG for climate change policies at different governance levels. The climate change policies of EU member states are strongly influenced by EU but also international climate change policies (e.g. Delbeke and Vis 2015). Even Norway, which is not an EU member state, is strongly influenced by EU climate change polices and has recently even joined the EU ETS (Chapter 13). When the EU offers exemplary leadership (e.g. on emissions trading and its attempted move towards a low-carbon economy) it is also closely watched by non-European states such as the US, China and India. Unsurprisingly the chapters on federal states (Germany, US and India) detailed a relatively large number of regional climate change initiatives (Chapters 8, 16 and 18). The same applies to Spain, which can be perceived as a quasi-federal state granting a significant degree of autonomy to its regions (Chapter 11). However, regional or local climate change initiatives and experiments also seem to have played an important role in Denmark and China (Chapters 6 and 7).

Schreurs and Tiberghien (2007: 25) have argued that 'multi-level governance has not just multiple veto points, it has created numerous leadership points where competitive leadership has been initiated'. The Chapters on EU institutional actors offer empirical examples for this claim (Chapters 3–5). The development of German climate change policy seems to show that a member state may find itself caught by a 'kind of "enforced leadership"' through EU institutions (e.g. the Commission) (Chapter 8). At least under certain circumstances (the precise nature of which requires further empirical research and analytical consideration) a 'multi-level reinforcement' mechanism seems to be at work (Jänicke 2014; Chapter 8).

The fact that with a few exceptions (see especially Chapters 2, 14 and 15) the above chapters have unearthed relatively few polycentric climate governance examples is perhaps not surprising as their main focus was on the international, EU and national level, while polycentric governance is likely to be found on the regional and local level. However, the novel bottom-up approach enshrined in the 2015 Paris Agreement may possibly necessitate the wider use of polycentric climate governance also within the EU.

Conclusion

Judging from the empirical evidence and the analysis put forward in the preceding chapters, EU institutional actors (Chapters 3–5), member states (Chapters 6–13) and societal actors (Chapters 14–15) have all tried, although to different degrees, to offer climate leadership while demanding from the EU that it takes on a leadership role in international climate politics. The differentiation of leadership into different *types* (structural, entrepreneurial, cognitive and exemplary) and *styles* (humdrum/transactional and heroic/transformational) of leadership has allowed for a more fine-grained analysis of what exactly climate leadership entails. The analytical use of different *types* of leadership has, for example, made it possible to explain why the EU is still able to offer leadership in international climate change politics, although its general influence in international politics has arguably been declining at least in relation to rapidly developing countries such as China and India (Chapter 2, 17 and 18). The 2016 Brexit decision is likely further to enhance this decline. To some degree the EU has been able to compensate its declining structural leadership capacities (or structural power) with the help of a combination of entrepreneurial and particularly cognitive climate leadership. For example, at crucial stages of the international climate change negotiations, the EU has managed to forge strategic alliances with (among others) developing countries (i.e. entrepreneurial leadership) while presenting the challenge of climate change not only as a threat but also an opportunity for a low-carbon economy which will generate economic growth and jobs (i.e. cognitive leadership). However, the Brexit decision is likely to negatively affect particularly the EU's structural and, although to a lesser degree, its entrepreneurial leadership capacity.

The differentiation of leadership into different styles of leadership has helped to assess how climate leaders and pioneers act. It also introduced a time dimension (i.e. short or long term) for assessing the actions of climate leaders and pioneers (see Liefferink and Wurzel 2016). An actor adopting a humdrum/transactional leadership style will normally adopt a short-term strategy which is likely to lead only to marginal adjustments of existing policies; adopting a heroic/transformational leadership style, in contrast, requires the adoption of long-term objectives, strong political coordination and the assertion of political will. Humdrum/transactional leadership is likely to lead to incremental change although it may also produce transformational change if pursued over a long period of time (Chapter 1). Heroic/transformational leadership, on the other hand, brings about history making or 'revolutionary' changes. Unsurprisingly the chapters in this book have unearthed relatively few examples of heroic/transformational leadership – arguably because it can usually be used only in exceptional circumstances and even then only sparingly (Liefferink and Wurzel 2016).

At first sight the differentiation into structural, entrepreneurial, cognitive leadership types seems to further deepen old theoretical divides because structural leadership is the main domain of realists/neorealists, entrepreneurial leadership is arguably most closely related to pluralist/neofunctionalist theory whereas cognitive leadership best fits constructivist approaches. However, our typology of different leadership types as applied in the preceding chapters, show that the relationship is not necessarily disjunctive. On the contrary different types of leadership can be mutually enhancing in facilitating integrative bargains. As we have argued above, more than one leadership type is usually required to achieve integrative bargaining success and thus also an enhanced likelihood of successful and longstanding problem-solving. Our leadership classification may therefore help to bridge the division of political science into international relations and comparative politics which seems to have created a kind of 'intellectual apartheid' (Bulmer 1994: 355). The analytical leadership framework put forward in Chapter 1 does not claim to constitute a new theory; rather, its more modest ambition was to encourage the chapter authors to generate new analytical insights and empirical findings while drawing on a reasonably robust analytical framework. Whether this attempt was successful is something which the reader must decide.

References

Bäckstrand, K. and Elgström, O. (2013) 'The EU's role in climate change negotiations: from leader to "leadiator"', *Journal of European Public Policy*, 20(10): 1369–86.

Bickerton, C.J., Hodson, D. and Puetter, U. (2015), 'The new intergovernmentalism: European integration in the post-Maastricht era', *Journal of Common Market Studies*, 53(4): 703–22.

Bulmer, S. (1994) 'The governance of the European Union: a new institutionalist approach', *Journal of European Public Policy*, 7(1): 351–80.

Burns, J.M. (1978) *Leadership*, New York: Harper & Row.

Damro, C. (2012) 'Market power Europe', *Journal of European Public Policy*, 19(5): 682–99.

Delbeke, J. and Vis, P. (2015) *EU Climate Policy Explained*, London: Routledge.

Grubb, M. and Gupta, J. (2000) 'Leadership', in J. Gupta and M. Grubb (eds), *Climate Change and European Leadership. A Sustainable Role for Europe?*, Dordrecht: Kluwer, 15–24.

Haverland, M. and Liefferink, D. (2012) 'Member State interest articulation in the Commission phase. Institutional preconditions for influencing "Brussels" ', *Journal of European Public Policy*, 19 (2): 179–97.

Hayward, J. (2008) *Leaderless Europe*, Oxford: Oxford University Press.

Jänicke, M. (1993) 'Über ökologische und politische Modernisierungen', *Zeitschrift für Umweltpolitik und Umweltrecht*, 16: 159–75.

Jänicke, M. (2011) 'German climate change policy: political and economic leadership', in R.K.W. Wurzel and J. Connelly (eds) (2011) *The European Union as a Leader in International Climate Change Politics*, London: Routledge, 129–46.

Jänicke, M. (2014) 'Multi-level reinforcement in climate governance', in A. Brunnengräber and M.A. Di Nucci (eds), *Im Hürdenlauf zur Energiewende*, Berlin: Springer, 35–47.

Jordan, A., Huitema, D., van Asselt, H., Rayner, T. and Berkhout, F. (2010) *Climate Change Policy in the European Union. Confronting Dilemmas of Mitigation and Adaptation?*, Cambridge: Cambridge University Press.

Liefferink, D. and Wurzel, R.K.W. (2016), 'Environmental leaders and pioneers: agents of change?', *Journal of European Public Policy*, DOI:10.1080/13501763.2016.1161657.

Manners, I. (2002) 'Normative power Europe: a contradiction in terms?, *Journal of Common Market Studies*, 40(2): 235–58.

Moravcsik, A. (1999) 'A new statecraft? supranational entrepreneurs and international cooperation', *International Organization*, 53(2): 267–306.

Nye, J. (2008) *The Powers to Lead*, Oxford: Oxford University Press.

Schmidt, J.R. (2008) 'Why Europe leads on climate change', *Survival*, 50(4): 83–96.

Schreurs, M. and Tiberghien, Y. (2007) 'Multi-level reinforcement: explaining European Union leadership in climate change mitigation', *Global Environmental Politics*, 7(4): 19–46.

Stern, N. (2007) *The Economics of Climate Change*, Cambridge: Cambridge University Press.

Torney, D. (2015) *European Climate Leadership in Question: Policies towards China and India*, Cambridge, MA: MIT Press.

Wurzel, R.K.W. (1996) 'The role of the EU Presidency in the environmental field: does it make a difference which member state runs the Presidency?', *Journal of European Public Policy*, 3(2): 272–91.

Wurzel, R.K.W. (2008) *The Politics of Emissions Trading in Britain and Germany*, London: Anglo-German Foundation.

Wurzel, R.K.W, Zito, A. and Jordan, A. (2013) *Environmental Governance in Europe*, Cheltenham: Edward Elgar.

Young. O.R. (1991), 'Political leadership and regime formation: on the development of institutions in international society', *International Organisation*, 45(3): 349–75.

Index